THE FIRECRACKER BOYS

The
FIRECRACKER BOYS

H-bombs, Inupiat Eskimos and the Roots
of the Environmental Movement

Dan O'Neill

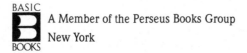

BASIC
B A Member of the Perseus Books Group
BOOKS New York

Books published by Basic Books are available at special discounts for bulk purchases in the United States by corporations, institutions, and other organizations. For more information, please contact the Special Markets Department at the Perseus Books Group, 2300 Chestnut Street, Suite 200, Philadelphia, PA 19103, or call (800) 255-1514, or e-mail special.markets@perseusbooks.com.

This book was previously published by St. Martin's Press in 1994.

Designed by Timm Bryson
Set in 10 point Weidemann Book

Library of Congress Cataloging-in-Publication Data

O'Neill, Dan (Daniel T.)
  The Firecracker Boys : H-bombs, Inupiat Eskimos and the roots of the environmental movement / Dan O'Neill.
      p. cm.
  Includes bibliographical references and index.
  ISBN-13: 978-0-465-00348-8 (pbk. : alk. paper)
  ISBN-10: 0-465-00348-6 (pbk. : alk. paper) 1. Project Chariot. 2. Hydrogen bomb—Testing—Alaska—Thompson, Cape (North Slope Borough) 3. Nuclear excavation. 4. Nuclear weapons—United States—Testing—History. 5. Antinuclear movement—United States. 6. Teller, Edward, 1908–2003. 7. Inupiat—Alaska—History—20th century. I. Title.

  UG1282.A8O39 2007
  621.48—dc22

                          2007030702

To the memory of my father
John Terence O'Neill

The author acknowledges with gratitude the support of the Alaska Humanities Forum, the National Endowment for the Humanities, and the Rasmuson Library, University of Alaska Fairbanks.

[There was] a general atmosphere and attitude that the American people could not be trusted with the uncertainties, and therefore the information was withheld from them. I think that there was concern that the American people, given the facts, would not make the right risk-benefit judgements.

—**Peter Libassi,** Chairman, Interagency Task Force
on the Health Effects of Ionizing Radiation

I know of no safe repository of the ultimate powers of society but the people themselves; and if we think them not enlightened enough to exercise their control with a wholesome discretion, the remedy is not to take it from them, but to inform their discretion with education.

—**Thomas Jefferson**

CONTENTS

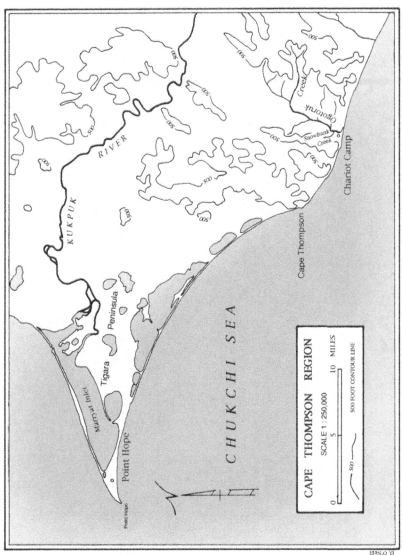

CAPE THOMPSON REGION

SCALE 1 : 250,000

0   5   10   MILES

500   500 FOOT CONTOUR LINE

CHUKCHI SEA

Point Hope

Tigara Peninsula

Marryat Inlet

Point Hope

KUKPUK

RIVER

500

500

500

500

500

500

500

500

500

500

500

Cape Thompson

Chariot Camp

Snowbank Creek

Ogotoruk Creek

D. O'Neill

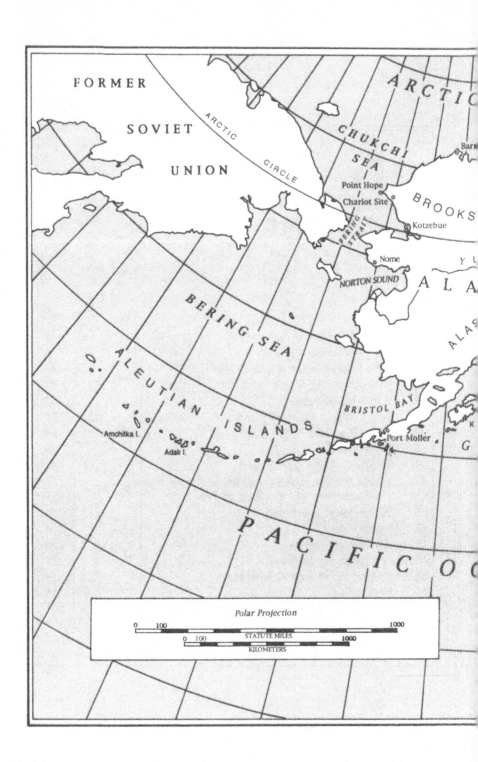

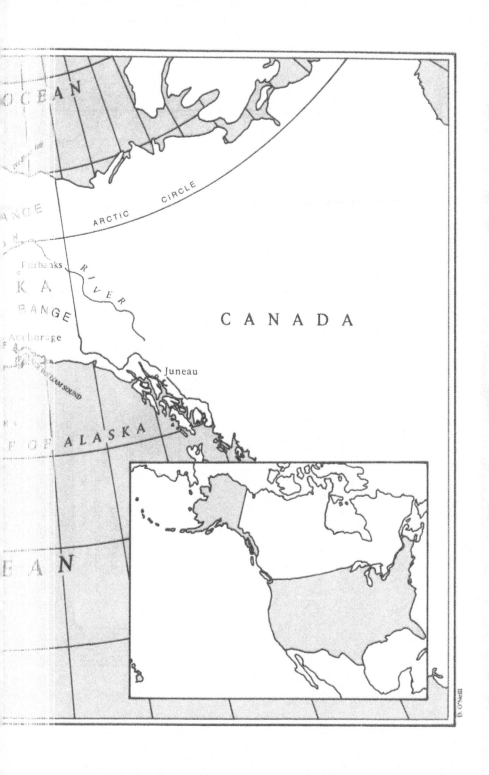

OCEAN

ARCTIC CIRCLE

RANGE

Fairbanks

K A

RANGE

Anchorage

PRINCE WILLIAM SOUND

RIVER

CANADA

Juneau

P OF

ALASKA

E A N

D. O'NEILL

1 THE SEA PEOPLE OF TIKIGAQ

No truly primitive group could exist
under such conditions. Only by means
of a highly complex technology and
through a highly developed knowledge of natural phenomena could human beings penetrate the Arctic.

—Helge Larsen and
Froelich Rainey,
Arctic archeologists

In the extreme northwest corner of the North American continent, in the
region of Bering Strait, prehistoric Inupiat hunters discovered a spit of land
jutting twenty miles into the Chukchi Sea. It pointed like a finger back
across the water to the people's ancestral homeland, Asia, just 160 miles
away. They called the place *Tikigaq,* the Inupiaq word for "forefinger."

The peninsula was an ideal place to intercept migrating sea mammals.
From here the Inupiat hunted walrus, seals, and polar bear. They hunted
the small white whale called beluga, and *ugruk,* the bearded seal. They
collected the eggs of nesting seabirds from the inland cliffs near a little
stream they called Ogotoruk. Eventually, they would load flint-pointed
harpoons into *ugruk*-skin boats and chase, strike, and land fifty-ton bowhead whales. At Tikigaq, they built semisubterranean houses employing

what few materials existed in a place a hundred miles from the nearest tree: whalebones and driftwood for the arching structural ribs, and slabs of sod stacked up igloo-fashion to form a covering. As wind and rain smoothed the contours, and sod and flowers sprouted again, the settlement looked like a hummocky patch of grassland, a green rejoinder to the surrounding blue, rolling sea.

*   *   *

That was the summer picture, anyway. In winter, the Chukchi Sea freezes. The spit stands only a few feet above the sea ice, and both are covered indiscriminately by hard-packed, windblown snow. It is hard to tell where the land leaves off and the sea begins. To the extent that Tikigaq looked like a village at all, it must have seemed to be one adrift on the sea ice, twenty miles from shore, unsheltered from the strong winter winds, more than a hundred miles north of the Arctic Circle.

Most oral history accounts agree that the early inhabitants of the region around Tikigaq organized themselves into a social system akin to a modern nation. They occupied a precisely defined territory, defended it, and, like the citizens of modern states, were not averse to expansionist forays. Sometimes the Tikirarmiut, as they called themselves, penetrated several hundred miles into neighboring dominions. In time, they became the largest and most powerful society in northern Alaska, "the aristocrats of the Arctic," as one writer has called them. By the late 1700s, their territory included a stretch of coast from Icy Cape, 200 miles to the north, to Kotzebue Sound, 200 miles to the south.

It was during this time, in 1778, that Capt. James Cook became the first European to sail up the Alaska coast, through Bering Strait and into the Chukchi Sea. But, though he explored all the way to Icy Cape, he missed the low Tikigaq spit. Shortly after Cook's voyage, around the turn of the nineteenth century, the Tikirarmiut lost a decisive battle with a people on their southern frontier and may have lost half their entire population. It is likely that many of the best hunters died and a period of starvation followed. Europeans first encountered Tikigaq during this period of instability. In 1820, Glieb Semenovich Shishmarev of the Russian navy saw the settlement from some miles offshore. Sailing by without landing, he named the point Cape Golovin in honor of a fellow captain in the Russian navy.

The next Europeans to glimpse Tikigaq reenacted the "discovery" and, for the second time, the most prominent features of the Eskimo's homeland were renamed, this time in honor of Englishmen. Capt. F. W. Beechey sailed HMS *Blossom* into the Chukchi in 1826 and closed on a high bluff, which he named Cape Thomson (later written "Thompson") after Deas Thomson, a commissioner of the British navy. Putting ashore, Beechey became the first outsider to encounter the Tikirarmiut when his party was "met upon the beach by some Esquimaux who eagerly sought an exchange of goods. Very few of their tribe understood better how to drive a bargain than these people." Beechey's account describes a "robust people above the average height of Esquimaux: the tallest man was five foot nine inches, the tallest woman five foot four inches. All the women have tattooed upon the chin three small lines . . . all the men had labrets [lip ornaments]." The Englishmen found the natives to be "very honest, extremely good natured, and friendly."

After visiting a small encampment near the base of Cape Thompson, Beechey ascended the promontory. From its vantage he "discovered low land jetting out from the coast to the W.N.W. as far as the eye could reach. As this point had never been placed in our charts," he wrote, "I named it Point Hope, in compliment to Sir William Johnstone Hope." The next morning Beechey sailed away to the west to trace the extent of the low point he had seen from Cape Thompson. On nearing its tip, he saw "a forest of stakes . . . and beneath them several round hillocks." The stakes were drying racks on which chunks of dark purple seal meat hung and skins lifted in the steady breeze. Other racks, made from the jawbones of whales, supported *umiaqs* (skin boats). The hillocks were the sod houses of Tikigaq.

A strong current kept Beechey's gig from landing, but inclusion on world maps had befallen the Tikigaq people. Significant contact with the white man would follow. Though industrial trade goods had filtered into northwest Alaska before 1849, Western culture had had very little influence on the Tikirarmiut. They subsisted on a harvest of animals obtained without the use of firearms. And, though their general health was not always good, they had not yet been ravaged by smallpox, syphilis, influenza, measles, diphtheria, and tuberculosis, nor had they been intro-duced to alcohol.

But, starting in 1849, each spring saw more than a hundred ships from the American whaling fleet pass northward through Bering Strait and re-turn with barrels of oil and bundles of baleen, the durable, elastic "whale-bone" used primarily for corset stays. In little more than a decade, Yankee whalers had nearly decimated the bowhead whale, not only a crucially important food for the local people, but the center of Inupiaq culture as well. The walrus was an alternate source of nutrition for the Eskimo, but it was also an alternate source of crude for the original oil men from the south. And if the 3,000-pound pinnipeds didn't have baleen, they did have ivory. Turning their technology on the walrus, the whalers, inside of fifteen years, all but wiped out those herds. As a consequence, they nearly wiped out the Eskimo as well. Anthropologists estimate that in scarcely more than a generation the Eskimo population of northwest Alaska was reduced by half, maybe two-thirds.

White men established their first shore-based whaling station in Alaska on the Tikigaq spit in 1887. The Tikirarmiut had steadfastly refused to al-low the station to be built in their village, and even refused to work there, but it took root five miles east. Inhabited by "American whalers, Hawai-ians, Negroes and men of all nations," Jabbertown, as it was called, accel-erated the introduction of disease and alcohol to the Tikirarmiut.

•   •   •

If the polar bear, because it spends so much of its life hunting on the ice offshore, is classified by biologists as a sea mammal rather than a terres-trial one, a similar logic might apply for the Tikirarmiut. Every hunter spent much of his life upon the crystalline ocean, the gnashing, churning, jumbled wilderness of sea ice. From November to April, he hunted seals during every hour of light, often standing perfectly still at a breathing hole while high winds compounded the chill of temperatures reaching twenty, thirty, and forty degrees below zero. Strong surface winds are more fre-quent here than in any other part of Alaska, and the winds of greatest ve-locity, both on average and in the number reaching sustained gale force, occur during the coldest months of the year. As one meteorologist suc-cinctly put it, the area is "one of the most uncomfortable in the world."

These winds also have the power to tear the ice pack from its grip on the land, holding it offshore and creating a lead often miles wide. If the lead opened up behind a hunter, he was cut off and could be carried far out to sea. Men disappeared this way. They drowned, froze to death, or were devoured by polar bear, which also hunted on the ice. A hunter was considered lucky if he was carried all the way to Siberia. At Tikigaq, a lost hunter's family always held out hope until the following winter, when the men sometimes returned over the frozen sea.

There seems to be no other explanation for Tikigaq's size, its permanence, and its location at the tip of a barren spit than the fact that the people who lived here were a sea people. They were also one of the few communities that practiced the most venerated occupation in Eskimo culture: They were whale hunters.

Before the white man's presence began significantly to influence the Tikirarmiut, Eskimo whale hunting had changed little over the centuries. In late March or early April, when the gulls and snow buntings appeared, whaling crews lashed their skin boats to sleds and headed miles out on the shore-fast ice until they encountered open water. Still farther out to sea was the floating ice pack. So long as an offshore wind blew, the open lead between the shore ice and the ice pack was maintained. Here at the edge of the lead the men sat day and night watching for migrating bowhead whales. They did not allow themselves sleeping bags; their only protection besides their clothes was a windbreak made of snow blocks. For days or weeks they watched, eating frozen meat and drinking water from a seal-flipper flask worn inside their parkas. They slept only intermittently.

At home the women observed prescribed rituals. Froelich Rainey, an archaeologist who spent the winter of 1940 at Tikigaq, noted that they refrained from any work during the whale hunt. The women did not wash or comb their hair, change their clothes, or scrub the floor. They believed that if they scrubbed the floor, the skin of the whale would be thin. A woman was to remain tranquil and "act like a sick person" so that the harpooned whale would be calm and easy to kill. Nor could women ever use knives—someone would even cut their food for them—because they might sever the harpoon line attached to the whale.

When a spout of water appeared, the men jumped for their skin boats, knowing the whale was offering itself. Each boat had six men at the paddles, a helmsman in the stern, and a harpooner in the bow. Paddling silently, they made for a spot where they thought the whale would surface. When they saw the broad back rise and a blast of vapor shoot into the air, they paddled rapidly toward it, the harpooner positioning himself in the bow. He threw with all his might and the stone-tipped point sliced through the tough, black skin into soft blubber. As the shaft fell away and the line tightened, the point pivoted like a toggle, spanning the entrance hole on the underside of the whale's skin. The walrus-hide line playing out from the bow of the boat had a series of inflated sealskins attached, each with a buoyancy sufficient to lift 200 to 300 pounds. Their drag tired the whale as it dived. Again, the crew paddled hard to follow the whale as other crews joined in the chase. When the whale was finally too exhausted to dive, the helmsman took the boat right up to its flukes and, working from the bow, the best lance man attempted to cut the tendon controlling the tail to hamstring the whale and prevent it from sounding again. Charles Brower, a twenty-one-year-old sailor who participated in a whale hunt at Tikigaq in 1885, describes what came next:

> Then the lance expert cut loose with all his skill. A stroke here, another there. Lightning speed! Deadly precision! Still the whale wouldn't die. The Eskimos had one last trick. Shoving the oomiak [the skin boat] along until part of it lay fairly across the whale's back put our expert in position to lance a large artery similar to the jugular vein in a man. Poised for a mighty effort, he jabbed deeply at this vital spot and a great fountain of blood spurted high and crimsoned the water all about. We backed away in a hurry. A dangerous moment! For in dying, the animal went into his death flurry with such a thrashing of fins and flukes that any boat would have been stove in and its crew injured or drowned.

If a boat capsized, a crew could die in minutes in the icy water. But if all went well, the jubilant crew sang traditional songs as they towed their prize to the ice edge nearest the village. There, thirty or forty men wielding flint knives with handles ten feet long worked through two days and nights butchering the carcass. Women and children likewise worked continuously, hauling the meat back to the village's underground caches.

The dances and feasting that followed featured the most prized food in the Tikirarmiut's diet: *muktuk,* the whale's black, three-quarter-inch-thick skin together with an inch or so of the adjacent blubber. (A non-Eskimo might be tempted to compare this chewy delicacy to a chunk of automobile tire bonded to a layer of greasy fat.) The meat of the whale would feed people and dogs throughout the year; the blubber would be rendered for food and fuel. The bone and baleen were also valuable raw materials in the Eskimo technology. Bone was used for house frames, cache and grave scaffolds, and tools; baleen produced nets, line, and baskets. And the whale hunt itself provided the framework within which the people of Tikigaq attained social stature, material wealth, and spiritual virtue.

●   ●   ●

A "forest of stakes . . . several round hillocks." For sixty years after Tikigaq's discovery, incurious white men added nothing to this prosaic summation. They did not know that they had visited what was probably the most ancient village site in the Arctic, one of the great capitals of the aboriginal North. "We now realize that Tikigaq was one of the largest settlements in the entire Eskimo-speaking world," writes anthropologist Ernest Burch. "Both culturally and socially its significance was thoroughly underestimated by the first European observers." During the 1800s, the population was probably 600 to 700 people in the winter, but the village was virtually abandoned in the summer when the white men visited. By then the people had moved off, traveling in skin boats and hauling tepee-like tents. They hunted caribou in the interior hills, fished along the Kukpuk River, and collected the eggs of migratory seabirds from the cliffs at Cape Thompson.

Tikigaq was an Arctic Machu Picchu, one that was still thriving, a fact that would not be understood until 1939 when archaeologists Louis Giddings and Helge Larsen visited the spit to excavate the Old Tikigaq site, adjacent to the modern village. When Giddings and Larsen arrived, they found that the local people had scavenged the area for artifacts to sell to the crews of government ships. "The mound," said a disheartened Giddings, "looked like it had been bombed repeatedly from the air." With their assistants, they explored other possibilities on the spit. One midnight, with shovels slung over their shoulders and anticipating their dinner of whale meat, Giddings and Larsen began their long trudge home. Their route rose and fell as they crossed the gentle undulations of ancient beach ridges. Stopping at one of these corrugations to rest and to look at the red midnight sun, they noticed something odd. The low-angle rays lit up in relief a network of slight depressions along the entire length of the ridge top. "Frost scars," explained Larsen. Such disturbances were common in the Arctic. But then they saw the same phenomenon on the next beach ridge. And there was something else. The outlines appeared to be square. The polygonal deformations formed by frost action always took irregular shapes. These, by contrast, had to be the work of man. "But no," Giddings wrote in a memoir, ". . . these could not be man-made because no village in the Arctic could have been so large. The markings appeared again on the third ridge and on the fourth. . . . I made a suggestion: I had nothing to lose, I said, by digging a hole in one of the squares to see what it might contain—if, that is, we could find them again when the sun shone high in the sky. Larsen thought a moment, then said, 'Let's all come and try.'"

What they found on the old beach ridges not far from the village and the Old Tikigaq site was "by far the most extensive and complete one-period site yet discovered and described in the entire circumpolar region." Ipiutak, as they called the ancient settlement, contained 600 or 700 houses. It revealed that occupation of the spit had been essentially uninterrupted for millennia. Tikigaq, today called Point Hope, is one of the oldest continuously occupied sites on the North American continent. It not only predates Plymouth Rock, it may predate the pyramids.

A peculiarity of the ocean currents is eliminating our chances of ever knowing just how old the village is. Currents that formed the peninsula seem now to have reversed themselves and are eroding the spit, which used to extend several miles farther out to sea and hook to the north. The most ancient sites, the ones farthest west, are gone, and successively more recent eastern sites are disappearing. Today, 99 percent of the Tikigaq people's archaeological heritage is under the sea.

•   •   •

If the artifacts of the Tikirarmiut's culture were washing away, the people nonetheless sustained a sense of themselves as a cohesive society, steeped in and sanctified by the traditions of forebears extending back thousands of years. They were able to incorporate those aspects of Western culture most compatible with this inheritance. They readily took to a plan introduced by the Episcopal missionaries in the 1890s that allowed for democratic participation in church government. As early as 1920, Point Hope had a village council, again established with the help of an Episcopal clergyman. The system was not precisely a model of representative democracy—the whaling captains directed how their extended families and crew voted in elections—but it did represent a modern structure for self-determination in local affairs. By 1940, the U.S. government had formally ratified Point Hope's constitution, authorizing it to form a legal community government within the federally administered territory.

Nowhere on the Arctic coast of Alaska was this tradition of cohesiveness and sovereignty more deeply incised than at Point Hope. And, it was this fact, rooted in a feeling, as one writer puts it, "of belonging inalienably where they were," that would save the Tikirarmiut in their brush with extinction in the atomic age.

2   THE FIRECRACKER BOYS

> It is not too much to expect that our
> children will enjoy in their homes
> electrical energy too cheap to meter.
>
> —Lewis Strauss,
> Chairman, U.S. Atomic Energy
> Commission

In the 1940s and 1950s, life at Point Hope went on pretty much as it always had. Nature—the cycles of light and wind and wave, the habits of animals, the character of the landform—didn't so much impinge on human experience as merge with it. But on another frontier, a technological one, change was occurring at an almost unbelievable speed. In radical contrast to the worldview of the Eskimo, who understood so well the matter of nature, the oracles of the atomic age were busily exploring the nature of matter and imagining man's unfettered control over nature.

Following the discovery in the late 1930s of nuclear fission (the splitting of an atom) and its release of enormous amounts of energy, scientists and observers of science declared the dawning of a golden age. Writing in *Collier's* magazine in 1940, physicist R. M. Langer of the California Institute of Technology described an atomic garden of Eden where people scooted about parklike landscapes in uranium-powered cars and airplanes. The abundant energy, "so cheap it isn't worth making a charge for it," would allow the countryside to remain uncluttered as daily activities—even agriculture—could take place underground. And while offering "unparalleled richness and opportunities for all," the advantages would not be purely material. Class distinctions would evaporate, and social ills would become "relics" as everyone enjoyed the good life. "War itself will become obsolete" because economic stresses would disappear. And this was not, said Langer,

> ... a promise of Utopia centuries away. It is a statement of facts that will profoundly change for the better the daily lives of you and yours. . . . This is not visionary. The foundations of the happy era have already been laid. . . . Reality is about to be handed from the scientists in their laboratories to the engineers in their factories to application to your daily life. It is a new form of power—atomic power.

Within a year of Langer's epiphany, the world was at war and it was the possibility of a nuclear bomb, not of a nuclear Shangri-la, that roused men's imaginations. In 1942, President Roosevelt, in a race to beat Germany in the development of the atom bomb, moved to establish the Manhattan Engineering District. One hundred and fifty thousand people went to work in facilities all over the country. Three new laboratories—secret cities, really—were built at Oak Ridge, Tennessee; Hanford, Washington; and Los Alamos, New Mexico. Oak Ridge enriched naturally occurring uranium to weapons grade, Hanford made plutonium, and Los Alamos designed and built "the gadget" itself.

These "atomic cities" bore little resemblance to Professor Langer's promised land. They were company towns where the government owned all the land and all the buildings. No one was permitted to live there, or even visit, without a government pass. Scientists accustomed to free intellectual exchange faced "compartmentalization" of information and military secrecy. Only a tiny percent of the workforce even knew that they were working on the atomic bomb. Until future Nobel Prize–winner Richard Feynman was permitted to tell them, technicians at the Los Alamos computing center had no idea of the purpose behind the laborious calculations their machines performed. To many, security regulations were maddeningly excessive. Security officers read and censored outgoing mail, bugged homes and offices, and monitored telephone calls, interrupting any conversation that headed off course. When they got a rare day off to go up to Santa Fe, scientists noticed that they were tailed by plainclothes intelligence agents. Hotel porters turned out to be counterespionage informants, and bartenders, FBI agents. One drollery had it that because physicist Henry D. Smyth was in charge of two departments simultaneously, he required official permission to talk to himself. Thus, the emerging nuclear culture was pervaded by secrecy, insularity, and a sense of exclusive ownership of the truth—a value system profoundly at odds with the openness of Eskimo society.

Working night and day under the direction of forty-year-old J. Robert Oppenheimer, the Manhattan Project scientists and engineers at Los Alamos achieved their goal on July 16, 1945, at 5:25:49 A.M. At that moment, in a desert called Jornada del Muerto (Journey of Death), while the people of nearby Alamogordo slept, the predawn darkness was ripped by a flash of light bright enough to be seen from another planet. Physicist I. I. Rabi was there, lying in the sand, shielding his eyes.

> Suddenly, there was an enormous flash of light, the brightest light I have ever seen or that I think anyone has ever seen. It blasted; it pounced; it bored its way right through you. It was a vision which was seen with more than the eye. It was seen to last forever. You would wish it would stop; altogether it lasted about two seconds. Finally it was over, diminishing, and we looked toward the place where the bomb had been; there was an enormous ball of fire which grew; it went up into the air, in yellow flashes and into scarlet and green. It looked menacing. It seemed to come toward one.

A millisecond after detonation atop a 100-foot tower, the ball of fire reached the ground with the pressure of 100 billion atmospheres, pounding out a crater nearly a quarter mile wide. The fire mass grew symmetrically, like an inverted bowl, pushing out in front of it along the ground a surge of billowing dust. At the center of the rapidly expanding inferno, temperatures reached 10,000 times that of the surface of the sun. It ascended into the darkening sky, rolling and boiling into convolutions like a fiery brain or a bloodred eye. From a base of churning black clouds and exploding gases, thousands of tons of radioactive earth were sucked up into a convection stem that trailed the fireball into the sky. At 15,000 feet, hitting warmer layers of air, its head mashed and stretched into a luminous canopy. With its stem rising eight miles above the desert floor, the flattening lobe spanned a mile and created a visual image that would thereafter become a symbol as deep-seated in the human psyche as the cross: the mushroom cloud.

Most who witnessed the world's first atomic explosion, code-named "Trinity," suffered pangs of doubt and spiritual compunction, "that we puny things were blasphemous to tamper with the forces heretofore reserved to the Almighty," as Gen. James Farrell wrote in his report. Oppenheimer recalled a line from the *Bhagavadgita,* "I am become death, the Shatterer of Worlds." More prosaically, test director Kenneth Bainbridge congratulated Oppenheimer with a handshake and the words: "Now we are all sons of bitches." But Edward Teller (later to be called "Father of the H-bomb") was sanguine: "I was looking, contrary to regulations, straight at the bomb. I put on welding glasses, suntan lotion, and gloves. I looked the beast in the eye, and I was impressed."

●   ●   ●

By 1944, many men from Point Hope were serving in the armed forces throughout the world, defending, they felt, the values of their great nation. That same year Gen. George Patton had taken Strasbourg from the retreating Germans. Among the first to enter the fallen city was Samuel Gouldsmit, a physicist whose specialty was sleuthing the status of German atomic research. Sifting through papers at the University of Strasbourg late into the night, Gouldsmit found a bundle of papers that proved conclusively that the German A-bomb project, generally assumed to be ahead of the Allied quest, was at least two years behind it. The Nazis had not yet built plants to produce U-235, or plutonium, the material needed for a chain reaction. Sharing his excitement and elation with a military officer who acted as liaison between the War Department and the atomic physicists, Gouldsmit remarked, "Isn't it wonderful that the Germans have no bomb? Now we won't have to use ours." The career soldier's reply finally awakened in Gouldsmit an understanding of how fundamentally antagonistic were the aims of science and the aims of war: "Of course you understand, Sam, that if we have such a weapon we are going to use it."

Though Germany was crumbling and the Japanese were in no position to build an atomic weapon, many scientists began to realize that the bomb was likely to be deployed anyway. The military was disinclined to use the atomic bomb merely as a threat that might bring the Japanese to their senses. Nor did it care to drop a bomb in an unpopulated area as a demonstration, even if such a display seemed likely to result in capitulation. Nor would it be reserved for a strictly military target. When the decision was made to drop the bomb on a city, the option of issuing a prior warning to spare wholesale loss of civilian life was rejected. The brilliant Hungarian physicist Leo Szilard, who first suggested American atomic bomb research in 1939, addressed this worry some years later:

> During 1943 and part of 1944 our greatest worry was the possibility that Germany would perfect an atomic bomb before the invasion of Europe. . . . In 1945, when we ceased worrying about what the Germans might do to us, we began to worry about what the government of the United States might do to other countries.

Many writers and scholars argue that Japan's defeat was imminent. Having deciphered the primary Japanese code, the Americans had known since Trinity that Japan sought a negotiated end to the war. Perhaps American policy makers, who had pressed Russia for assistance in defeating the Japanese, saw a way to end the war quickly without opening up Japan to the expansionist Soviets. Perhaps by 1945 the unremitting effort to deploy the bomb had simply reached a degree of "bureaucratic momentum" that it could not be reined in. Or perhaps military planners and politicians were aware of the public relations value of the bomb's use. After all, how might the American people react if they learned of a $2 billion crash program to build an atomic bomb and the war ended without its use? By contrast, if the bomb could be employed as the apparent cause of a swift surrender, wouldn't a "battle of the laboratories" narrative better serve the military and political leadership?

Just three weeks after Trinity, on August 6, 1945, two lone B-29s flew over the Japanese city of Hiroshima. One plane carried the bomb, the other, instruments to measure the destruction. Not suspecting an air raid from only two planes, the people below did not move into the bomb shelters. At 8:16 A.M. a Los Alamos-built uranium bomb nicknamed "Little Boy" exploded 2,000 feet above the city. No one knows how many died outright in the firestorm, perhaps 70,000, with a like number left dying of injuries and severe radiation poisoning. Five years later, deaths from the Hiroshima bomb were estimated to have reached 200,000.

Seventy-five hours later, barely time for the Japanese government in Tokyo to confirm that the destruction had resulted from a single bomb—a nuclear bomb—and to digest that fact, Nagasaki was destroyed by an even larger blast. Two days later Japan surrendered. The plutonium bomb was a twin to the Trinity device, and likewise had a yield equivalent to 21,000 tons of TNT. The death toll at Nagasaki reached 70,000 by the end of 1945, and 140,000 within five years.

The magnitude of the technical accomplishment astounded the world. Harry Truman, who had been sworn in following President Roosevelt's death less than four months before Hiroshima, praised the participants in the Manhattan Project for having accomplished "the greatest achievement of organized science in history." The American people were fascinated with details of the secret project that had so spectacularly ended the war, and the press abetted the frenzy with a flood of sensationalistic ink. At the same time, however, the atom became the focus of serious political discussions. When Truman announced sixteen hours after Hiroshima the world's first use of nuclear weapons in warfare, he also told the American people that he would ask Congress to "consider promptly the establishment of an appropriate commission to control the production and use of atomic power." The administration's recommendations emerged in the form of the May-Johnson bill, which held that the central concern of the new commission should be weapons production, and that control should reside mainly with the military.

But the scientific community became alarmed at the prospect of military censorship of physics. The public, too, was tired of military authority and opposed the extension of wartime powers into peacetime governance. When a public outcry accompanied the War Department's attempt to rush the bill through Congress, Truman withdrew his support. Sen. Brien McMahon introduced a substitute that offered two principal concessions. First, it showed some deference to the civilian purposes to which atomic energy might be applied, and second, it stipulated that the five-member Atomic Energy Commission (AEC) it proposed to establish would be composed entirely of civilians.

Still, the bill set up a government monopoly on the ownership of fissionable materials and introduced new classification and secrecy regulations, the violation of which could carry the death penalty. Notwithstanding any clash with traditional notions of American free enterprise—and with First Amendment freedoms—the compromise was successful and Truman signed into law the Atomic Energy Act of 1946.

As a practical matter, though, civilian control of the commission proved illusory. The statute had organized the AEC into divisions, one of which was the Division of Military Applications (DMA), to be headed by a member of the military. During its first fifteen years, DMA dominated the commission's activities, commanding fully 70 percent of the AEC's expenditures. And, though the law required commissioners to be civilians, Lewis Strauss, the powerful third chairman of the AEC, was a reserve admiral in the navy, and three of the first five general managers of the AEC were military men who simply stepped down from active duty to accept the job. The act permitted the military to control the AEC and to keep nuclear weapons development the agency's first priority for the next thirty years.

In 1954, the Atomic Energy Act was revised in a way that fostered commercial development of atomic energy while giving the military full control of the country's weapons policy. Furthermore, the Department of Defense was given authority to prohibit access to data that might have permitted scrutiny of such policy. Conspicuously absent from either the original bill or the revision was any evidence of concern for public health and safety. No safety standards were set for power plants, and the health effects of atmospheric testing of nuclear weapons were ignored completely.

•   •   •

After the war, most of the atomic scientists drifted back to their universities. But Edward Teller pressed for a continuation of the national effort to develop nuclear bombs. Even before the scientists at Los Alamos had built the world's first atomic bomb, Teller saw its construction as inevitable and insistently pursued the fusion, or hydrogen bomb, "the Super" as it was called then. Cornell University's great physicist, Hans Bethe, Teller's long-time friend, was, as leader of the Manhattan Project's theoretical physics division, also Teller's boss. Bethe, who had also fled Germany when Hitler came to power, and who would later win a Nobel Prize for working out the nuclear fusion process that occurs in stars, remembers the difficulty he encountered in trying to get Teller to focus on the job at hand:

> I hoped to rely very heavily on him to help our work in theoretical physics. It turned out that he did not want to cooperate. He did not want to work on the agreed line of research. . . . He always suggested new things, new deviations. So that in the end there was no choice but to relieve him of any work in the general line of the development of Los Alamos, and to permit him to pursue his own ideas entirely unrelated to the World War II work.

Teller's lack of cooperation may have stemmed, in part, from his particular type of creative genius. Friends referred to his intense maverick streak, his disinclination toward routine, method, and discipline: "What Edward can't carry in his head and solve in his head, he doesn't want to bother with." He liked to be working on concepts that were ten years ahead of their time, not grinding out proofs to problems he regarded as solved, theoretically. He worked in a scattershot way, producing great numbers of ideas, many of which he himself would quickly pronounce idiotic. As Bethe put it, "nine out of ten of Teller's ideas are useless. He needs men with more judgment, even if they be less gifted, to select the tenth idea which is often a stroke of genius."

Teller's upbringing may also have had something to do with his monomania about the H-bomb and his emergence as one of the nation's most hawkish scientists. Born into a family of prosperous, assimilated Jews in 1908 in Budapest, Hungary, Teller felt the effects of political upheavals in the aftermath of World War I, and as an impressionable adolescent, saw the breakdown of the social order in his homeland. When he was ten, the Austro-Hungarian monarchy fell to a popular revolution that demanded such democratic reforms as freedom of the press and women's suffrage. This new Republic of Hungary lasted just four months and then, oddly, transmogrified into a repressive communist regime under Béla Kun. The communists nationalized the economy, suspended domestic law, cracked down on dissent, and even appropriated a portion of the Tellers' apartment for the billeting of two soldiers. Violence broke out in the streets of Budapest as a counterrevolutionary movement rose, and Teller, if he didn't see them, at least heard about the corpses of "traitors" hanging from lampposts. When the communists were ousted, a violent fascist government took power and the killings multiplied tenfold.

In 1926, Teller's parents sent him off to school in Germany, where the arts and sciences were flourishing. At Leipzig, he earned a Ph.D. in theoretical physics under Werner Heisenberg, one of the pioneers of quantum mechanics. But again the rise of a repressive political regime intervened. With respect to political awareness, however, the young Teller was not precocious. He recalls that he remained apathetic even in 1933 when the Nazis seized power: "In day-to-day politics I still was hardly interested. Except that I wanted to continue my work." Worried that anti-Semitism might impede that work, twenty-six-year-old Edward Teller left Germany in 1934 and made his way to America.

If science was Teller's refuge—providing stability and order, as well as bringing acclaim—it was also a weapon to combat communism. In 1939, Teller accompanied fellow Hungarian expatriate, Leo Szilard, on a historic visit to Albert Einstein. Szilard persuaded Einstein to sign the letter to President Roosevelt that resulted in the epochal decision to build the atomic bomb. When Teller made up his mind to join the Manhattan Project and work on bombs, he entered the scientific and political inner circle of nuclear weapons development. He would remain there for more than half a century.

●   ●   ●

The nucleus of U-235, an unwieldy glob of 92 protons and 143 neutrons, can barely hold itself together. If it absorbs one more neutron it will shudder wildly for a millionth of a millionth of a second, then burst apart into two nuclei with an appreciable release of energy: nuclear fission. Along with the energy release, the original nucleus will also let go some of its 143 neutrons. These shoot off, colliding with and being absorbed by other uranium nuclei, which also shudder, split, and release energy and more neutrons. Because fission is initiated by neutrons and is responsible for the release of neutrons, the process may sustain itself, like a fire, so long as fuel is supplied. Each fission releases the binding energy that had held the atom together, and the explosive chain reaction will not stop until a great deal of energy has been released.

The atom bomb consisted of a nut-sized neutron source (the initiator), surrounded with a quantity of enriched uranium (the core), enclosed in another layer of naturally occurring uranium (the tamper), and all jacketed by blocks of ordinary high explosive (TNT, essentially). These blocks were shaped like pyramids with the points cut off and arranged side by side with the tapered ends inward in a sort of spherical arch. Finally, electronic detonators were installed around the outside of the sphere atop each lens. When the detonators fired, and the TNT ignited, most of the energy went outward, but some went inward in the form of shock waves. As the explosive reaction progressed inward from the bomb's outer edge, the inward-moving shock waves moved in concave shapes. In two dimensions, it would look like the ripples from many pebbles dropped simultaneously in a circle around a target floating on the surface of a pond. The bomb makers were able to improve the shape of this implosion by imbedding within each lens a bubble of slow-burning explosive. This had the effect of retarding the leading edge of the concave wave until it became convex. Now the inward-moving shock waves looked like the ripples from one pebble shown in retrograde motion, converging on a point. And this symmetrical implosion compressed the bomb's innards.

The squeezed initiator released millions of neutrons into the core, which was itself compressed to a "critical mass" where fission was more likely. The tamper of dense metal served to reflect the escaping neutrons back into the core to ensure that some of them started the chain reaction. The tamper also held back the explosion for an instant, making it that much more powerful.

If a conventional explosive such as TNT could serve as the trigger for an A-bomb, what sort of bomb might be possible if an *atomic* explosion served as a trigger? Teller and others thought such a wallop might so compress and so heat a core material that atoms as light as those of hydrogen might fuse together. And this fusion bomb, or hydrogen bomb, might release 1,000 times more energy than the one dropped at Hiroshima. Many people had worked on the H-bomb idea at Los Alamos—even before Los Alamos—but for Teller the quest was all-consuming. After the war, but before a sound theoretical basis existed upon which to direct develop-ment, Teller urged that Los Alamos be placed essentially at his disposal to pursue a crash program to build the H-bomb. He continued to pressure laboratory officials even after the calculations upon which his proposed model was based were shown to be wrong.

Eventually, a new method, unrelated to Teller's earlier model, oc-curred to mathematician Stanislaw Ulam, and Teller quickly refined and extended it. The laboratory moved to direct research on the new ther-monuclear concept. But it didn't move fast enough for Teller. As he was known to do during the Manhattan Project, the maverick physicist agi-tated for sweeping changes: The people assigned to the work were not competent; the laboratory needed complete reorganization; in fact, a new laboratory should be set up to do the work that he was suggesting.

Teller quit Los Alamos and began lobbying vigorously and successfully at the Pentagon for a new laboratory, which he said was essential to the national defense. When the air force began to show interest in building the lab, the AEC stepped in to protect its monopoly, designating the site of the University of California accelerator planned for Livermore, California (fifty miles east of Berkeley), as the newest of AEC's national laboratories. The University of California Radiation Laboratory opened in July of 1952 under the leadership of Teller, Berkeley physicist Ernest O. Lawrence, and Lawrence's recent graduate student Herbert F. York. (In 1958, upon Lawrence's death, it was renamed Lawrence Radiation Laboratory [LRL]. In the 1970s, during the period of antiwar protests, the AEC wanted the word *radiation* removed from the name, and the complex became known as Lawrence Livermore National Laboratory, or LLNL.)

For the first year, according to York, Teller alone was given personal veto authority over all elements of the scientific program. Needless to say, the development of thermonuclear weapons became the laboratory's paramount priority. Meanwhile, as Livermore organized, the Los Alamos laboratory went on to test the prototype H-bomb successfully in October of 1952.

•   •   •

After World War II, with nuclear bombs piling up in U.S. arsenals, atomic weaponeers joined the savants and futurists in imagining peacetime uses for the apocalyptic technology. The mainstream media gave sensationalistic play to the stories, which were eagerly consumed by an atom-crazed public. Many of the speculations were fruitful, such as the suggested use of radioisotopes in medicine, but often they were championed without an equally spirited investigation of possible environmental or health and safety repercussions. At times, the suggestions were wildly fantastic and arguably crazy. William L. Laurence, *The New York Times* science writer who had left his job to become a consultant with the Manhattan Project, wrote in 1948 of "a chance to enter a new Eden." Laurence did not foresee nuclear technologies as creating environmental contamination or mass apprehension and foreboding. On the contrary, he anticipated the opposite effect, seeing them as "abolishing disease and poverty, anxiety and fear." With any luck, man would "learn to control weather and heredity," he wrote, even "find the key to the riddle of old age."

"Heat will be so plentiful that it will even be used to melt snow as it falls," predicted the chancellor of the University of Chicago, adding that "utilities will be so cheap that their cost can hardly be reckoned." The science editor of the Scripps-Howard newspapers confidently declared, "The day is gone when nations will fight for oil."

Atomic airplanes seemed a natural extension of the new technology. In 1948, the AEC put forty scientists and engineers at the Massachusetts Institute of Technology to work on a project to develop nuclear-powered aircraft. The planes were expected to be capable of circling the globe several times without refueling. *Flying* magazine speculated that a nuclear airplane "will revolutionize flight with its unlimited range and endurance. On less than one pound of uranium, an A-plane will be able to fly 100,000 miles. In the future it can eliminate our need for a worldwide chain of military air bases." Thirteen years and $1 billion later, the program was scrapped as military officials admitted the plane would necessarily spew radioactive fission products across the skies. No aircraft reactor was ever built, and by the 1960s the project was considered one of the AEC's "most notable failures."

In the 1950s and 1960s, atomic rockets were seen as a way to close the gap with the Soviets in space technology. But, as the journal *Nucleonics* complained, neither the army nor the navy had yet requested such a program. Reasoning backwards, the editor advised that "an easier flow of development dollars" would follow if "a clear cut military requirement" could be demonstrated. Research did begin on the rocket reactor, and development dollars eventually flowed at the rate of $100 million a year.

In the late 1950s, the Advanced Research Projects Agency put $1 million into Project Orion, which imagined a unique spacecraft propulsion system whereby the vehicle would dispense and detonate nuclear bombs then ride out the shock wave like an interplanetary surfer. Orion designers expected to visit Saturn in a vehicle as tall as a sixteen-story building and 135 feet in diameter. Its most prominent feature would be a gigantic pusher plate to receive the shock of blasts up to twenty kilotons. The spaceship would store 2,000 nuclear bombs—NASA called the explosives "pulse units," the air force called them "charge propellant systems." One at a time as needed, the bombs would be ejected, traveling down a chute and, with a blast of compressed gas, out a hole at the bottom of the ship. At a distance as close as a few hundred feet, the bomb would detonate.

Because the mechanical process of dispensing the bombs reminded the scientists of a soft-drink vending machine, they consulted with Coca-Cola Company on the design of Coke machines. Eventually, even the interplanetary nuclear Coke machine struck Orion scientists as insufficiently imaginative. One afternoon, for reasons not entirely clear, they calculated the weight of Chicago, and what it would take to fly it through space.

*   *   *

In his famous "Atoms for Peace" speech before the United Nations in 1953, President Eisenhower declared that "this greatest of destructive forces can be developed into a great boon for the benefit of all mankind." He advocated the creation of an international "atomic pool" into which all the nuclear nations would deposit fissionable material. Any nation could then draw out quantities of the substances to be used for peaceful purposes. The pool would make the "deserts flourish." In fact, it would account for a string of benefactions. Putting a nuclear spin on the corporal works of mercy, promoters claimed the split atom would help "to warm the cold, to feed the hungry, to alleviate the misery of the world." A massive public relations blitz launched by the White House spread the good news. Two hundred thousand copies of the speech were printed in ten languages. The American press responded as requested with headlines like FORESTRY EXPERT PREDICTS ATOMIC RAYS WILL CUT LUMBER INSTEAD OF SAWS and ATOMIC LOCOMO-

TIVE DESIGNED. Unfortunately, other nations were disinclined to deposit fissionable material, and the vision of an atomic pool dissolved.

Forging ahead with the policy alone, the United States initiated nuclear power research projects in Japan, Brazil, Venezuela, Spain, Italy, and elsewhere. The program was a public relations success with the American people. But, in its zeal to promote the dream of an idyllic nuclear future, Atoms for Peace boosters dispersed hardware, know-how, and the rare fissionable material that could be used to make nuclear bombs to developing countries throughout the world. In 1974, India detonated an atomic bomb using plutonium produced in a reactor obtained from Canada under the Atoms for Peace program. Besides India and the original five nuclear powers (the United States, Russia, Britain, France, and China), other nations thought, or known, to have nuclear weapons, or the wherewithal to make them, now include Israel, Belarus, Kazakhstan, Ukraine, India, Pakistan, North Korea, Japan, South Africa, Germany, Argentina, Brazil, Czechoslovakia, Belgium, Canada, Sweden, Italy, Iran, the Netherlands, and Australia.

David Lilienthal, the first chairman of the AEC and former head of the Tennessee Valley Authority, is one of the few central figures in the early AEC who has written of these times with critical self-examination.

> I had a hand in formulating and popularizing that hope of peaceful potentials . . . . The basic cause, I think, was a conviction, and one that I shared fully, and tried to inculcate in others, that somehow or other the discovery that had produced so terrible a weapon simply *had* to have an important peaceful use.

If Lilienthal and his successors at the AEC were attempting to atone for the new technology by promoting nuclear *power*, Edward Teller and his followers at the new AEC laboratory at Livermore were even more ambitious. They sought to redeem the nuclear bomb itself.

●   ●   ●

In the fall of 1956, following the withdrawal of British troops from Egypt, President Nasser nationalized the Suez Canal. That action prompted

Britain and France to launch an attack on Egypt to try to retake the strategic waterway. Before fighting in the "Suez Crisis" was halted by a UN cease-fire, sunken ships lay strewn across the canal and the transportation link was closed to all traffic for months.

Theoretical physicists at Livermore found the crisis thought-provoking. The north end of the Red Sea forks as it nears the Mediterranean. The western branch is the Gulf of Suez, where the canal was cut; to the east is the Gulf of Aqaba. Suppose a new canal were cut from the Gulf of Aqaba to the Mediterranean through Israeli territory. Gerald Johnson was a young physicist at Livermore at the time:

> We looked at the topography of Israel, and what we thought we could do with nuclear explosions, and concluded we could dig a sea level canal—from a purely technical point of view—all the way across Israel, from the Gulf of Aqaba to the Mediterranean. Of course, we never talked to the Israelis about this . . . it was just an exercise for the student at the blackboard . . . it looked like a feasible thing at least to think about.

What they were thinking about was burying a string of nuclear bombs from gulf to sea and touching them off simultaneously to excavate an enormous ditch. The Suez Crisis faded, but the idea of the peaceful use of nuclear explosions had taken hold at Livermore.

The idea may have had some earlier currency as well. In 1949, the year the Soviets developed the atom bomb, that country announced that they were employing nuclear explosions for economic development projects: "We are razing mountains. We are irrigating the deserts. We are cutting through the jungle and the tundra." Edward Teller had heard the claim, and whether he ascribed the boast to propaganda or took it seriously, he certainly shared with the Russians a vision of the future where nuclear explosions played a peaceful role: "I wish that this statement described our program," he lamented in 1962.

In any event, the seed planted by the Suez Crisis flourished in the fertile Livermore valley. Predictably, the next thing to occur to the "the firecracker boys," as Alaskans would later call them, was the possibility of a sea-level canal in their own hemisphere, across the Central American isthmus. The bomb of choice would be thermonuclear. The fusion fuels (such as deuterium, which is found in water) are much cheaper and more abundant than fission fuels. And, for a given yield, the thermonuclear explosion produced less radiation than a fission one. The H-bomb did produce some radioactive fallout, of course, because a fission detonation served as the trigger for the thermonuclear reaction. But if Livermore could develop the so-called clean bomb—where the fission contribution of a thermonuclear explosion was kept small—then fallout could be reduced even more.

Shortly after the Suez Canal speculations, three Livermore scientists, Edward Teller, Harold Brown, and Gerald Johnson, wrote a letter to the AEC in Washington outlining a proposal to use nuclear explosions literally to reshape the earth. They sought funding for a new AEC program that, in alliance with private industry, would employ atomic blasts for a host of nonmilitary purposes.

Gerald Johnson remembers the startup plan for the innovative program:

> Our thought was that we would generate the technology, publish it, and then as industry or other agencies got interested in a project, we would . . . provid[e] the explosion and safety analysis and so on, which they would pay for.

The AEC gave a provisional nod to the idea, but it cautioned that work on peaceful applications was not to interfere with weapons development. In November of 1956, the AEC agreed to fund a classified conference at Livermore called the First Symposium on the Industrial Use of Nuclear Explosives. Livermore's Harold Brown was the organizer. Though still in his twenties, Brown had been head of Livermore's A-Division (thermonuclear weapons design) for four years. Twenty years later he would become secretary of defense under Jimmy Carter.

The secret meetings occurred from February 6 to 8, 1957, at Livermore. Participants came from the AEC's laboratories at Livermore and Los Alamos, and from the Rand Corporation, Aerojet-General Nucleonics, Princeton University, and Sandia Laboratory. Sandia was a private corporation created to do the ordnance engineering that transformed an explosive device into a deliverable weapon. Teller opened the proceedings with a few remarks, among them a characteristically brash proposal. It was the kind of suggestion AEC chairman James Schlesinger may have had in mind when he reviewed nuclear excavation some years later and admitted to having "some questions with regard to the environmental aesthetics of the program, if I can put it that way." Succinctly put, Teller proposed to nuke the moon:

> One will probably not long resist the temptation to shoot at the moon. The device might be set off relatively close to the moon and one would then look for the fluorescence coming off the lunar surface, or one might actually shoot right at the moon, try to observe what kind of disturbance it might cause.

Symposium participants went on to consider how nuclear explosions might facilitate power production, oil and mining procedures, even the feasibility of creating diamonds by smashing carbon. But by far, the bulk of the program dealt with "landscaping" projects, specifically the blasting of a new sea-level isthmian canal. The existing forty-four-mile Panama Canal required ships to pass through a series of locks in order to lift the vessels eighty-five feet over the highest elevation along the route. With a sea-level canal, ships would no longer have to queue up waiting for the locks to fill and empty; they could steam through nonstop in both directions. L. J. Vortman of Sandia, who presented the canal paper, predicted that the existing facility would be unable to handle traffic demands after 1961. (Actually, by 1971 the canal was operating at only 80 percent capacity, and that year the AEC projected that "by 1990 a new canal will be imperative.") Vortman reviewed several likely routes in Central America, including one through Mexico's Isthmus of Tehuantepec and another through Lake Nicaragua, one of the largest natural lakes in the world. If the latter route were technically possible, Vortman allowed, the resulting drainage of Lake Nicaragua would be "politically prohibitive." The more preferred alignments required, on average, twenty-six nuclear devices totaling 16.7 megatons. Later the estimates were revised upward: by itself, one alignment through Colombia would have required 262 nuclear bombs with an aggregate yield of 270.9 megatons. Radiation hazard was not considered an insurmountable problem; workers would use conventional protection measures, and local residents would be temporarily relocated.

As might be expected, the conference reported enthusiastically on the prospects for peaceful uses of nuclear explosions. Harold Brown's concluding remarks called for "a group of some number, such as 10 or 20 people, thinking about it for 6 months or a year, picking out the good ideas, and working them out in some detail." He also noted in the unclassified version of the conference proceedings that "there is some kind of public relations problem here." Apparently mystified by worldwide apprehension over atmospheric testing, Brown groused, "In the past 12 years all kinds of phobic public reactions have been built about nuclear bombs." Peaceful use of the explosions "could provide a fine opportunity for people to gain a more rational viewpoint," and he suggested that those in the AEC with public relations responsibilities take note. He may well have sharpened this point in internal, classified accounts of the symposium.

Project Plowshare was to be the name of the program. Harold Brown came up with it. He had been talking to the physicist I. I. Rabi about his current interest when Rabi quipped, "So you want to beat your old atomic bombs into plowshares?" The reference was to the well-known passage from the Bible: "And they shall beat their swords into plowshares, and their spears into pruning hooks; nation shall not lift up sword against nation, neither shall they learn war any more" (Isaiah 2:4). The Atomic Energy Commission approved the program to investigate nonmilitary uses of nuclear explosives on June 27, 1957. Still leery that the weapons program might suffer neglect, the AEC limited first-year Plowshare funding to $100,000. Though it was first and always a Livermore project, administratively Plowshare was part of the AEC's Division of Military Applications, which was headquartered in Washington.

AEC chairman Lewis Strauss was delighted to showcase the sunny side of nuclear technologies during a cold war arms race. In a striking revelation at an AEC meeting in 1958, he acknowledged that Plowshare's public works value was subordinate to its public relations value. Strauss stated explicitly what Harold Brown had only implied: Plowshare was intended to "highlight the peaceful application of nuclear explosive devices and thereby create a climate of world opinion that is more favorable to weapons development and tests."

•   •   •

Project Plowshare was formally established at Livermore in July of 1957. In the spring of 1958, Edward Teller became director of the laboratory, and Plowshare grew as he pressed the AEC for increased funding. By 1959, the program was receiving $3 million annually, and by 1960, $6 million. Funding had doubled again by 1964 to $12 million per year. At its height Plowshare employed 290 people and spent $18 million annually to recast the bomb as a peacetime tool. Physicists became public works engineers and set about to correct "a slightly flawed planet." "To remedy nature's oversights," detonations would gouge out canals and "instant harbors." They would slice passes through mountain ranges. Edward Teller spoke of "a new and important discipline: Geographical engineering. We will change the earth's surface to suit us." One ardent Plowshare partisan, military author Ralph Sanders, saw wholesale reconstruction of the landscape as man's birthright, and denounced naysayers who would try to "stem the surging tide of technical progress."

> Should one from our generation visit the earth a hundred or two hundred years from now, most likely he would find many of our up-to-now abiding geographical landmarks grossly altered. . . . In all probability, the transformation will be more radical than we think. New technologies (and nuclear explosives seem destined to be one of them) will enable men to remold many landscapes.

Nuclear explosions would dam straits to deflect ocean currents along more preferred routes. Rivers would be caused to flow upstream. Atmospheric bursts could modify climate and weather, either by seeding rain clouds or by blocking solar radiation with dust. Even earthquakes could be tamed by a series of small preemptive, underground shots along fault zones. To rescue the Arctic from its stigma as "the refrigerator of the world," atomic scientists would "simply blast the ice pack, greatly roughening its surface and increasing its absorption of solar radiation." The resultant rise in sea level, Plowshare enthusiasts had to concede, would be disastrous, especially for the residents of the world's great coastal cities.

If the Arctic struck nuclear landscapers as too icy, the planet's deserts were too dry. A classic suggestion attributed to Edward Teller involved closing the Strait of Gibraltar with a nuclear bomb. As the Mediterranean Sea gradually rose and freshened, he speculated, it could be used to irrigate the Sahara. Unfortunately, there would be a price to pay for the miracle of a verdant North Africa: Venice, for instance, and other sea-level cities would be lost to flooding.

On a more local scale, ice-locked harbors and channels might be cleared by underwater detonations that would churn up the water column, bringing warmer water to the surface, melting sea ice and opening shipping lanes. Mining would be facilitated by the fracturing of ore bodies, petroleum forced out of low-grade deposits, and trapped natural gas released. And, should the need arise, the earth could be defended against asteroids bearing down on it from outer space.

It was as though the liberation of the atom's energy had triggered a release of technological hubris as well. Plowshare scientists seemed like the "militant tinkerers" writer Wendell Berry spoke of, who "believe that the ability to do something is the reason to do it." Or as humanist author Konrad Lorenz has written, it became "a 'must' of the technocratic religion: everything that it is at all possible to do should by all possible means be done."

And the first thing to do, it seemed, was to excavate a new Panama Canal. Here the Livermore scientists were fortunate to have an important ally in Gen. Alfred Starbird, director of the AEC's Division of Military Applications. As an Army Corps of Engineers officer, improving the canal was of interest to Starbird. He suggested to Livermore's Gerald Johnson, who had been put in charge of the program, that the corps ought to be involved. As a result, a unit of army engineers was assigned to Livermore to learn the nuclear cratering business. "The Panama Canal Company interest, and the Corps of Engineers' interest," said Johnson, "is really what put the heavy focus on excavation. Because they, in a sense, were a potential customer. We had a target and a high level target." Along with Edward Teller, Johnson even briefed President Eisenhower and the cabinet on the potential for the nuclear-excavated canal and other Plowshare projects.

Plowshare planners knew from hydrogen bomb tests at Eniwetok Atoll in the South Pacific that the blasts left big holes. Some of the smaller bombs—ones that didn't vaporize the islands on which they sat—left substantial water-filled craters in the coral sand. Except for the lack of an entrance channel, they looked like harbors. But these tests had been surface detonations. And the physicists knew from their experience with nonnuclear explosives that to produce the biggest craters for a given size explosion they had to bury the bombs. Exact scaling laws for nuclear explosions, however, could only be guessed at without experimentation. Also, there was the question of slope stability. Cuts 1,000 feet deep were required along some of the possible canal alignments. Would the loose material along the slope of the cut be prone to slumping, especially in a wet environment? Would it fill the ditch back in, or make construction of facilities along the bank risky? And finally, what about radioactivity? Compared with a surface burst, a buried explosive mixed the radioactivity it produced with much more material. How fast and in what quantities or concentrations and with what effect would the radioactivity invade ecological cycles? Answers to these questions were not known as Livermore physicists began to consider where in the world to conduct the first Plowshare experiment.

Some supporters, like Edward Teller, felt strongly that "Plowshare should be demonstrated at home before it is exported to others." Because radiation hazard was one of the unknowns that the experiment was designed to address, what foreign country would want to host it? As Ralph Sanders summarized:

> It is naive to assume the Congolese or Guatemalans would permit nuclear earth moving in their countries before we prove out the technology here. Words alone will fail to persuade. The Panamanian, who may fear these nuclear explosions may kill or sterilize him, may justifiably criticize any attempt to use his country as a "test tube." One successful demonstration in the United States would induce acceptance where many words would fail.

Even if a foreign country was receptive to the idea, testing abroad would surely require an international agreement—an added complication. So the Plowshare scientists began to consider where in U.S. territory they could find a site that met their requirements. It should be remote and sparsely populated, a necessity given the radiation danger. And it should be coastal so the experiment could test the effects of water on the stability of crater slopes. "And that ended up with Alaska," said Gerald Johnson. "None of us had ever been in Alaska." Alaska was certainly remote and thinly populated. It had more coastline than all the rest of the United States put together. And, though not yet a state, Alaska was U.S.

territory. "We looked at the maps, you know, where the settlements were, what the topography was, and so on. And so it looked to us like any area from Nome to Point Barrow would be fine," remembers Johnson. "And then is when we decided, well, if we're going to do this, we'll produce a small harbor." It would be like the craters left behind on the Pacific atolls, except this time they would add an entrance channel.

For more information on Alaska, a delegation from Livermore was dispatched to the U.S. Geological Survey offices across the bay in Menlo Park, California. One of the scientists they spoke with at USGS in the spring of 1958 was David M. Hopkins. The men did not seem, according to Hopkins, "particularly conscious of the closeness of the Soviet Union," which at Bering Strait is only fifty-five miles away. He worried about the cold war implications of fallout drifting over that country and raised the issue with the Livermore physicists:

> They said, "Oh, well, that's very easy. That's no problem at all. We're going to blow this off when the wind is blowing the other way." In [1958] the ability to predict Alaskan weather was much poorer than it is now. It's only improved because of the existence of satellites. So I was sort of shocked at the casual way these guys dealt with some really heavy problems.

Hopkins had taken his geology Ph.D. at Harvard, and his labors in the trenches of the Alaska Branch at USGS would eventually result in two books that became the definitive formulation of the Bering land-bridge theory. The Livermore physicists impressed him as "bright young guys looking for a project. . . . They had a lot of money, it would be something fun to do, and they wanted to do it." He thought Livermore must be an "autonomous and irresponsible bureaucracy."

Once before, Hopkins had been asked to write a classified study identifying Alaska lands that had geological substrates analogous to those of Moscow and Leningrad. The purpose was to find places where atomic bomb tests might reveal the effect of a nuclear attack on the Soviet Union's two largest cities. Eventually Hopkins came to regard his participation in the assignment as "one of the great sins of my life." He was leery of military-oriented science, but the harbor scheme at least had some constructive value. Despite misgivings, he joined two colleagues and began a survey of the literature and to assemble aerial photographs and maps looking for geologic and oceanographic factors that might indicate a likely harbor site. Originally they were asked to look at the entire northwest coast of Alaska, which stretches 1,600 miles from Nome to Barrow. But a short time later, Livermore requested the USGS to limit their survey to a twenty-mile strip of coast between Cape Thompson and Cape Seppings. Most of the shoreline between those points, as Hopkins and his colleagues noted in their April 1958 report, consists of high bedrock ridges. Only three sites offered low-lying terrain where a harbor might be built. One of them is thirty-one miles southeast of Point Hope, in the vicinity of Cape Thompson, where a little creek the Eskimos call Ogotoruk enters the Chukchi Sea.

*   *   *

Four months after the USGS report was issued, a stocky young Eskimo whaling captain named Daniel Lisbourne crunched through loose gravel, lugging a ten-horse outboard motor down the beach at Point Hope to his *umiaq.* The boat's frame of expertly shaped and lashed driftwood was covered with the sun-bleached skins of six *ugruk,* or bearded seals. From this boat on an April day a few months earlier, Lisbourne and his crew had taken their first whale. But on this calm mid-August afternoon it was time to look for caribou. Daniel's young nephew and another hunter, Peniluke Omnik, had the *umiaq* loaded by the time he arrived: a tent, sleeping bags, a gasoline camp stove, food, and rifles.

Perhaps Lisbourne asked Peniluke if he thought they would see caribou at Ogotoruk Creek. If so, Peniluke probably made no answer, except maybe to lift the edge of his rain parka and fish out a cigarette. He may have looked over to the hills that rose twenty miles inland from the Point Hope spit, following them southeast along the skyline to the cliffs at Cape Thompson. Beyond the cliffs, out of sight, lay the Ogotoruk Creek valley. In every month of the year, since before memory, the Tikirarmiut took caribou at Ogotoruk. Using an inflated seal skin as a roller under the keel, the two men and the boy pushed the *umiaq* into the sea. Daniel yanked on the engine's starting cord and the light craft sped away toward Cape Thompson.

Two nights later the hunting party returned to Point Hope. At home Daniel passed on bits of news to his wife. Yesterday at Kisimilok they saw caribou, he said, but they were too far off. On their way home today they shot two caribou at Imikrak. Not far from there they saw six grizzly bears together moving along a ridge. And oh, yes, at Ogotoruk there were a bunch of tents and *nalaukmiut.* (*Nalaukmiut* is an Inupiaq term for white men and is derived from the word for a bleached white seal skin.) Jimmy Hawley from Kivalina was working with them; he said they were surveyors. Daniel Lisbourne, president of the Point Hope Village Council, wondered what in the world surveyors were doing at Ogotoruk.

"WE LOOKED AT THE WHOLE WORLD . . ."

[He] told how the navy had searched the world for a test site and had determined that Bikini was the best.

—Robert C. Kistte, on how the Marshall Islands' Military Governor convinced Bikinians of the need for weapons tests there

On July 14, 1958, Edward Teller, now director of the Lawrence Radiation Laboratory, and his entourage from Livermore landed unannounced in Juneau, the seat of the Alaska territorial government and, according to guidebooks, perhaps the most beautiful of American capital cities. Coming from the flat pastureland and dry rolling hills of the Livermore valley, they must have been impressed with this place where a rain forest of mammoth conifers straddles saltwater channels. The waterfront may have reminded them of San Francisco, with its racket of swooping gulls and smells of fish and creosote and diesel; with its wet streets, lined with curio shops and bars, rising to Victorian-era wood-frame houses stuck into the hillside. Most people visiting Juneau for the first time would pick up on the West Coast flavor of the place, but would be unprepared for the scale of things. The hillside rose to a mountainside so steep that in places long flights of stairs serve as streets. And the gusts here were not the mild breezes off San

Francisco Bay, but instead carried down through the dank cedar and spruce the cold bite of the Taku Glacier thousands of feet above.

If Teller and his physicists from Livermore were not prepared for Alaska, neither were territorial officials prepared for them. For a start, no one knew they were coming. The legislature, not then in session, had dispersed throughout the territory, and Gov. Mike Stepovich was in the Lower Forty-eight at the time. Half of Teller's group scrambled to set up a news conference, imposing on the Department of Health to round up business leaders and people from government agencies. They also set up a presentation to the Rotary Club for the next day. The other half of the party flew on to Anchorage to address the chamber of commerce there.

At his Juneau press conference, Teller said he had come to unveil a proposal known inside the AEC as "Project Chariot." The idea was to create an instant deepwater harbor at Cape Thompson in northwest Alaska by simultaneously detonating several thermonuclear bombs. Seventy million cubic yards of earth-moving would be accomplished instantly, as the sea rushed in to fill a keyhole-shaped crater. "We looked at the whole world—almost the whole world," he said, not mentioning his own criterion that the site must be in U.S. territory, "and tried to pick a spot where we could most effectively demonstrate the peaceful uses of [nuclear] energy." On the basis of a preliminary study, Teller said, "the excavation of a harbor in Alaska, to open an area to possible great development, would do the job."

The audiences Teller addressed in Juneau, and later in Anchorage and Fairbanks, liked the economic development angle of the peculiar project. Alaskans were troubled by a postwar economic decline and hopeful that statehood, which had just passed Congress and was to take effect the following January, might bring an increased flow of federal dollars to the new state. Federal spending—most of it military spending—had accounted for 60 percent of all jobs in Alaska prior to 1957. After that date, however, spending for military purposes declined dramatically. Mindful of Alaskans' concerns, Teller described a range of economic advantages that would attend the project. He said that the AEC would be spending $5 million on the harbor, "two-thirds in labor and things we will buy here." Moreover, the employment of many Alaskans in construction activities was only the short-term benefit. The real payoff would come when the harbor began to export "the largest deposits of highest quality proven coal deposits in Alaska," which he said were adjacent to Cape Thompson. Alaska's "black diamonds can pay off better than its gold ever did or will," said Teller. The fishing industry would also profit. According to AEC press handouts, commercial fishing in the Chukchi Sea had "been impeded by the lack of a safe haven."

But the Alaska businessmen and government officials, confirmed developers though they were, could not see the utility of a harbor at Cape Thompson some 300 miles north of Nome. They pointed out that the harbor would be ice-locked by the frozen Chukchi Sea for nine months of the year. Still, as men who had spent long years beating back the wilderness, they weren't about to turn down a free offer to move dirt. They suggested alternatives: a canal excavated through the Alaska Peninsula near Port Moller would shorten by 400 miles the shipping lanes to Bristol Bay, the world's richest fishing grounds. If the AEC and Livermore had to have a harbor, there was a bona fide need for one in Norton Sound to serve Nome. And even with the short shipping season on the Arctic coast, a harbor that could serve Umiat, where oil was known to exist, would be more useful than one at Cape Thompson.

If Teller was caught off guard by the Alaskans' strong opinions, he adjusted immediately. "I'm delighted. This is just the type of suggestion and objection we are looking for," he said. "We came here to be partners with you, and because we want suggestions." Teller downplayed his interest in performing an experiment and emphasized his interest in doing useful public works: "We want not just a hole in the ground, but something that will be used. If this project isn't feasible, what else can we undertake?" Everywhere they spoke, Teller and his colleagues stressed that whatever project the AEC selected, it "must stand on its own economic feet over a long range period." He reassured his audiences that "the blast will not be performed unless it can be economically justified."

Before leaving Juneau, however, Teller tried once more to salvage the economic argument for a harbor at Cape Thompson. On the morning of July 15, 1958, he tried to recruit Dr. George Rogers, a longtime Alaskan and Harvard-trained economist, as an adviser for the project.

> [Teller] invited me to breakfast . . . he gave me the pitch
> again. . . . Then I said well, the Native people, they depend on

the sea mammals and the caribou. He said well, they're going
to have to change their way of life. I said what are they going
to do? Well, he said, when we have the harbor we can create
coal mines in the Arctic, and they can become coal miners.

Rogers was not eager to advise Eskimos to abandon an economy and
culture rooted in gathering food from the land and sea for low-wage work
and the expectation of black lung disease. Nor was he likely to be com-
forted by assurances that the AEC had a great deal of experience in relo-
cating people. The tragic experience of the Marshall Islanders, whom the
AEC had uprooted for atomic testing in the Pacific, had been widely reported in the popular press. Rogers shifted the focus of the discussion to
economics, mentioning the short shipping season, and remembers Teller's
reaction: "He said we'll just have to create warehouses and tank farms to
retain these things until the season opens." When Rogers noted that the
coal deposits the physicists seemed to be referring to were on the back
side of the Brooks Range, hundreds of miles from the proposed harbor
site, Teller said it was not a problem, a railroad would be built. Agape,
Rogers asked if Teller had any idea what these things would cost to construct in the Arctic. "By this point he realized I was not the economist he
wanted, so he changed the subject totally and said time is running out,
where would you advise me to buy my souvenirs?"

Though Teller managed to convey his ignorance of such things as the
length of the shipping season, the location of mineral deposits, and construction problems in the Arctic, these issues had been covered in an economic study Livermore had commissioned from the E. J. Longyear
Company some months earlier. Longyear, a mining concern that held Livermore contracts for drilling at the Nevada Test Site, was hired to analyze
the "economical mineral potential of Alaska between Point Barrow and
Nome." Its report, prepared in just nine weeks' time and without the benefit of travel to Alaska, concluded:

> Most of the valuable mineral resources will be in the form of
> bulky material, namely oil and coal. These, in order to be eco-
> nomic, must be produced in large tonnages, put aboard rela-
> tively large ships, which in turn require a large harbor and
> facilities.

The Longyear report suggested a harbor with a "minimum basin area of 6,000 by 3,000 feet." It also cautioned that "the effect of ice on navigation to the north of Icy Cape is considerably greater than the effect to the south." In the course of their study, therefore, Longyear researchers had decided that the harbor should be far enough north to be near known North Slope oil deposits (though the massive Prudhoe Bay oil field, nearly 500 miles northeast of Cape Thompson, had not yet been discovered), and near some North Slope coal deposits, but that it should be south of Icy Cape. Increasingly, they had begun to zero in on the area around Cape Thompson, and LRL had advised Dave Hopkins and his group at USGS to restrict their study to the stretch of coast between Cape Thompson and Cape Seppings.

*   *   *

While Teller was still in Juneau, two of his associates spoke to a chamber of commerce-sponsored meeting in Anchorage and told the audience there that the harbor would export millions of dollars in coal and oil—the Longyear report estimated $176 million annually in twenty-five years. But, they cautioned, the AEC would only pay for the excavation. Private industry or some other government agency would have to put up money for port facilities. And these costs, said Teller's associates, might be between $50 million and $100 million: "We have to be reasonably certain that someone will have money for this development before we dig the hole." The Anchorage businessmen, though generally interested in applications of the new technology, were taken aback by this sum. One man pointedly reminded the physicists that "economic feasibility and scientific feasibility don't go hand in hand."

Teller had to do some nimble damage control when he arrived in Anchorage the next day. While the costs his subordinates had quoted came right out of the Livermore-commissioned study, Teller categorically disavowed them, saying: "figures of 50 to 100 million dollars are greatly exaggerated." An *Anchorage Daily Times* writer said Teller dismissed the usefulness of such estimates, believing that it was "foolish to give figures now." He himself had considered the cost of port and warehouse facilities to be more like $5 million. It was not an inconsequential issue. In fact, though he didn't say so, Teller knew that a directive by the AEC commissioners had made their approval of the harbor plan "contingent upon receiving outside sponsorship for the development of harbor facilities."

Juneau and Anchorage businessmen had given the Cape Thompson harbor idea a cool reception, and Teller had responded by conveying the impression that he was flexible. By the time his group reached Fairbanks, the excavation project appeared wide open—the AEC might construct dams, canals, or harbors, whatever Alaskans favored. About forty civic leaders and members of the science faculty showed up at the faculty lounge on the University of Alaska campus to hear Teller say that all sorts of projects might be considered and that suggestions were welcome.

Albert Johnson, a faculty member who heard Teller's presentation, remembers Teller as charming and persuasive, as speaking matter-of-factly and thoughtfully. According to Johnson, there was little reaction from the audience, though the boomtown businessmen gladly contributed a few suggestions of their own for a Plowshare project to boost the Alaskan—and maybe the local—economy. "Many of these were of the wild variety," wrote Al Johnson in a 1961 letter to a colleague, "I can't recall anything very sensible being said." The businessmen thought the AEC might consider damming the Yukon River with a strategic detonation at Rampart, which could create an impoundment the size of Lake Erie; or damming the half-mile-wide Copper River, which carried its load of glacial silt like a great gray conveyor belt out of the Wrangell Mountains; or the mighty Susitna, which drained practically the whole south side of the Alaska Range. Besides a harbor at Cape Thompson and another near Nome, one might be built at Katalla on the Gulf of Alaska near promising coal deposits. And a canal across the Alaska Peninsula still looked promising.

By contrast, the science faculty, particularly the biologists, were generally skeptical, even mildly alarmed. Teller insisted that radiation hazard was a nonissue. "We have learned to use these powers with safety," he said, claiming the detonation would add even less radiation than the background amount to which all persons are constantly exposed due to cosmic rays. That statistic made Chariot's radiation hazard sound insignificant, but the science faculty could see that the statement was just a tricky way of saying that the harbor project might nearly double the amount of radiation to which people were already exposed. Chariot might only add an amount that was just under the background level, but because it would be *added* to the already existing background level, it could account for a near doubling. And no one really knew for certain if that amount might do significant harm. Those scientists who were in the best position to know—the world's geneticists—were unanimously convinced that *any* increase in radiation exposure would result in increased genetic damage. Privately, the University of Alaska biologists discounted Teller's salesmanship, though publicly they said very little. They took heart at his insistence that the project would have to be justifiable from an economic standpoint. In another letter, Al Johnson recalled the faculty mood after Teller's presentation:

> The statement was made by someone . . . that the shot would not be made unless it could be shown to be economically useful. I remember breathing a sigh of relief at that news and several of us discussed the project in light of that news, being sure that nothing could come of it.

Teller was an effective speaker, by turns flattering and swaggering, soft-soaping the crowd, saying that Alaska had been chosen to host the visionary technology because "you have the fewest people and the most reasonable people," and that Alaska was "a big state with big people." A bold project like Chariot was to be seen as of a piece with Alaskans' pioneering spirit and intrepidity. Permitting himself a bit of frontier bravado, Teller boasted that the AEC could control a nuclear excavation so precisely as to "dig a harbor in the shape of a polar bear if desired."

C. W. Snedden's *Fairbanks Daily News-Miner* covered the speech, framing the story as a lucky break for a flagging economy. Under a headline that said ALASKA CAN HAVE MASSIVE NUCLEAR ENGINEERING JOB, the story's lead read: "An opportunity for Alaska to have a $5 million earth-moving job done at no cost to the new state was outlined here yesterday." The question of whether or not Alaskans wanted the project was not raised by the *News-Miner.* Publisher Snedden was a tough campaigner for every brand of resource development and economic growth. In the week following Teller's visit, the *Daily News-Miner's* editorial page gave Project Chariot an unqualified endorsement, the onetime mining-camp newspaper fairly groveling with gratitude: "Alaska has been invited by the high echelon of nuclear scientists of our nation to furnish the site. . . . Dr. Teller flattered Alaskans. . . . We say to Dr. Teller and his associates:

Come ahead, Alaska will be proud to be the scene. . . . Alaska welcomes you, tell us how we can help."

The piece was written by Snedden's right-hand man, editorial page editor George Sundborg, who apparently construed the massive nuclear explosion as some sort of pyrotechnic fanfare—the sort of fireworks that might befit Alaska's entry into the union and herald the new state's brilliant destiny: "We think the holding of a huge nuclear blast in Alaska would be a fitting overture to the new era which is opening for our state," he enthused. Sundborg predicted the blast "would center world scientific and economic attention on Alaska just at the time when we are moving into statehood and inviting development." Any worry over the hazards posed by radiation was "nonsense." "The fallout," Sundborg insisted, "would occur where the scientists foresee and prepare for it."

The editorial concluded that those who had heard "the famous physicist" speak "appeared to emerge of one mind that having a nuclear experiment of the kind outlined would be a good thing." With respect to the Fairbanks businessmen and civic leaders in attendance, the *News-Miner* was probably right.

•   •   •

While Teller tried to promote the Cape Thompson excavation, he had stressed throughout Alaska that the AEC was open to alternatives. "If this project is not feasible, what else can we undertake?" he had said. And Livermore did seem prepared to shift from its preferred site, as evidenced by its quiet studies of Katalla and Port Moller, areas that had been suggested by Alaskans. But what Teller had not said was that some two months earlier his laboratory had asked the AEC to initiate a classified application with the Department of the Interior to withdraw from the public domain a chunk of the Cape Thompson hinterlands nearly the size of Delaware. Even as Teller spoke in Fairbanks, encouraging his audiences to suggest alternative projects, his associates landed at Ogotoruk Creek and began determining where the bombs might be placed.

From July 17 to 19, Gerald W. Johnson, test division leader at Lawrence Radiation Laboratory, and a small team of consultants held a field conference at Ogotoruk Creek. The group joined a USGS field party already camped there. The USGS group had spent ten days studying three alternative harbor sites between Cape Thompson and Cape Seppings. It was at this meeting, after listening to the geologists review each site, that Gerry Johnson selected the mouth of Ogotoruk Creek as ground zero for the detonation. He showed the geologists where the holes for the bombs would likely be drilled. The USGS team then spent the remainder of the season, until August 25, in the area evaluating geologic and oceanographic factors that might bear on harbor construction.

The matter of site selection wasn't the only card Teller was playing close to the vest. He had decided not to release the 1958 Longyear report to the public during the promotional swing through Alaska, preferring instead to summarize its contents. It was eventually released more than a year later. George Rogers, the Alaskan economist whom Teller had tried unsuccessfully to recruit for the project, had an opportunity to review it:

> There is no analysis nor comment as to actual economic feasibility. . . . There is no attempt to analyze probable demand. . . . There is rather strong circumstantial evidence that the volume [of exported minerals] may have been arrived at by working backward from the contemplated size of the harbor, through an estimate of the number of ships which could be moved in and out within the shipping season, thence to a volume figure.

The report suggested that a railroad 50 to 100 miles long would bring coal to the new harbor. But Rogers said that such a line would reach only small deposits along the coast. To reach deposits "of any interest in the Arctic" would require a railroad 250 miles long and cost over $100 million to build. Rogers concluded that the Longyear report *under*estimated the costs to develop the mining ventures, *over*estimated the value of the output, and worst of all, ignored completely the fact that there was no market for the high-cost minerals in view of more accessible reserves elsewhere. As for the notion that fishing had been impeded for lack of a safe haven—another of the AEC's claims—the fact was there simply were no fish stocks of commercial interest in the Cape Thompson region.

What Livermore really wanted, of course, was to get information about cratering phenomena and radiation hazard in order to proceed with the Panama Canal project. The claims of wanting "not just a hole in the ground," but something to boost the Alaskan economy, had little basis in reality and seemed in the end to be rhetoric aimed at securing public acceptance of the proposal. In an interview conducted in 1989, Gerald Johnson was presented with George Rogers's criticisms of the economic justification for Project Chariot: "He could very well be correct. I mean, it didn't matter to me because we just wanted to do an experiment. We didn't care if there was a harbor there or not."

Teller's claim that the economic viability of the harbor was a sine qua non proved to be specious. Likewise, the assertion that Plowshare projects were strictly peaceful in nature and had no military connection has been undermined by documents from Livermore declassified as a result of this research. Writing to Gen. Alfred Starbird, the director of the AEC's Division of Military Applications, Teller revealed that the military value of Project Chariot was, in fact, a subject of interest. He wrote Starbird that he had discussed the harbor idea with groups in Alaska, but had found no one who could justify, on a "military basis," the need for the harbor. The August 1958 letter also reveals that Teller discussed the possible strategic value of other nuclear excavations with Alaska-based Gen. Frank Armstrong.

The AEC confirmed the military value of Plowshare projects in statements contained in the previously classified minutes of the Atomic Energy Commission meeting of May 22, 1959. Until then, Plowshare had resided, administratively, within the AEC's Division of Military Applications. Some of the commissioners worried about the "unfavorable implications" of having an ostensibly peaceful program supervised by "an organization which bore a military connotation." These members proposed the creation of a new Office of Civilian Applications. Interestingly, those at the meeting who opposed the suggested restructuring took the position that Plowshare was so alloyed with weapons development that segregating the programs would be inefficient. According to the minutes, the AEC's general manager, Gen. A. R. Luedecke, reasoned that "the circumstances created by an integration of work performed to produce explosive nuclear devices, whether for peaceful application or for weapons, makes it very impractical to separate these functions." And Chairman John McCone "expressed his belief that the basic development of the devices, irrespective of their ultimate use, is [so] interwoven in the weapons program, that organizational separation would seriously impair the program." In addition, he continued, "it is clear that Dr. Teller believes the detonations carried out in the near future under the Plowshare program will be related to our weapons program; therefore, the Commission must exercise care in any arrangements made for conducting nuclear shots under this program."

*   *   *

Among Edward Teller's many gifts was the ability to concoct aphorisms extemporaneously. He invariably supplied the strategic sound bite, the quotable quote. When he said that the AEC had refined the science of crater production to such a degree that they could "dig a harbor in the shape of a polar bear if desired," the reporters were happy to perk up their stories with the colorful image and, in the process, to reassure the public with Teller's cocksure style. But three months later, in a more sober tone, Teller wrote in a document classified "Secret" that the cratering dynamics of buried shots in media other than Nevada soil and Pacific coral were simply not known. He said the Alaska blast would produce helpful data "without endangering the project even if the actual crater was off by *30 per cent* or more from the calculated size" (emphasis added). In a second classified letter a few months later, Teller again referred to "considerable uncertainty" about cratering laws and said there was "even more uncertainty" with respect to how deeply to bury a bomb so as to contain radiation.

Had Alaskans, as the nation's "most reasonable people," been given this information, they might have concluded that a 30 percent margin of error amounted to something less than sculptural precision, and that a hypothetical polar bear harbor might stand a fair chance of resembling a big ragged hole in the ground.

A related issue also left unaddressed by Teller or the newspapers concerned the operational details of the shot and the effects of the blast. What sort of explosive yield was being considered for the Cape Thompson shot? How many bombs would be used? How deeply would they be buried? How much of what kind of radiation would be released? What area would be contaminated, and for how long? Was the AEC proposing to evacuate the people of Point Hope? And if so, where would they go and when could they return?

The *News-Miner* had reported Teller's figures for the dimensions of the proposed harbor: the entrance channel would be more than a mile long by nearly a quarter mile wide (6,000 feet by 1,200 feet), leading to an oval "turning basin" more than a mile by a half mile (6,000 feet by 3,000 feet). The basin's dimensions were exactly as recommended in the Longyear report. Later, the harbor's size varied slightly almost every time it was mentioned in internal documents and press accounts, but the important fact was that an excavation this enormous would require a huge explosion—in the megaton range—and therefore a thermonuclear, rather than a fission, one. Writing in a classified letter to the AEC a few months later, Teller revealed the configuration he originally had in mind: a 2.4-megaton detonation consisting of four 100-kiloton and two 1-megaton devices to excavate the entrance channel and turning basin, respectively.

Nuclear weapons designers commonly refer to a bomb test as a "shot." But the word seems inadequate when applied to a detonation on the scale of Project Chariot: the equivalent of 2.4 million *tons* of TNT. If that amount of TNT were transported by one-ton flatbed trucks loaded to their legal limit and driving literally bumper-to-bumper, the convoy would stretch from Fairbanks, Alaska, to somewhere in southern Argentina.

The enormity of the Hiroshima explosion seemed to mark permanently all who saw it. The blast Teller envisioned for Alaska would be 160 times larger than Hiroshima. In one instant, the salvo at Ogotoruk Creek would discharge firepower equal to 40 percent of all the explosive energy expended in the whole of World War II.

As to the danger from radiation—both the immediate effects and the long-term contamination from fallout—any assessment of the hazard relied on knowing the depth of the bomb's burial and the percent of the bomb's yield attributable to fission. The fission trigger of the thermonuclear device contributed the most dangerous radionuclides, so the larger the fission component, the "dirtier" the bomb. But the AEC never revealed what percent of the yield would be attributable to fission. In fact, it is likely, as the following evidence suggests, that the original Project Chariot concept was sold to Alaskans before any engineering design—drawings and calculations showing number of devices, yield, depth of burial, percent of vented radiation, crater size, etc.—was begun.

The report of the USGS field party who visited Ogotoruk Creek obliquely suggests that the nuclear devices were to be buried at a depth of some hundreds of feet. This squares with the recollection of Gerald Johnson, who was technical director of the Alaska project. But other evidence suggests that Livermore intended that the explosion would have been a surface or near-surface burst, which could still accomplish excavation. In 1958, Livermore asked the U.S. Weather Bureau to analyze wind records and to speculate as to possible fallout patterns of the Chariot shot. The bureau based its calculations on the assumption that the blast would be in the megaton range and that 100 percent of the radioactivity would vent. In other words, the bureau understood that Chariot designers at Livermore contemplated a surface or near-surface thermonuclear burst.

Also in 1958, the same year that Teller and his subordinates toured Alaska, the laboratory produced a film entitled *Industrial Applications of Nuclear Explosives.* It shows an image of the Ogotoruk Creek valley apparently adapted from the aerial snapshots taken by LRL visitors to the site. Like the photos, the film shows the green, treeless hills and the broad valley that meets the sea between high limestone cliffs. Graphics superimposed on this view indicate the placement of the explosives. Four 100-kiloton explosives are positioned to excavate a channel that leads to a basin to be dug by a single one-megaton device. Though the depiction is far short of the sort of technical drawings that would accompany an engineering design, the schematic is certainly a simplified representation of the Chariot experiment envisioned by the Livermore physicists. (Speaking at Point Hope in 1960, the AEC's Russell Ball confirmed that the film represents Project Chariot, although, as he said, "it is not mentioned by that name.") The only difference between the film's portrayal and the original Chariot configuration is the elimination in the film of one of the one-megaton bombs. The burial depth is shown as 30 meters for the smaller bombs and 50 meters for the larger one, or 98 and 164 feet respectively. Such near-surface bursts allow nearly 100 percent of the radiation to vent into the atmosphere, and the potential destruction from the air blast (the powerful shock wave traveling through the air) is nearly the same as for a surface burst. With respect to the radiation hazard, the shallow burial actually produces a dirtier explosion than a surface burst because a larger crater is formed, hence more dirt is mixed with the fireball and rendered radioactive. In the case of Chariot, hundreds of billions of pounds of earth would be pulverized, irradiated, and blasted as high as the stratosphere.

The documentary film was shown at the Second International Conference on Peaceful Uses of Atomic Energy in Geneva in September of 1958. Because the physicists showcased the film to an international audience of their peers and the press, it can be taken as illustrating Livermore's state-of-the-art techniques in nuclear cratering. The film proves that in 1958 Livermore physicists did in fact believe that a harbor could be safely excavated in a remote coastal area with a megaton-range blast where the bombs were buried at relatively shallow depths.

"I don't see how those numbers got in the game or where they came from," said Gerald Johnson in an interview. Considering the shallow depth of burial, he said, "That would have been pretty damn dirty. . . . That design could not have lasted long within the laboratory." The design did not last long. By January 1959, LRL had downscaled the blast to a mere 19 percent of the size originally proposed. Johnson also noted that a 2.4-megaton blast "could not have been safely carried out in Nevada for two reasons. I would have worried about the seismic [implications] as well as the radioactivity . . . we would have been ruled out on both counts in Nevada." Las Vegas is the nearest community of significant size that would have been affected by such a blast in Nevada. Yet Point Hope is closer to Ogotoruk Creek than is Las Vegas to the Nevada Test Site.

Gary Higgins, who succeeded Johnson as the head of Plowshare projects at Livermore in 1960, corroborates the notion that the Chariot scheme, as initially proposed, would have been shallow and, therefore, dirty. "I think the original concept would have been for a surface or near-surface detonation," says Higgins. In an interview the same year, the radiochemist allowed that the Project Chariot concept was promoted in 1958, before any realistic consideration was given as to the possible consequences of the blast to the residents of Point Hope:

> I know there was no real serious thought of air blast. That's another phenomenon that's very important and precludes essentially using megaton-level explosions around any kind of human activity within hundreds of miles. You break things and knock them down and do all kinds of damage.

That's what the experts say today—that hundreds of miles ought to separate human activity from such shots. But in the 1958 film, the Livermore scientists, perhaps allowing their optimism for the new technology to get the better of their scientific training, claimed that "overpressures and seismic effects would be acceptably small beyond a radius of 30 kilometers," or about nineteen miles.

Higgins considered the original sketches of the nuclear excavated harbor as "sales charts that they used to take to the Congress and to the AEC to say this is what we have in mind." He called the animated film and the mock-up charts "cartoons":

> I don't think that, in the original cartoons of Chariot (I think that's what they really were), anyone thought about the off-site effects and the environmental impact. You know, I really mean they really hadn't even thought about them, let alone dismissed them.

When asked, "What could have been the effect from air blast at Point Hope thirty-five miles away or so, from that kind of shot as a near-surface burst?" Higgins said:

> Well, it would have wiped out everything. It would only take someone to sit down and start to do the details of an engineering design to realize you couldn't do that experiment. You just couldn't do it. . . . For a one megaton burst under those circumstances, you probably want seventy or eighty miles separation, at a kind of minimum.

Seventy or eighty miles' minimum exclusion for a one-megaton burst. Point Hope was less than half that distance from ground zero, and the shot was to be more than twice that size. Like Gerald Johnson, Gary Higgins believes that any plan for a megaton-range blast thirty-odd miles from Point Hope would never have survived a proper engineering analysis. Nevertheless, it was precisely that plan that was promoted in Alaska by the "highest echelon" of American scientists as the safe use of a proven technology. And those Alaskans who, as the *News-Miner* put it, feared exposure "to the effects of a powerful blast and to radiation" were said to be talking "nonsense."

●   ●   ●

Teller made many claims in Alaska in July of 1958. He said that the AEC had looked at nearly the whole world for a demonstration site; that a harbor at Cape Thompson was economically justified, and that economic feasibility was important to Livermore; that high-grade coal existed in abundance in the area; that commercial fishing had been impeded for lack of a harbor there; that the cost of developing harbor facilities would be much less than quoted by his colleagues; that LRL was not particularly set on Cape Thompson as the site for the excavation project; that the physicists could control the crater dimensions of the explosion to a highly refined degree; and that the proposed blast presented no danger to local people from radiation or from air blast. Though these remarks played well with the newspaper editors and with many civic boosters, nearly every single material claim Edward Teller made in Alaska seems to have been untrue. As Teller and his retinue headed back to Livermore, work was quietly begun at the mouth of Ogotoruk Creek to ready the site for a detonation.

●   ●   ●

Holmes & Narver (H & N), Livermore's engineering and construction contractor, moved quickly to establish a camp at Ogotoruk to support topographic and offshore survey crews. The USGS field party had built a 700-foot airstrip, and by August 7, 1958, Wien airlines was flying in H & N personnel in a Cessna 185, and B & R Tug and Barge Company steamed up from Kotzebue to land supplies from barges. Of course, planning for construction of a camp during the short Arctic summer had, by necessity, begun much earlier. The AEC had authorized field studies in April 1958, and H & N had received in mid-May, two months before Teller's visit to Alaska, its "scope of work" order outlining the company's contractual responsibilities.

The USGS party had set up their camp, a half-dozen white canvas wall tents clustered on the benchland above the beach. H & N personnel, experienced in running AEC projects in the Pacific and at Nevada, had been flown up from Los Angeles and were ready to get the work under way. They quickly selected a site on a sandbar in the lee of high ground near the mouth of the clear-running Ogotoruk Creek. Tents, sleeping bags, camp chairs, and messing facilities sufficient to handle twenty-five field-workers were all in place by August 12. Fourteen surveyors from Philleo Engineering Company of Fairbanks arrived that day just as the weather changed. It was an ordinary Arctic summer storm: It raged. Tent flies billowed and flapped wildly, straining at their snub lines. The rain pelted down and gushed in rivulets along the tundra-covered hillsides. Shortly, the tents, the sleeping bags, the camp chairs, and tables of H & N's camp were leaving wakes in the numbingly cold waters of a rising Ogotoruk Creek.

There was nothing for H & N supervisors to do but to order, in the midst of the storm, the relocation of the camp to higher ground. That's when the wind kicked up in earnest. The higher ground was more exposed, and tents caught gusts up to sixty-five miles per hour. Canvas parted and tents collapsed as soaked and shivering men struggled in the chaos of a country very different from Las Vegas or Eniwetok, learning much about it even before their formal investigations began.

The blow lasted four days. Then it was calm. Calm enough for Dan Lisbourne, his nephew, and Peniluke Omnik to launch their skin boat from Point Hope and hunt for caribou around Cape Thompson. As Lisbourne would later recount to his wife, when the hunters motored around Crowbill Point they saw tents at Ogotoruk and turned the *umiaq* toward shore. Jimmy Hawley, an Eskimo from Kivalina working at the camp, recognized Lisbourne in the boat. He waved, and walked down to the beach to say hello.

4   A BOMB ON THE WORLD STAGE

> Teller and the AEC scientists and their friends were determined to continue testing. Their device was through "peaceful" explosions.
>
> —Stephen E. Ambrose,
> historian

By 1957, the nuclear nations of the world had been exploding atomic bombs in the atmosphere for more than a decade. As radioactive fallout gradually sifted down from the stratosphere and covered the earth, worldwide anxiety mounted. Most worrisome was the fact that some of the radionuclides chemically resembled nutrients, and these were finding their way into the tissues of living organisms. Strontium 90, for example, was readily taken up with calcium. Though the element did not exist naturally, tiny quantities could now be found in every plant and animal in the world, and in the bones of every human being. And strontium 90 was an almost unimaginably toxic substance. A single teaspoon of it, if distributed evenly among all the people on earth, would kill every one of them within a few years.

That year, 1957, Nobel Peace Prize-winner Albert Schweitzer called for a ban on weapons tests. In the U.S. Congress, hearings focused public concern on the cancer hazard from exposure to radiation in the minute concentrations associated with fallout. And Linus Pauling, who had won the Nobel in chemistry, gathered 9,000 signatures from concerned scientists worldwide who advocated halting further testing. (By 1962, Pauling's accurate warnings about the danger of fallout prompted the Nobel Committee to take the extraordinary step of awarding him a second Nobel Prize, this time for peace.)

President Dwight Eisenhower took the public's apprehension seriously. Already deeply troubled personally by the arms race, Eisenhower also recognized that the rising pressure of unfavorable world opinion would damage the United States politically as it sought to advance its foreign policy and generally to present itself in favorable comparison with the Soviet Union. When the Russians suddenly proposed a two- to three-year moratorium on nuclear testing, the president was inclined to embrace the opportunity. He told a press conference on June 19, 1957, that he "would be perfectly delighted to make some satisfactory arrangement for temporary suspension of tests."

Edward Teller, however, would not have been delighted. He emphatically opposed any freeze on nuclear weapons testing, as did Lewis Strauss, the chairman of the Atomic Energy Commission. A few days after Eisenhower's comments to the press, Strauss brought Teller and two other AEC physicists into the Oval Office to dissuade the president from a test ban. Chairman Strauss had briefed Teller as to Eisenhower's long-cherished vision of the peaceful use of atomic energy. And he knew the president would be vulnerable to the argument that testing was necessary to realize that dream.

In his conference with Eisenhower, Teller spoke of great possibilities that only continued testing would allow: changing the courses of rivers, cutting through mountain barriers, even the chance "to modify the weather on a broad basis by changing the dust content of the air." He told the president that the development of "clean" bombs for such projects—that is, bombs free of radioactive fallout—would come within six or seven years. Before the Joint Congressional Committee on Atomic Energy, Teller testified that it would be "a crime against humanity" if even a temporary moratorium on nuclear testing took place, because it would slow the development of the clean bomb.

Despite Eisenhower's personal conviction that the arms race was a disastrous course, he deferred to the technical consultants from the AEC and reversed his support for what appeared to be a promising avenue of negotiation with the Soviets. Speaking to the press, the president said he understood from AEC scientists that with continued testing "an absolutely clean bomb" could be developed in a few years. He called his physicist advisers "the most eminent scientists in this field," and apparently believed them when they said that the safe, peaceful use of nuclear explosions was near at hand. Secretary of State John Foster Dulles remembers that on the day after the meeting with Teller the president was convinced that "the real peaceful use of atomic science depends on their developing clean weapons."

David Lilienthal, a former AEC commissioner, penned an acid entry into his journal upon reading the newspaper coverage of the meeting: "Teller is reported as saying, rather pathetically (for he is so transparent a salesman), that 'of course' they can't produce this wonderful new little bomb . . . 'if we can't keep on working,' i.e., if his Livermore laboratory is told to suspend weapon testing."

Writing in a classified memo within weeks of Teller's claims, the AEC's Gen. Alfred Starbird warned that there was "a great danger" that the agency was deluding the public regarding the early expectation of a clean bomb. In fact, no absolutely clean bomb appeared in the next six or seven years, as Teller had claimed, nor was one ever developed. And the reduced-radiation bombs that were eventually built proved to be little more than engineering novelties, being larger, more costly, and more complicated than their counterparts. Furthermore, contrary to Teller's assurances, the actual direction of nuclear weapons development over the years has been toward explosives that release more, rather than less, radiation. Nevertheless, by selling the clean bomb and its use in Plowshare civil works projects as "well on its way to success," Teller helped to kill the comprehensive test ban and, by extension, to continue the nuclear arms race.

*   *   *

The test ban issue came to a head in 1958, the year Teller first visited Alaska. In March, the Soviets finished an extensive series of thermonuclear tests, whereupon Premier Khrushchev promptly announced a unilateral suspension of testing. He added, disingenuously, that if the United

States persisted in testing, the Russians might resume as well. Outmaneuvered by Khrushchev, Eisenhower permitted the AEC to proceed with the previously scheduled nuclear test series code named "Hardtack," thereby handing the Russians a major propaganda victory in the cold war.

But impetus toward a cessation of testing emerged from a scientific advisory panel led by Hans Bethe. The group reported that Soviet compliance with a test ban could be verified by establishing listening posts around the world and recommended that the United States stop testing after the Hardtack series. Further support for the Bethe panel's report came from James Killian, the president's new science adviser. But the AEC's Strauss, who had heretofore insulated Eisenhower from all but carefully selected hard-line experts like Teller, argued against any cessation of testing, saying the AEC scientists disagreed that such verification was possible. Overruling objections from the Pentagon and the AEC, President Eisenhower invited the Soviets to send a delegation of scientists to Geneva to work out a system of detecting violations during a test ban. Khrushchev quickly agreed to the talks, which began July 1, 1958.

While Teller toured Alaska that same month, the negotiators in Geneva gradually narrowed their differences. By mid-August, they had reached an agreement that 180 control posts located throughout the world could detect all nuclear explosions greater than five kilotons. With the important verification issue resolved, the next step was to negotiate the terms of the test ban itself. And the president, supported by his science adviser, decided to take that step.

Eisenhower set to work on a public statement calling for test ban talks to commence October 31, 1958. As a good-faith gesture, he also wanted to announce that the United States would, with the start of negotiations, stop all testing for one year if the Russians would do likewise. Learning of the impending announcement, John McCone, Strauss's successor at the AEC, quickly scheduled a meeting with the president to say that the five-member Atomic Energy Commission was unanimously opposed to any moratorium. Eisenhower replied testily that the AEC "was not concerned with the world political position."

McCone would not give up. He decided to play what was becoming the AEC's ace in the hole—Plowshare. As historian Stephen Ambrose describes:

McCone made one last, desperate effort. Fully aware of
Eisenhower's great interest in using atomic power for peaceful purposes, he told Eisenhower that if only the AEC were
allowed to continue underground testing, it would soon be
able to use atomic energy to extract oil from deep deposits,
blast tunnels through mountains, and achieve other goals.

For the second time in thirteen months, the AEC had attempted to de-
rail arms control negotiations by conjuring up dazzling images of the
peaceful use of atomic blasts and arguing that such marvels would be possible only with the continued improvement of nuclear weapons. This
time, however, Eisenhower resisted the AEC's vision of atomic utopia; he
was going ahead with his announcement "no matter what our military
might say." On August 22, 1958, to a roar of protest from Teller and others as reported in the media, Eisenhower issued his statement announcing American willingness "to proceed promptly to negotiate an agreement
. . . for the suspension of nuclear weapons tests." He said that once negotiations began, the United States would withhold testing for a year, provided the Soviets would do the same. To mollify the AEC, however, he
also inserted a statement suggesting that any agreement should "deal
with the problem of detonations for peaceful purposes, as distinct from
weapons tests."

But before Khrushchev accepted the proposed moratorium a few days
later, McCone was back in the Oval Office to plead for "one more test,"
saying he required a decision "immediately." Annoyed and exasperated,
and with the clamor of hawkish dissent still ringing in his ears, the weary
president relented, saying that "he supposed that the AEC might as well
go ahead."

The AEC launched Operation Hardtack II within two days of obtaining
the president's approval. Within two weeks, the first detonation took place
at the Nevada Test Site. The AEC had only two months until the moratorium on testing would take effect on October 31, 1958, and they planned
to make the most of it. The press dubbed the hasty series "Operation
Deadline." Altogether, Hardtack II accounted for the detonation of thirty-
seven nuclear bombs in Nevada during September and October 1958, at
least twenty-five of which were atmospheric, rather than underground.

The Soviets also conducted a flurry of tests, exploding more bombs in October alone than they had in all of 1957. Levels of radioactive fallout were at an all-time high. But in November of 1958, the explosions stopped. Because of President Eisenhower's resolute determination to address the nuclear arms race, and the influence of men like Hans Bethe, James Killian, and Isador Rabi, his new group of scientific advisers, there would be no atomic testing for the next three years.

•   •   •

Not only did the AEC attempt to forestall the moratorium by invoking Plowshare miracles, but *during* the moratorium the agency argued that the peaceful explosions should be considered as exempt from the treaty's provisions. From the AEC's point of view, if there had to be a moratorium, then an exemption for Plowshare detonations could serve their interests very nicely. Eisenhower recognized that fact in a meeting with John McCone in 1960, where the AEC chairman and representatives from the military strongly objected to any test ban treaty, saying the Russians would surely cheat. As historian Stephen Ambrose writes:

> Eisenhower calmly reminded them that the United States was preparing a program, code named Plowshare, for exploding nuclear weapons for peaceful purposes, such as building tunnels. He did not need to remind every man in the room that the United States expected to get new military information from the blasts, although publicly insisting that they were for peaceful uses only. In essence, the United States was already cheating.

Edward Teller specifically assured the president and his cabinet that Plowshare projects would provide a great deal of information of interest to the military. He also said that moratorium restrictions "can be evaded with complete safety by us." Again, Teller played the Plowshare card in an attempt to head off disarmament talks with the Soviets. He told a cabinet meeting that he wanted to go forward with Project Chariot as well as the sea-level isthmian canal and a variety of fantastic geographical engineering projects. But, he said, these opportunities depended upon the continuation of nuclear testing.

In a follow-up letter to the president two weeks later, Teller singled out the Chariot shot as one of the most important preliminary steps toward realizing the Plowshare dream. Eisenhower gave permission for Plowshare planning to proceed, but insisted that "final authorization for the actual detonation would be reserved for action by the President." He did not want AEC explosions such as Chariot, 180 miles from the Soviet Union, scuttling his arms control negotiations at a delicate stage.

●   ●   ●

One of the last-minute tests conducted at the Nevada Test Site as part of Operation Hardtack II in October of 1958 was to bring about the radical redesign of Project Chariot. The test was called Neptune, and it indicated that the Alaska harbor blast, at 2.4 megatons, was excessive and should be reduced more than 80 percent. But a month before Neptune was fired, the AEC and the Lawrence Radiation Laboratory showcased the full-scale plan to a world audience.

Teller, as director of LRL, and Gerry Johnson, his associate director, attended the Second International Conference on Peaceful Uses of Atomic Energy in Geneva, Switzerland, from September 1 to 13, 1958. The conference was a grand public relations extravaganza for atomic industries, with more than 6,000 people participating, including 911 accredited members of the world news media. Former AEC chairman Lewis Strauss led an American delegation that presented technical papers, exhibits, and films. Nearly half of the massive exhibition hall was taken up by displays from the United States. A full-scale cutaway model of the Shippingport reactor core towered forty feet above conference-goers "in the center of a rotunda where the story of the U.S. atomic energy program was told with numerous large, color transparencies," according to an AEC report. Over the first five days of the conference, a team from the AEC's Argonne National Laboratory assembled, in full view of visitors, an operating nuclear reactor. In a special ceremony on the sixth day, former AEC chairman Strauss brought the ten-kilowatt reactor to criticality by inserting a radioactive "wand" into a control device.

Scientists from forty-eight nations submitted more than 2,000 papers, and Livermore's Teller and Johnson were among those who presented them. Teller's talk, titled "Peaceful Uses of Fusion," concluded that a controlled thermonuclear reaction for power production was not a practical possibility, but that thermonuclear *explosions* could perform useful industrial tasks. Gerry Johnson's paper, "Non-Military Uses of Nuclear and Thermonuclear Explosions," said that the most important of those industrial tasks were excavations, such as the creation of harbors.

On hearing Johnson's advocacy of Plowshare projects, the session chairman, V. S. Emelyanov, head of the Soviet atomic energy program, denounced Johnson from the floor. He charged that Plowshare was merely a disguise for weapons tests and an attempt to dodge the nuclear testing moratorium, which was about to go into effect. The Russians contended that there really was no such thing as a strictly "peaceful" or nonmilitary nuclear explosion. Regardless of the possible civil works benefits, any continuing experience with nuclear explosions necessarily had military value. Experts would continue to be recruited and trained to fabricate and fire steadily improving nuclear devices. If the moratorium broke down, the designers of peaceful explosions could be quickly transformed into weapons designers. In addition, the shots would take place in different geologic media, yielding important data on the effects of bombs. And the seismic, air blast, and radiological information from Plowshare explosions would also give weaponeers useful information as to the bomb's effects. As it turned out, Plowshare detonations in the years to come did have military "add-on" experiments. And Glenn Seaborg, AEC chairman from 1961 to 1971, acknowledged in 1981 that military and national security interests within the United States and the Soviet Union saw peaceful nuclear explosions as a potential way to escape the provisions of a comprehensive test ban treaty, should one be negotiated. In short, Project Plowshare was inextricably intertwined with military interests from the beginning.

For all Comrade Emelyanov's protestations, however, he overlooked the fact that peaceful uses of nuclear explosions were also a declared intention of the Soviet nuclear program from the outset. On the occasion of the announcement of the first successful Soviet test in 1949, the Russian foreign minister declared that the Soviet program was already "spreading life, happiness, prosperity and welfare in places wherein the human foot-steps have not been seen for a thousand years."

Within days of Emelyanov's complaints, the Soviet embassy in Washington delivered to the U.S. State Department a bluntly worded statement protesting Plowshare and Project Chariot. The Russians expressed "deep concern" over the plan to detonate nuclear explosions "under the pretext of building . . . a harbor" when all the world called for an end to testing:

> The plans announced in the USA to conduct tests with nuclear explosions in Alaska cannot be evaluated otherwise than as an attempt to evade permanently the cessation of such experiments and as an effort to set up a legal camouflage for the continuation of nuclear tests. . . .
>
> The Soviet Government firmly objects to the plans being made to conduct nuclear explosions in Alaska . . . such explosions would create a serious threat of contaminating . . . the Soviet Union.

The open hostility of the Russians toward Plowshare was matched by privately expressed skepticism on the part of Western scientists attending the conference. As the press reported these objections, and the potential for the "peaceful use" program to stall test ban negotiations, the U.S. delegation spent much of its time at the conference attempting to meet the objections to Project Plowshare. At one point, the AEC even seemed to back away from Project Chariot. Willard Libby, one of the five AEC commissioners, was reported in *The New York Times* as disclosing that the AEC was "about to drop plans to excavate a harbor in northern Alaska with nuclear energy because nobody seems to want the harbor."

But in another part of the exhibition hall, Project Chariot was being touted in the short film called *Industrial Applications of Nuclear Explosives*. With a sound track available in four languages via headphones, the film was requested and shown more than 100 times, making it the most popular film at the conference. When Teller and Johnson left Geneva for Livermore in September 1958, it was with a distinctly mixed review of Chariot's feasibility in particular, and Plowshare's potentialities generally. If they were to salvage the program from skeptics, they were going to have to do some serious repackaging and hard selling of the Alaska project.

•   •   •

The Neptune shot of October 14, 1958, at the Nevada Test Site was unusual for several reasons. In an oral history interview conducted by Livermore's archivist, Dr. James Carothers, with Ralph Chase, who worked for Holmes & Narver at the Nevada Test Site, Chase remembered a shot in tunnel U-12-C, which he called "a small fiasco." The explosive device was placed at the end of a horizontal tunnel dug into the side of a mesa. Chase said the problem was that "whoever planned the tunnel failed to take into account that the earth cover over the shot point was a shorter distance than from the shot point to the mouth of the tunnel."

> CHASE: Do you remember that?
> CAROTHERS: Now, Mr. Chase, I think the shot you are referring to is Neptune, and we don't call that a fiasco around here. We call it the first cratering experiment.
> CHASE: Yes it was. It did come out of the side of the mountain.
> CAROTHERS: It just depends on how you look at it.
> CHASE: That's right it was a definite cratering experiment. We weren't far away from it at the time, and we got a good close look at it. 'Look at that! Wow!' Claude was there and we were leaning on the hood of the car, watching it. The local county policeman . . . he got one look at that and he just came unglued. He started to tell everybody to get away, but I think we all realized we shouldn't stay too close.

Actually, Neptune was not a cratering experiment at all. Instrumentation for measurement of cratering characteristics had not been set up because Neptune's crater and release of radiation were entirely accidental. Even the movie camera positioned to document any accidental venting of gases was pointed at the tunnel mouth, while the site of the eventual crater was nearly out of the frame. The fact is that Neptune was designed to be a completely contained test of safety features associated with the XW-47 Polaris missile warhead. But, instead, the side of the mesa ballooned into a hemispheric dome until escaping gases pierced its thinning shell and blew it apart. Spectacular plumes of dust billowed 1,000 feet into the sky. Cannonading boulders whistled aloft in wide arcs, leaving hundreds of smoky contrails hanging in the desert air. A rock slide smashed into a trailer in the staging area near the tunnel entrance. When test officials returned to the site—after beating a hasty retreat—they were surprised to see that the accidentally formed crater was so large (33,000 cubic yards), considering the depth of burial (98 feet) and the low yield (.09 kilotons). When they returned to their desks, they were able to overcome any initial chagrin they may have felt and wrote up the results as if the blowout had been part of the plan. Their report was titled: "The Neptune Event—A Nuclear Explosive Cratering Experiment."

Neptune's results, if serendipitously obtained, proved significant on three counts. First, Neptune showed that the published accepted curves used to calculate the depth of burst for maximum cratering "were in considerable error," according to Livermore reports. The physicists learned that with a deeper burial than was indicated by the previous formula, "it was possible to produce much larger craters for the same size charge than had been previously thought possible." The second and even more interesting discovery was that "in the Neptune event less than 1–2% of the gross radioactivity of the explosion escaped into the atmosphere even though almost maximum crater dimensions were realized." As Teller wrote in a letter to the manager of the AEC's San Francisco office a few months later, "The significance of the Neptune results is clear as applied to excavation. It may be possible through burial at appropriate depths to produce excavation without penalty and at the same time to reduce the escape of radioactivity to the surface to very low levels."

In other words, up to a point, crater size increased with depth of burial. After that point crater size decreased. Meanwhile, vented radiation decreased steadily with depth. Putting these two realities together, the physicists saw that the ideal depth of burial for a given charge was deeper than they had imagined. A deeper burial might sacrifice little or no cratering efficiency while containing a great deal of the harmful radiation underground.

Finally, the "blast effects" such as flash, air shock, and sound seemed to be greatly reduced compared with previous shots less deeply buried.

The Neptune data not only suggested that Chariot ought to be buried more deeply, but also that the explosion should come from fission rather than fusion. The logic went this way: The fusion, or thermonuclear, configuration probably had been preferred originally for several reasons. From a cost standpoint, there was an economy of scale at work. A large excavation could be accomplished much more cheaply by nuclear means, the AEC said. And nuclear excavation compared even more favorably with excavation by conventional means as the excavated volume—and hence, the explosions—got larger. All of this argued for the use of the large, thermonuclear explosives in Plowshare projects. Livermore had gone so far as to publish a price sheet for nuclear explosions that listed shots of "a few kilotons" at $500,000, and shots of "several megatons" at $1 million. So for just two times the money, a customer could buy 1,000 times the energy, making large excavations even more cost effective compared with traditional digging techniques.

Secondly, the larger, thermonuclear shot would be an impressive display; the harbor would be enormous. Coming on the heels of the Soviet coup in launching Sputnik 1, the world's first artificial satellite, a major U.S. technological response may have been viewed as desirable. Especially if it was conducted on the USSR's doorstep—the Chariot cloud would be visible from Siberia. And the final incentive to employ thermonuclear devices for the Chariot harbor was that they produce much less radiation than fission bombs.

But Neptune threw the accepted wisdom into doubt. Though Neptune had been a fission bomb, it had excavated a large hole relative to its yield, and had released comparatively little radiation. If the same results could be achieved at the 100-kiloton level, the fission bomb might prove a more useful technology for Plowshare projects, especially when used near populated areas. In any event, whether fission or fusion bombs were employed, Neptune showed that burial at an appropriate depth could both dig a substantial crater *and* contain a large percentage of radiation.

The upshot was that the Chariot design as promoted in Alaska was now obsolete. Within a few weeks of the Geneva conference, where Soviet and Western scientists alike expressed opposition to Plowshare and where Commissioner Libby seemed to announce Project Chariot's demise, the scheme was found to be *technically* questionable as well. Nevertheless, Teller did not intend to abandon the test opportunity, which presented him with a dilemma. On the one hand, he needed immediately to lobby Alaskans to embrace Chariot before the AEC in Washington acted on Libby's prediction and canceled the project for lack of support. On the other hand, he knew that the concept, as he had been promoting it, ought to be scrapped. Chariot needed to be completely redesigned in order to verify and extend the Neptune data. Though the two objectives seemed at odds with one another, Teller decided to pursue both at once. He sent two of his scientists, Harry Keller and Vay Shelton, on a fence-mending mission to Alaska. Their instructions were to hit the chambers of commerce, the newspaper editors, and "opinion leaders," wherever they might be found. They had to drum up a tangible expression of support for Project Chariot that Teller could use to deflect opposition from the AEC in Washington. At the same time, others at LRL were set to work analyzing the Neptune data and radically altering both the purpose and the engineering design of the very project Keller and Shelton were boosting in Alaska.

By January 1959, Teller had settled on the new Chariot configuration. In a classified letter to the AEC's San Francisco Operations Office, he described a fission blast employing two large bombs at 200 kilotons each, and three smaller ones at 20 kilotons each. At 460 kilotons, it was less than a fifth the size of the original 2.4-megaton thermonuclear design. Teller made it plain that the revised plan would yield experimental information suggested by Neptune—the cratering characteristics of large buried explosions fired simultaneously, the effectiveness of containment of radioactivity, and the suppression of blast effects—but the experiment he proposed had nothing to do with creating a public works project that was economically viable.

5 FRACAS IN FAIRBANKS

> To be fair, they were completely un-
> suspecting and we sandbagged them
> without warning.
>
> —Professor Albert Johnson, on a
> meeting with Livermore scientists
> at the University of Alaska

Six hundred and fifty miles of mountains fan out like a dam in an im-
mense parabolic arc across the bottom of Alaska. The mountains define
and isolate Alaska's vast Interior from everything southern, including the
climate-moderating influences of the Pacific Ocean. Here the first snow
can come in late August, and gardeners must wait until June before it is
safe to set out plants. Mid-winter temperatures can fall to seventy degrees
below zero Fahrenheit and colder; mid-summer, they can reach ninety,
even a hundred above. It is a measure of Alaskans' disinclination to take
the elements for granted that they often insert the qualifier *above* after
temperatures that would be almost unimaginable *below* zero.

The Interior is dog-team country, where the dry cold and the relatively
light snowfall permit a trail system connecting remote camps and villages.
And it is steamboat country, where packets nearly as grand as those on
the Mississippi once worked the broad Yukon and Tanana rivers. It is the
heart of Alaska, and the heart of the heart is Fairbanks—the only settle-

ment in the whole of this Texas-sized hinterland with a plausible claim to city status. The fact that it thrived is a story in itself, given the bedlam attending its wilderness birth.

In 1901, a trader named E. T. Barnette had talked Captain Adams of the *Lavelle Young* into venturing up the Tanana River farther than any sizable steamer had gone before. With his load of supplies, Barnette was bent on establishing the "Chicago of Alaska" in a region of gold, copper, and agriculture prospects. But after traveling 700 miles up the Yukon and about 200 up the Tanana, while still 200 miles short of Barnette's goal, the *Lavelle Young* hit braided shoals. None of the channels had enough water for passage. Captain Adams had a written contract with Barnette that if he could get no farther "Barnette would get off with his goods wherever that happened to be." But Barnette, who had captained a steamer himself, persuaded Adams to drop back down the Tanana a few miles and try going up the Chena River, which he'd heard from an Indian might have a slough that entered the Tanana above the shoals. The captain agreed to give it a try, but not far up the narrow Chena the *Lavelle Young* scraped bottom again and Adams called Barnette to the pilothouse.

For an hour they argued as perhaps only two rivermen can, Barnette protesting that Adams couldn't put him out in the middle of nowhere, that if he did Barnette would be stuck there for at least a year, that an Indian had told him of another route, that Adams could at least take him back to the Tanana. But Adams knew he could get himself stuck for days going downstream with a full load. He held to the terms of the contract, saying to hell with Barnette's Indian stories. He put the trader ashore on the south bank of the Chena, a pretty but unremarkable little river, ten miles above its confluence with the Tanana. As the *Lavelle Young* pulled away from shore, and while his wife sat crying, Barnette laid his ax into trees for a stockade. And with extreme reluctance on the part of its founders, Fairbanks was born.

Barnette's luck changed not long after when a prospector emerged from the woods across the river. From a hill fifteen miles to the north, Felix Pedro had seen the smoke from the stern-wheeler's stack and hustled down to the river "anxious to buy anything eatable that was for sale." He also reported that other miners were scattered throughout the creeks. Making the best of the bleak situation, Barnette said he would operate as a trading post until the following summer, when he would move his goods farther up the Tanana on a steamer with shallower draft. But the next summer, before Barnette could pack up and leave, Felix Pedro discovered gold just twelve miles from the spot where Barnette had landed. Miners stampeded from a declining Klondike, and within two years, Barnette's lonesome trading post had become the largest log-cabin town in the world, with four hotels, two stores, a newspaper, and a row of waterfront saloons spilling strands of honky-tonk music out over the Chena River. To please the judge of the U.S. District Court, a loyal Republican, Barnette saw to it that the town was named in honor of the influential Republican senator from Indiana, Charles W. Fairbanks, who would later become vice president. It wouldn't hurt, Barnette figured, to have a friend in the nation's capital.

Even today, Fairbanks retains more than a hint of its frontier outpost beginnings. Though shopping malls and box stores proliferate, the center of town probably still contains more log cabins than concrete or metal-clad structures. It still seems—especially in winter—alone in the wilderness, hunkered down in the cold beneath rising white plumes from tin chimneys. As the temperature falls, wood smoke drifts horizontally in stratified air. The buckling lids of cooling oil drums bang out over the town like gunshots. At forty and fifty degrees below zero, the heavy air transmits sound with such disorienting clarity you can hear the footfall of a man feeding dogs a quarter mile away.

December days at sixty-five degrees north latitude are hardly days at all. At the winter solstice, the shortest day of the year, there are just three hours and forty-two minutes of possible sunlight. At high noon, the sun barely scales the Alaska Range on the southern horizon. It slides laterally a bit along the skyline, a yellow smudge beneath blue-gray clouds, a continual sunrise-sunset.

•   •   •

In 1959, the University of Alaska sat apart from greater Fairbanks (population 12,500), though the homesteads along the four-mile connecting road were beginning to metamorphize into subdivisions. The campus included 2,250 acres of boreal forest atop a low spur of the Yukon-Tanana uplands. From the doorstep of the main building, you could look out over the broad Tanana River valley, across ninety miles of spruce-covered lowlands to the mountains. On a good day, you could make out 20,320-foot Mount McKinley, 160 miles away, anchoring the Alaska Range on the southwestern horizon.

At this school, wilderness was more than an abstraction. Students walked by moose browsing in the gardens of professors' houses on campus and shooed black bears from picnic baskets at Smith Lake. Dormitory residents could keep firearms in their rooms, though they were asked to register them with the dean of students.

Like Fairbanks, the campus buildings at the university in 1959 were an easygoing collection of mismatched architectural styles—like a hodgepodge of fashions at a family gathering. Ensconced on the best spot on the hill was the original wood-frame edifice built during the First World War, a dignified old-timer known as "Old Main." Nearby, the stolid and pragmatic middle generation gathered—blocky concrete structures painted in industrial hues, looking as though they had just been discharged from military service. And a few boxy youngsters in steel and glass (poorly clad for a subarctic climate) struck Bauhaus poses around the quad.

A scattering of trailers with attached shacks called "wannigans" backed up to the base of the hill and served as married students' housing. In the woods beyond, more adventurous students—and professors, too— lived in log cabins they'd built themselves, some commuting to school by dogsled, staking out the teams in the center of campus. As the moon rose in the darkening afternoon, the young scholars could hear a cacophony of low, dolorous howls and rasping brays mixed contrapuntally with insistent yelps until, as if thirty or forty dogs followed one conductor, the entire racket stopped on a beat. The concerts were finally silenced when the administration—in an attack of late-fifties modernity—banished dog teams from the center of campus while classes were in session.

President Ernest Patty, a former mining engineer and successful independent gold miner, presided over what he liked to call "the northernmost star in the intellectual firmament." In the early 1920s, he had held short courses for local prospectors on mining geology and mineral identification. Indicative of Patty's practical approach to frontier education was the teaching aid he devised to appeal to the rough collection of miners who were drawn to his class: He invented "rock poker." The men were dealt a pile of rocks and the game hinged on the proper identification of

the anted minerals. At times the distinction between classroom and sa-
loon blurred, as Patty wrote in his memoirs: "As the game progressed, the
air grew blue with tobacco smoke and tempers flared, there were mo-
ments when I had to move fast to break up fights."

Albert Johnson, who would eventually become the vice president of
San Diego State University, started his academic career at the University
of Alaska and remembers the college under President Patty in the late
1950s: "There was no tenure at the University of Alaska. Faculty were
paid what President Patty thought they were worth. The only fringe was
a discount at the Fairbanks Medical Clinic." With a faculty of around 80
and a total student body of about 800, Johnson says the young school
shared many attributes with its gold miner president: "It was unsophisti-
cated, impatient, independent, unpretentious and friendly."

In one way, it was an improbable setting for a historic confrontation
between academic scientists and the United States nuclear establishment.
In another way, a clash between two such groups of opinionated, pioneering people was, perhaps, inevitable.

•   •   •

Friday, January 9, 1959, dawned a crisp forty-three degrees below zero as
Professor Al Johnson went about his duties at the university. Actually, it
was the warmest day of the week; on Tuesday and Wednesday it had hit
fifty below. Johnson heard from a colleague that a Dr. Keller and a Dr.
Shelton were visiting Fairbanks from the Lawrence Radiation Laboratory
(LRL) in Livermore, California. They were apparently giving interviews to
the *Fairbanks Daily News-Miner* and making presentations to service
groups—the Rotary and Kiwanis—about the harbor the AEC hoped to
blast at Cape Thompson. Johnson, a tall, young scientist with jug ears and
a crew cut, thought it must be Dr. Teller who was in town because he had
never heard of a Dr. Keller. In any event, because Johnson served as president of the local chapter of Sigma Xi, a research society, and was always
on the lookout for program speakers, he wondered if the visiting scientists
might be willing to make a presentation on campus. As President Patty
was out of town, Johnson prevailed upon Vice President Robert Wiegman
to invite the physicists to address the university, and Keller and Shelton
agreed to come that very afternoon.

Keller was test-group director in charge of field operations at Livermore, and Shelton worked at the laboratory's test division and specialized in the effects of nuclear explosions. Before visiting the campus, the men would first give a two-hour presentation to the board of directors of the Fairbanks Chamber of Commerce. They were keeping a busy schedule in Alaska. The day before, in Anchorage, they had managed to address both the staff of the State Department of Health and the Anchorage area legislators. That night they would be interviewed on Fairbanks television.

Faculty members threw on parkas and hustled along webs of snowy trails to Constitution Hall. It was in this building three years earlier that fifty-five delegates from all over the territory—fishermen and truck drivers, merchants and homesteaders, independents, misanthropes, and near-anarchists—had hashed out a constitution for Alaska, which the National Municipal League would call "one of the best, if not the best, state constitutions ever written." Stomping the snow from their boots, the professors tromped up the stairs to the second-floor faculty lounge. Because the program had been organized hastily, there were no reporters or city officials present. When everyone was settled, Harry Keller, sporting a gaudy tie some years out of fashion, launched into what was becoming his stump speech on Project Chariot.

Newspaper and other accounts of Keller and Shelton's several other engagements in Alaska showed that each presentation covered the same ground: They had come to explain the project and—they didn't mind saying—to seek support for it. Since Edward Teller's initial visit to Alaska the previous summer, Keller began, Alaskans had been strangely silent on the subject of Project Chariot. He wondered why there had been no official statement of support for the harbor. "The people in Washington," he said, "are presently in the notion that the people of Alaska are not very enthusiastic about the project." And that, Keller explained, "is one of the reasons we came up here."

The AEC had an image problem, Keller told his listeners. It "has done a good job of selling the destructive effects of nuclear devices," but now the big job was to convince the public that there was "a whole gamut of engineering problems where use of fission and fusion devices is completely economical and feasible."

As Keller put it, bluntly, "We need an expression of favorable opinion from Alaskans themselves if the project is to be completed by our target date." He suggested that Alaskans write their senators and congressmen. Dr. Shelton added that if people expressed support "right away," the experiment could be done "as early as November."

The atmosphere in the faculty lounge grew tense. Keller and Shelton were contract employees of the federal government, and they were using public money not to inform the public, but to lobby it in support of work that would pay their salaries. One professor wanted to know if the men thought Alaskans should also write their representatives if they *didn't* want the blast. The physicists admitted that they probably should.

Another of Keller and Shelton's pronouncements that alarmed some of the university people was that the AEC had shifted the official rationale for the Cape Thompson project. The summer before, Dr. Teller had been adamant that the "blast will not be performed unless it can be economically justified—it must stand on its own economic feet." "We want not just a hole in the ground," Teller had said, "but something that will be used." Before the AEC would do the excavation, Teller and other LRL officials had said, private groups or other agencies would need to commit to build the necessary harbor facilities.

Now it seemed the AEC had abandoned the economic criterion completely. The new rationale held that the primary value of the harbor at Ogotoruk Creek was to demonstrate conclusively to the public that nuclear power can be applied beneficially to large-scale construction jobs. But even as the men from Livermore described the need for such a demonstration project, they seemed to want to have it both ways. They claimed—especially to the chambers of commerce—that the "mineral potential of the Cape is large." In its coverage of the chamber meeting, the *Fairbanks Daily News-Miner* allowed the men to work both sides of the street: "The officials pointed out a harbor at Cape Thompson could open up vast mineral reserves to future development, but presently the project was not being proposed as a matter of economics."

The truth was, Livermore had already dropped the economic-advantage argument. Teller had written some five months earlier in a classified letter to Gen. Alfred Starbird of the AEC that "in discussing the usefulness of such a harbor with groups in Alaska we found no one who could justify the harbor on an economic or military basis." Nevertheless, Keller and Shelton were sent north to "sell" Chariot as a demonstration justifiable in its own right, while evoking the prospect of an economic shot in the arm.

The boon would come both from the 3 million federal construction dollars to be spent in the state and from the eventual exportation of "vast mineral reserves."

Shelton continued to reassure the group that the AEC expected no fallout problem. The blast would be made when favorable winds would carry the radioactive dust away from land areas. He told the gathered professors that fallout from the Cape Thompson shot would produce "no changes" in the animal and plant populations there. But Shelton was not an expert in biology, nor in botany, nor in genetics—all subjects that Al Johnson taught. Johnson asked on what scientific authority Shelton made his claim. Radiation was known to cause genetic damage, and fallout landing on plants would undoubtedly cause genetic changes. Only then did Shelton concede that damage would occur but asserted that it would be, for the most part, undetectable. Johnson agreed that this was probably so.

The tendentious nature of the physicists' presentation was beginning to rub the Alaskans, particularly the biologists, the wrong way. How was it that physical scientists, with no competence in radiobiology—let alone radiobiology in an Arctic ecosystem—could give such categorical assurances? Furthermore, Keller and Shelton were claiming that the demonstration project would yield important bioenvironmental information that would be applied to future excavation projects. But they did not propose pre-shot studies of the biological environment on a scale anywhere near what would be required to draw useful conclusions. Why, Al Johnson asked, was Livermore not planning comprehensive environmental studies as President Patty had suggested in his letter to the AEC last fall? A letter from Patty? Keller and Shelton said they knew nothing about it. Dr. Wiegman, the vice president, ran over to Patty's office and found the letter. He returned with verifaxed copies, which Keller and Shelton immediately read.

On October 7, 1958, President Patty had sent a letter addressed simply to *The Atomic Energy Commission, Washington, D.C.* He stated the university's support of Project Chariot and suggested that environmental studies of the area be made prior to any blast. Three weeks before that, on September 19, Patty had let Gerald Johnson at LRL preview the letter. Johnson wrote back that the laboratory did "not see how we could improve the letter in any way and, therefore, suggest that you mail it in its present form."

What Patty told the AEC was that the university was "intrigued by the imaginative thinking," but that "some of our people expressed apprehension that such a blast might do lasting damage to the abundant marine life in the Arctic Ocean." He said he had discussed the matter with fifteen of his scientists and that they "all felt that detailed field studies of the wildlife, both land and water types, should be carefully studied before and after the blast; and that a complete program should be planned for this." Patty concluded by saying, "this background information is submitted for what value and interest it may have to you and, also, to assure you of the University's interest in your tentative ideas for research here."

Keller and Shelton were astonished. Here was the only expression of interest in Chariot to have come out of Alaska, and it had fallen through the cracks in Washington. Moreover, the only thing that seemed to stand in the way of the university's support for the project was some research contracts. Keller and Shelton suggested straight away that the scientists might submit funding proposals for the studies they thought necessary. But Tom English and Al Johnson insisted that the University of Alaska group was obviously not qualified to outline an experimental design to measure the effects of radiation. That was a job for the AEC's own Division of Biology and Medicine. After such a program was conceived, if the Alaskans could help by virtue of their special knowledge of the Arctic, they would be happy to do so.

It wasn't that the biologists weren't eager to land a federal contract to do basic research in their area of specialty. They just weren't going to be bought off with half measures. And if the case for environmental studies could only be made by pointing out the AEC's discredited positions on the safety of fallout, well, Tom English for one was prepared to do it.

English was a young fisheries instructor with a wicked sense of humor. He spoke with an engaging stutter that often disarmed the soon-to-be victims of his deviltry. Rising from his seat in the faculty lounge, English cited the dangers of radiation, emotionally evoking the specter of the birth of crippled babies. Keller and Shelton trotted out the now famous arguments that, even then, had been refuted by the geneticist Linus Pauling in debates with Teller: that radiation from fallout was less harmful than a luminous dial wristwatch, than living at high versus low altitudes, in brick houses versus wood houses, than occasional X-rays.

At one point in the meeting, English said that the faculty wouldn't have had too much trouble with the men's presentation if it weren't for the AEC's well-known reputation for "mendacity." Keller and Shelton suggested a brief adjournment, then quickly ducked into drama professor Lee Salisbury's office, where they headed straight for the dictionary to look up *mendacity*. When they read, "untruthfulness—See Synonyms at *dishonest*," they were furious.

Recalling the meeting in a letter written in 1961, Al Johnson wrote, "We harassed them rather unmercifully on the subject of fallout—actually we all said some ridiculous things. . . . To be fair, they were completely unsuspecting and we sandbagged them without warning."

Finally, Keller and Shelton thought it wise to adjourn the meeting until the following day, Saturday. Wrapping up, Keller tried to coax from the faculty some expression of enthusiasm for the atomic harbor. He said they had heard all the complaints against the project, wasn't anyone for it? There was dead silence. Then the dean of the School of Mines jumped up to say that there certainly was and all he wanted to know was when do we push the button!

Things did not go much better for Keller and Shelton on Saturday. Overnight, Tom English had prepared a written statement that did not mince words. Nearly everyone in the Biological Sciences and Wildlife departments had signed it. Standing before the assembled group in the faculty lounge, English read his statement to Keller and Shelton. In part, it said:

> We are told that the results will be used in planning further explosions. It therefore seems especially necessary to gain as much biological information as possible, and we feel embarrassed that it seems necessary to reiterate that useful conclusions are most likely to follow from a carefully considered experimental design. We have been given no assurance that such a study is contemplated on a realistic scale. We feel neither competent nor obliged to outline in detail an experimental design. Rather, we urge that employees and consultants of the A.E.C. who are familiar with such problems be assigned to prepare the experimental design.

English went on to urge that henceforth the AEC accompany its scientific claims with citations to the literature. He denounced the AEC's ongoing public relations efforts as propaganda masquerading as science, saying, "We have been subjected to conflicting and crudely misleading statements about the possible biological effects of this explosion and the entire weapons testing program of the A.E.C."

Al Johnson remembers that Vice President Wiegman "was obviously stumped during the meeting and didn't know what to say or do. He obviously thought we were nasties to say all of those things. But later when we got that big juicy contract, he confided to me that 'we had all done some good work.'" English gave Keller and Shelton a copy of the statement and mailed another off to their boss, Gerald Johnson, at LRL. He emphasized in his cover letter that "the University is willing and even anxious to have its people help you with their special knowledge of Alaska." Five days later, Gerald Johnson wrote English back, saying that Livermore "was most anxious to join with you in the necessary biological studies. . . . We will need and use all the help we can get, and particularly from those of you who have devoted time and effort specifically to Alaskan problems."

But English wasn't finished yet. After the Fairbanks Chamber of Commerce unanimously adopted a resolution in support of the Cape Thompson blast on January 12, English sent them a copy of the statement together with a letter. He urged the chamber to reverse its decision to act as boosters for the project, writing, "I am personally of the opinion you were sold a bill of goods." He sent a similar letter to the Anchorage Chamber of Commerce. When the *Daily News-Miner*'s Saturday editorial asked people to phone the chamber of commerce to cast a "vote" for Project Chariot, English called and talked for thirty minutes. It was a memorable performance according to Al Johnson, who was listening in on an extension. In a letter a few years later, he recalled English's bantering: "He was at his witty and vituperative best—which is very good. He called the Livermore chaps 'the firecracker boys.'"

*   *   *

At the same time that the University of Alaska scientists were trying to dampen enthusiasm for Project Chariot while holding out for environmental studies, the press and the chambers of commerce responded to the AEC pitch with unquestioning support. Under the headline, ATOMIC HARBOR O.K. VITAL, C. W. Snedden's *Fairbanks Daily News-Miner* certified without further investigation the claims of the Livermore scientists and dismissed probing questions out of hand as unreasonable. Affecting complete bafflement, editor Clifford Cernick wondered aloud in his January 10, 1959, editorial "what objections could there possibly be to this large-scale atomic harbor-blasting project?" Expressing no trace of professional skepticism, Cernick presented the AEC position in tones of trusting credulity:

> Scientists who have studied the entire matter carefully have given assurances that the project has been so carefully planned [that] blasts will have a minimum of fallout. Such fallout as results will be dissipated in a way that will not be injurious to human beings or animals.

Besides, Cernick noted, as the AEC often claimed, "The project is located in the wilderness, far away from any human habitation." Apparently he was unaware that the Eskimo village of Point Hope was only thirty-one miles from ground zero, and that for hundreds of years Point Hopers had seasonally occupied the Cape Thompson-Ogotoruk Creek area—ground zero itself—to hunt and to collect the eggs of cliff-nesting birds.

The *News-Miner* said Alaskans had been "handed the opportunity" to be in the forefront of an emerging technology and exhorted those who favored the experiment to "'vote' on the matter through their Chambers of Commerce." In a windup that may have struck even the chamber as melodramatic, Cernick declared, "Alaska's 'yes' vote on this vital project will result in incalculable benefits to all mankind."

Down at the *Anchorage Daily Times,* editor and publisher Robert Atwood clambered aboard the bandwagon. Even more than Cernick, Atwood argued that ordinary citizens had no business questioning the statements of legitimate experts. He presented a patronizing parable that seemed to bear little resemblance either to the rugged individualism so often ballyhooed in his newspaper or to the American democratic process:

> Asking Alaskans for a decision on this proposed atom experiment is like a doctor asking his patient whether he wants

an operation. The easiest answer is 'no.' The doctor usually tells the patient he must have one for his own good, and the patient does as the doctor says. The atom scientists are the doctor in this case. If they say that adequate safeguards for life and property have been provided, how can laymen say otherwise?

Atwood also proved a match for the *News-Miner* editors when it came to inflating the significance of the detonation. He was certain Alaskans would not want to thwart a potentially "glorious discovery" that might "open up a new era in the world" and "spur development of the vast resources" in Arctic Alaska. It "could be the greatest achievement of the century from many points of view," Atwood wrote, apparently ranking excavation by nuclear explosion alongside such recent human advances as space flight, antibiotics, and the transistor, to consider only science.

The Fairbanks Chamber of Commerce and the state chamber in Anchorage both formally endorsed Project Chariot in short order. A few months later, the Arctic Circle Chamber of Commerce, which represented both Nome and Kotzebue, conferred its benediction in a ringing testimonial authored by a Jesuit priest named William McIntyre. In his letter to AEC chairman John McCone, McIntyre employed the florid style and frontier imagery of the Alaskan newspaper editorials: "Alaskans, who are all pioneers, if not in fact then in heart, should feel proud that our state has been singled out to play yet another leading part in the Drama of the Century."

Pulling out all stops, McIntyre concluded his letter with lines from Klondike poet Robert Service's "The Law of the Yukon." With unintended irony, the priest assumed the role of procurer, offering the bosom of Alaska to the stalwarts from the AEC:

> *Long have I waited lonely*
> *Shunned as a thing accursed*
> *Monstrous, moody pathetic*
> *The last of the lands and the first*
> *And I wait for the man that will win me—*
> *And I will not be won in a day!*

And I will not be won by weaklings,
subtle, suave and mild.
But by men with hearts of Vikings
And the simple faith of a child.

●   ●   ●

It is a fact of small-town life in Alaska that any visible rallying of support for one cause serves also to muster troops in opposition. Besides the biologists from the University of Alaska, several members of the Fairbanks community began to speak out against the crusading newspapers and the chambers of commerce. One of them was Virginia "Ginny" Hill Wood, a bush pilot and the operator of a wilderness lodge near Mount McKinley. In an interview in 1988, she said she had some familiarity with the northwest Arctic region: "I'd been a bush pilot and gotten out into the Arctic quite a bit before I became a mother and stopped doing that. . . . I had been at Kotzebue and Nome and gotten weathered in for a couple of weeks at a time." She also knew something of the local people and their way of life: "[I'd] gotten to know the Eskimos and so forth, and realized how much they did depend on subsistence living. . . . That's not the place to have a commercial development for trade with the Orient. . . . I mean that's ridiculous."

Ginny Wood's typewriter was sizzling on January 11, 1959, the day after the *News-Miner*'s editorial requesting readers to "vote" through the chamber of commerce. Because serious treatment of opposing viewpoints was absent from editorials and news articles, Wood turned to the only forum available, the "Letters to the Editor" column. In her letter, she chided the downtown businessmen for being unable to spot the flimflam: "It seems to me that we may well be buying the proverbial 'pig in a poke' to accept unquestioningly the 'gift' of a harbor at Cape Thompson by atomic energy blasting." She objected to the AEC men's "proselyting" and "subtle salesmanship." She speculated that Chariot's real purpose must be "a defense measure disguised as a civilian project," perhaps a "submarine base," or that it was a "guise to continue testing" in light of the moratorium on weapons testing then in place. Sounding every bit the crusty sourdough disinclined to suffer flatlander foolishness, she attacked the atomic physicists' arguments head on: "Dr. Keller points out that the blast would be conducted when the wind conditions were right to blow the particles away from any animal or human life. Just where is that? Out to sea, there to contaminate sea mammals, which constitute the most important resources for the natives?"

A few days later Professor Al Johnson sent the newspaper a letter pointing out that "the *News-Miner* and the AEC have presented only one side of the story." He objected that "glowing accounts of future financial benefit to Alaskans seem to provide sufficient justification for everyone to jump on the atomic bandwagon." But even if economic advantage were guaranteed, which it was not, he said, there were "other scientific, social and political reasons why one may be opposed to this project without being called 'backward,' 'against progress,' or 'unpatriotic.'"

Johnson itemized the AEC program of baseless claims in which the newspaper, wittingly or not, participated:

> The AEC scientists tell us that we have nothing to fear from the amount of radioactive fallout we are receiving or will receive from continued testing of various devices. However, are the people of Fairbanks aware that substantially all the geneticists of the world believe that the opposite is true? Are they aware that the geneticists can demonstrate that radiation damage produces such conditions as leukemia and bone tumors in man? Are they aware of the incidence of leukemia in survivors of the Hiroshima blast? . . . I have not encountered this information in the local news media.

Two weeks later, a prominent Fairbanksan named Irving Reed added his name to the list of skeptical correspondents to the *News-Miner.* Reed had grown up in Nome, and was a fifty-five-year resident of Fairbanks. He'd started out his career as a mining engineer and had served as the territorial highway engineer; he'd been on the Fairbanks city council and on the territorial game commission. Reed was a builder. He *liked* public works projects, especially ones aimed at developing mineral resources. But with nuclear excavation, he temporized. He said he would rather see the money spent on road improvements and thought "Alaskans should be careful before we give our unqualified endorsement of Operation Plowshare and ask the AEC for more data and assurance than we have been so far given."

"I cannot refrain from adding my protest," wrote Olaus J. Murie, former Alaskan and then-director of the Idaho-based conservation group, the Wilderness Society. Murie noted that "the scientists who are defending the atomic blasts are those who are employees of the agencies which would carry them out." He challenged the mythology that saw the North as suitable proving ground for hazardous technologies: "many people who have spent very little time in the Arctic look upon the country as a wasteland, as a dumping ground, to do with as they please."

A man named Robert Needham wrote to inform readers of the formation of a group calling itself the Committee for the Study of Atomic Testing in Alaska. He said the committee objected to "the manner in which the A.E.C. endeavored to sway popular opinion through propaganda." The *News-Miner* reported that fifty Fairbanksans showed up at the committee's first meeting on January 30, 1959. "The consensus of the meeting," reported the paper, "was that the AEC was holding back information about the project that Alaskans should know." The citizens' group quickly drafted a letter to the AEC requesting answers to a list of specific questions.

As public opposition coalesced, University of Alaska president Patty continued to exchange cordial letters with Gerald Johnson, who was Keller and Shelton's boss and Teller's right-hand man for Plowshare at LRL. Patty continued to report the view of his faculty, which was that environmental studies were appropriate. On January 20, shortly after Keller and Shelton had reported back to Livermore, Gerry Johnson wrote Patty that LRL was receptive to the idea of environmental studies. He said that the AEC's San Francisco Operations Office would be in touch soon. For the first time, Johnson revealed the plan to downsize Chariot in response to the Neptune test, saying he and his colleagues at Livermore had "revised our concept somewhat of the Cape Thompson experiment . . . to do a smaller scale experiment than was proposed last summer." He said that he himself would be coming to Alaska in late February.

On February 18, Gerald Johnson telephoned Patty with news that caused the obviously elated president to dash off a memo to key staff people. Johnson would be in Fairbanks in nine days, "ready to proceed on the spot with the awarding of contracts." He would be accompanied by Elison Shute of the San Francisco office, who, Patty wrote, "will want to discuss with our scientific personnel the immediate awarding of contracts for research in health and safety features."

Though they had no idea what the AEC meant by "health and safety features," the university scientists were interested to see what sort of planned approach the AEC had to the proposed research, and they were hopeful about the prospects of obtaining funding. But when they went home that night, they read in the newspaper a report that AEC chairman John McCone had said the agency was now looking for someplace more suitable than Alaska for Project Chariot. The reason, he said, was that private developers willing to invest in harbor facilities had not come forward. He had made the statement the day before to the Joint Committee on Atomic Energy (JCAE) in Congress.

Obviously, Livermore was not acting in tight concert with the AEC in Washington. While Johnson was talking contracts, McCone was telling the press that the project was shelved for lack of interest in Alaska. This wasn't the first time Livermore had seemed to ignore headquarters while pursuing its own agenda. Keller and Shelton's trip north to lobby Alaskans had followed a public statement by AEC commissioner Willard Libby that Chariot might be canceled. Now, as opposition to Chariot sprouted in Alaska, McCone was clearly withdrawing the agency's backing of the controversial Livermore brainchild—"votes" cast with the chambers of commerce notwithstanding.

Things looked rather hopeful for the Chariot opponents. In the worst case, by having stood their ground, the Alaskans seemed to have forced Livermore to accept a full-scale environmental study before any detonation could be approved. In the best circumstance, if the AEC in Washington (as distinct from LRL) had its way, Chariot might be shelved. But did the AEC have any real control over the near-autonomous Livermore laboratory and its director, Edward Teller? Years later, Gerald Johnson put a sharper point on the issue of the laboratory's independence from the AEC:

> The thing that people need to realize is that in those days in the Laboratory we could go directly to the top. We didn't go to the president, but we'd go to the Joint Committee [on Atomic Energy]. . . . We didn't feel we had to work with anybody in between. . . . It wasn't a normal bureaucratic structure. Now the structure resented that. The people down the

line—well, people like John Kelly [director of AEC's Division of Peaceful Nuclear Explosives] felt they were running the Plowshare program. We never felt he had very much to do with the Plowshare program in terms of running it, and we'd ignore him if he'd try to tell us to do something we didn't want to do. . . . We had unlimited priority, unlimited money . . . I was rarely asked how much money I spent and no one cared.

From the Eskimos at Point Hope to the Fairbanks Chamber of Commerce, Alaskans on both sides of the Chariot issue waited to see if it would be the AEC administrators in Washington, or the factious Edward Teller and his cadre of young physicists at Livermore who had the final word on Project Chariot.

6   TENT CAMP AT OGOTORUK CREEK

A good field biologist will do damn
near anything to get back in the field
again.

—William O. Pruitt, Jr.,
biology professor,
University of Alaska

If you put your finger on a map at the Rocky Mountains near Denver and trace the keel of the North American continent through Wyoming and Montana, taking in the Bighorns and the Bitterroot; if you continue along the Canadian Rockies in Alberta and British Columbia, picking up the Selkirk and Cassiar mountains; if you reach up and track the McKenzies through the Yukon and Northwest Territories all the way to Alaska, you find that the northernmost and westernmost extension of the Continental Divide is the Brooks Range of Arctic Alaska. And if your finger follows the Brooks Range west until it runs into the sea, you end up descending the De Long Mountains and pointing to a spot on the Chukchi Sea coast very near Ogotoruk Creek.

The Inupiaq word *ogotoruk* is said to mean "poke," a bag in which game might be carried, and is thought to refer to the area's productivity. For centuries the Inupiaq people have hunted caribou in the uplands here and climbed the towering limestone cliffs to gather the eggs of nesting seabirds.

Ogotoruk Creek heads at 800 feet above sea level, six and a half miles up a southwest-trending valley two to three miles wide. With its several tributary rivulets, it drains some thirty-eight square miles, sliding undramatically into the sea at a two-mile-wide gap between 800-foot-high bluffs. Like many of the smaller streams in this region, the creek's outflow is periodically blocked. A big offshore storm will pile up beach gravel into a barrier bar across the creek mouth. For months or years, the stream fills a lake formed behind the bar, the fresh water slowly percolating through the gravel to the sea. But eventually, a big inland storm will come along, soaking the hills, swelling the creek with runoff, and filling the lake faster than the water can transfuse. When the impounded water finally spills over the bar, the gravel rapidly erodes away and the stream runs freely into the sea again, until the next big offshore storm reestablishes the gravel bar, and the cycle repeats.

There are no trees along Ogotoruk Creek, none on the hillsides. There are no trees at all along the Arctic coast here. Broad meadows on the valley floor look inviting from a distance but prove to be cotton grass tussocks with a foot of standing water and muck in between. Other tundra plants climb the gentler hillsides to talus slopes and rock outcrops—wildflowers, grasses, sedges, and shrubs: Arctic lupine and dock, dwarf birch, willow, Labrador tea, bog rosemary, blueberries, and bearberries. The valley also supports spongy carpets of sphagnum mosses and various lichen that take unusual forms in delicate shades of pale yellow, gray-green, charcoal, off-white, and pink.

On the morning of July 5, 1959, a tugboat pushing a heavily loaded barge pulled out of Kotzebue heading north. It steamed past Cape Krusenstern, past the little village of Kivalina, and past Cape Seppings. When it reached Ogotoruk Creek, it throttled back and slowly chugged into the beach.

●   ●   ●

Chariot was not dead, not by a long shot. In the face of AEC chairman McCone and Commissioner Libby's published statements regarding Chariot's imminent demise, Edward Teller and Gerry Johnson at Livermore had moved quickly to mollify the opposition in Alaska by offering to fund a major program of environmental studies. By January 1959, the strength of the Alaskan opposition had been made clear to LRL. Johnson had by then received President Patty's polite letters backing up the Alaska faculty's call for "detailed investigations"; Keller and Shelton had reported back on the grilling they had received in the faculty lounge; Tom English had forwarded his statement charging that studies had not been "contemplated on a realistic scale"; Ginny Wood, Al Johnson, and others had written outspoken and critical letters to the *News-Miner;* and fifty Fairbanksans had formed the Committee for the Study of Atomic Testing in Alaska.

By the time Gerry Johnson made his own visit to Alaska in late February, he was talking about an environmental program that might last more than one season. He didn't seem to like to use the word *environmental,* however, preferring to term the program "health and safety" studies. But, as one AEC official wrote in a revealing historical review of Chariot: "The first parties of AEC and Livermore representatives to visit Alaska were noticeably impressed by the demands of the Alaskans for environmental studies as part of Chariot. Gerry Johnson responded by demanding that AEC undertake an environmental program!" Even though it is clear from this statement that the original idea for such studies came from the University of Alaska biologists, AEC and LRL officials would later insist that the investigations were always a part of their Chariot program and were not developed in response to any external demands.

Before Gerry Johnson and his traveling companions, Duane Sewell of LRL and Elison Shute of AEC's San Francisco office, arrived in Fairbanks on the evening of Friday, February 27, they stopped in Juneau and Anchorage. While in Juneau, Johnson's party concentrated on legislators. In Anchorage, they gave a breakfast presentation to the chamber of commerce and conferred with Public Health Service officials. Shute, the man who was to have issued contracts "on the spot" at the university, did not accompany the others on to Fairbanks. Perhaps this was because Shute had found himself involved in a sales campaign that seemed at odds with the recently published statements of his boss, AEC chairman John McCone. One member of his staff at the San Francisco office said that Shute "was very discomfited to read some of the statements by LRL people in Alaska," particularly the solicitation of suggestions from Alaskans on "where to blow holes." He reportedly felt that he was going from one surprise to another while he was traveling with them.

In any event, Elison Shute was not on hand for the two meetings at the University of Alaska on Saturday, February 28, 1959. An informational meeting took place that morning in the small auditorium in the mining building. First, Gerry Johnson described the October 1958 Neptune test in Nevada. He said that from the small, ninety-ton (i.e., a little less than a tenth of a kiloton) Neptune shot, the AEC had developed new theories about the cratering characteristics and radiation release of nuclear explosions. But extrapolation of the theories to larger explosions needed to be confirmed with a test in the kiloton range.

As Johnson spoke, it gradually dawned on his audience that the Chariot design had undergone radical revision in the six weeks since Keller and Shelton's visit, and that LRL was now quietly revealing its second major policy shift in seven months. Justification for the blast had devolved from being a grand "economic development" opportunity requiring private financing, to a "demonstration project" underwritten by the government to showcase the beneficial uses of the new technology, to a scaled-down scientific "experiment" designed to answer more prosaic questions such as how big a hole in the ground would be made by a given charge buried at a given depth.

To corroborate the Neptune findings, the size of the shot need not be as large as the earlier model for Chariot, Johnson said. He revealed it had been scaled down to 460 kilotons, just 19 percent of the original 2.4-megaton blast. Furthermore, he said, because the new design would not require the wallop of a hydrogen bomb, a fission-type device would be employed. But the AEC's downsizing of the shot was small comfort to those Alaskans who cared about contamination of the environment. Gerry Johnson had acknowledged that the fission bomb was much "dirtier," producing 95 percent more radioactivity than a hydrogen bomb of the same yield. Gone, apparently, was the "clean bomb" rhetoric Teller had used in 1957 to convince President Eisenhower that a panoply of spectacular Plowshare projects depended on clean bombs. Now, a little more than a year later, when new data suggested Chariot should be a smaller—and therefore a fission—shot, Livermore's logic shifted.

*   *   *

In the time between Gerry Johnson's call to President Patty on February 18, 1959, in which he referred to "on the spot" contracts and his arrival in Alaska on the twenty-fifth, a welter of letters, phone calls, and miscommunications passed between Livermore, Fairbanks, and Washington. Professor Al Johnson questioned Alaska senator Bob Bartlett on his support of Chariot. Bartlett replied that while he had tentatively endorsed the project, he had reached "no fixed, unalterable conclusion on the subject." Within a few hours, he was writing Johnson again. Bartlett had heard that AEC chairman McCone made a statement before the Joint Committee on Atomic Energy intimating that Chariot was canceled. Bartlett confessed to Johnson that "my state at the moment is one of high confusion."

Al Johnson answered, stating that the university people were also quite bewildered: "We were informed on the one hand that the project is 'off,' and on the other that we should be thinking about contracts." Could the senator help clarify the situation before the Livermore scientists were due in Fairbanks the following week? Bartlett's terse reply showed his patience with the AEC had run out: "Quite candidly, the last thing I am capable of is clarifying anything having to do with the Atomic Energy Commission." The senator enclosed a carbon copy of a letter he had just sent to Gerald Johnson at Livermore in which he tartly withdrew his support for Project Chariot. Al Johnson received his copy in Fairbanks a couple days later, but when the original arrived at Livermore, Gerald Johnson was already in Alaska. So it came to pass that Professor Johnson had the pleasure of reading Sen. Bartlett's letter to his Livermore namesake at a public meeting in Fairbanks. In part it said:

> I had better tell you sooner rather than late that I am absolutely discouraged about this whole situation. . . . I had been convinced by you and others that this proposal was defensible and had in this respect taken a position opposite from that of some of my constituents. To have Chairman McCone announce as he did was anticlimactic and I must say my interest in the whole matter promptly evaporated.

Those in the audience who were on Bartlett's mailing list knew that the senator had been even testier in his recent newsletter: "I have now become completely disillusioned. If there is such absolute lack of coordination within the commission in planning, goodness knows what would happen when the trigger were pulled. For one, I hope the AEC does its blasting elsewhere."

But Gerry Johnson, who was as unflappable as he was affable (years later he would serve as a negotiator on SALT II and comprehensive test ban talks), laughed off the letter as unimportant. He told his Fairbanks audience that he had first heard of McCone's statement while in Juneau and had called Livermore to find out what was going on. It was only a misunderstanding, he said, and it had all been cleared up.

For the afternoon session, the university's scientific people convened in the faculty lounge. Gerry Johnson said he wanted to talk about the university's participation in "health and safety" studies. But only in a general way, as it turned out. As Al Johnson recalled in a 1961 letter, "everyone was getting very weary of the whole business." When asked about radioactivity and biology, Gerry Johnson and Sewell claimed to know nothing about it, except that biological risks would be minimized. Al Johnson did not remember any talk about contracts for the biologists.

President Patty's hasty memo announcing "on the spot" contracts was apparently premature. This latest AEC tour seemed purely informational—a chance to establish the new rationale for the project and to hobnob with those the AEC called "community opinion leaders." Meanwhile, on March 9, 1959, the Alaska legislature responded just as the Livermore physicists had requested: with a statement of support, which nudged Chariot closer to reality. House Joint Resolution #9 declared that, whereas the AEC's interest in the proposed nuclear excavation had been "effectively illustrated by the visit in Alaska of certain atomic scientists . . . the Cape Thompson project of the Atomic Energy Commission is hereby endorsed."

•   •   •

Dr. John N. Wolfe of the Atomic Energy Commission in Washington was known to be an urbane sophisticate brimming with wit and savoir faire, as comfortable quoting from literary classics as he was at explicating current scientific theories. He had earned his Ph.D. in ecology at Ohio State University, where he'd been strongly influenced by Edgar Nelson, one of the nation's pioneering ecologists. Wolfe had left a faculty appointment at Ohio State to accept a position as chief of the AEC's Environmental Sciences Branch in the Division of Biology and Medicine (DBM). In March of 1959, the AEC selected him to coordinate a program of biological studies

prior to the Chariot detonation. It was to be a choice of some consequence. As an academic scientist who had done notable research in the
then all-but-unknown field of ecology, Wolfe felt that "it [was] scarcely too
early in the Atomic Age to give considerable attention to the environment
which supports man on this planet." In a paper presented to the Second
Plowshare Symposium in San Francisco two months after he had visited
Alaska, Wolfe showed a knowledge and appreciation of the Arctic tundra
biome that was in striking contrast to the popular mythology:

> Not infrequently it is described as remote, barren, and climat-
> ically rigorous. Probably none of these adjectives is accurate,
> and very possibly they are all misleading. . . . It is not *remote*
> to the Eskimo, the Arctic fox, the ptarmigan; the flowering
> plants blooming by the thousands per acre in brilliant colors
> during the early period of wet earth belie its *barren*-ness and
> *desolation,* and represent the showiest natural floral display
> of any vegetation in the world; and the tundra climate is most
> salubrious to indigenous dwellers among which are some of
> the majestic animals of the earth, not excluding man.

In addition to his growing understanding of Arctic ecology, Dr. Wolfe
clearly saw the need to establish a new philosophy of development, one
that incorporated bioenvironmental knowledge. At the symposium, he
told the assembled would-be practitioners of "geographic engineering"
that it was "inconceivable that undertakings of this technical magnitude
should ignore the living aspects of geography." His paper, "The Ecological
Aspects of Project Chariot," seemed nearly a manifesto, articulating, possibly for the first time, the philosophic underpinnings of what would
someday be called "environmental impact assessment." He said that
while biologists were not competent to judge technical feasibility, "they
do raise the question of biological cost and can provide pertinent data as
to when, where, and under what biological conditions such an effort
could best be accomplished from a total environmental point of view."

On March 13, when Wolfe and his entourage touched down in Fairbanks, the *News-Miner*'s Albro Gregory launched into paroxysms of adu-
lation. Gregory was an ardent Chariot partisan whose news articles—not
"opinion" or "analysis" pieces—would regularly dub AEC scientists "bril-

liant" and "eminent," while labeling Chariot critics as "misinformed." After interviewing the scientists, Gregory's page-one story announced that "a team composed of some of the world's leading scientific minds is in Fairbanks today."

Wolfe had told the *News-Miner* that the group was formulating a program of "environmental" (rather than "health and safety") studies. He introduced his traveling companions as the nucleus of a still to-be-formed committee on environmental studies very much like the one the University of Alaska biologists had advocated should be established to oversee the investigations. Besides Wolfe, the other members of the Planning Committee on Environmental Studies, as they called themselves, were likewise AEC-affiliated scientists: two men from Livermore and two from other laboratories funded by the AEC.

Wolfe said that environmental studies would commence that summer, and that University of Alaska scientists would be working on some of them. He said that the studies would focus on food chains and on the movement of radioactivity through the environment, either by biological or physical transport mechanisms. "We want to know what is there," he said. If the danger was found to be too great, Wolfe told the newspaper, the blast would be called off.

In the faculty lounge at the University of Alaska, Wolfe was ready to get down to business. He wanted to know whom the university had, which parts of the envisioned study could be done by them, and what research would need to be contracted from other universities or agencies. But he did not yet have official authorization to spend money, and the university scientists pointed out that time was running out. It was March 13. The snow would be melting and the brief Arctic summer would be at hand. The logistics of a three-month stint of Arctic fieldwork was no small thing. There were people to hire, as well as tents, food, and gear to order and ship up from "outside," and all this to be transported to a spot on the tundra 560 roadless miles from Fairbanks.

Nevertheless, John Wolfe went to the comptroller's office and made statements that were so reassuring to the university administration that they were willing to authorize the hiring of additional biologists immediately. Because the university people had not yet prepared research proposals, Wolfe, according to one account, "essentially signed off to the university without having seen anything on paper, he essentially made a pledge . . . that AEC would supply funds and would supply contract money for university scientists to work in the Arctic the summer of '59."

In fairly short order, the University of Alaska biologists, who had led public opposition to Project Chariot, became contract employees of the Atomic Energy Commission. Not that anyone had any illusions that the university scientists, for all their protests, were entirely free of opportunism. As William Pruitt, a biologist hired for the Chariot studies, allowed years later:

> There [were] a number of people who, over a period of quite a number of years, had been trying to do Arctic research— field biology—living from hand to mouth, trying to pick up grants here and there. So that when this rather magnificent grant, or contract, came through, of course they jumped at it. Yeah, I did too. Sure. You know, a good field biologist will do damn near anything to get back in the field again.

Nor, says Dr. Pruitt, did the biologists fail to appreciate that with the funding came expectations of cooperation. There was always a feeling, says Pruitt, "that we were the tame biologists that they had bought."

Shortly after John Wolfe left Alaska, labor unions joined the newspaper publishers, the chambers of commerce, the legislature, and even some clergymen, by weighing in with support for Project Chariot. The Anchorage Building and Construction Trades Council, representing union workers hard hit by the postwar decline in construction, wrote to Wolfe rooting for results that would permit construction work: "We sincerely hope your findings will support the feasibility of the project."

·   ·   ·

William Pruitt had come to Alaska in 1953, a year after he had earned his doctorate at the University of Michigan. He was not a member of the University of Alaska faculty but rather a field biologist who lived near the campus in a log cabin without electricity or plumbing. During his first three years in Alaska, he had traveled all over the northern part of the territory doing small-mammal research under the auspices of the Arctic Aeromedical Laboratory. In 1957 and 1958, he took a contract with the

Canadian Wildlife Service to follow the caribou migrations throughout the year in northern Manitoba, Saskatchewan, and the Northwest Territories. With his wife and baby, Pruitt literally followed the herd for a year. His family stayed at an old abandoned Hudson's Bay post at Duck Lake in northern Manitoba. Later they moved on to Stoney Rapids, then to Yellowknife and Fawn Lake in the Territories. At Fawn Lake, Pruitt's family was with him in the field part of the time, living in a wall tent outfitted with a small wood-burning stove. Pruitt kept track of the animals with his dog team or on foot, and at night in the tent he examined caribou guts and blood samples by the light of a gas lantern. Pruitt's wife, Erna, had a master's degree in zoology, and when the baby was asleep, she assisted in the work, probing intestines and writing up the notes. The Canadian government provided good air support; while conducting aerial surveys of the migrating caribou, the bush pilots also flew in food, fuel, and scientific supplies. As the animals moved, Pruitt moved—sometimes by airplane, sometimes by dog team, sometimes by canoe.

According to Brina Kessel, then acting head of the Department of Biological Sciences at the University of Alaska, John Wolfe "demanded we hire Pruitt if we were going to get the contract." Wolfe thought the university team needed the strength of Pruitt's field experience. And, as Kessel conceded, "Bill Pruitt did Arctic field biology like nobody had ever done it before or since." As a consequence, Pruitt was hired as associate professor to teach biology and head the university's mammalian investigations at Cape Thompson.

Meanwhile, Al Johnson had begun to consider who else the university might hire to work that summer at Ogotoruk Creek, and he began to draft a formal research proposal for submission to the AEC. The proposal, which was ready by the twenty-fifth of March, 1959, sought over $107,000 from the AEC to support ecological investigations of the flora and fauna of the Cape Thompson region. The University of Alaska would stick to its areas of special competence—the vegetation, cliff-nesting birds, the mammals, including the peripatetic caribou—and leave other investigations to various other agencies. Al Johnson would head up the botanical research, cliff birds would be studied by L. Gerard Swartz, and Pruitt would lead the mammalian investigations. Written two months before John Wolfe's manifesto articulating the philosophy of impact assessment, the U of A proposal documents the biologists' hope for the ambitious study and foreshadows the scope of the "environmental impact statement," which would not be formally invented for ten more years:

> We should understand much of the basic ecology of the area, the interrelationships of all the plants and animals with their environment; the structure of the biotic communities; the dynamics of population fluctuations; the food chain—energy flow factors; and much more.

By May 22, Wolfe had secured AEC approval of the full $107,327 requested by the university for the first year's work. Besides the University of Alaska, which would conduct six studies, other institutions undertaking research included the University of Washington, Ohio State University, the U.S. Geological Survey, the Army Corps of Engineers, the General Electric Company's Hanford facility, the U.S. Fish and Wildlife Service, the U.S. Weather Bureau, and the Arctic Health Research Center. Forty-two separate investigations would include terrestrial, freshwater, and marine biology; oceanography; meteorology; geology and geophysics; archaeology; human geography; and radioecology.

One hundred and seven thousand dollars was big money to the University of Alaska in 1959. Aside from the affiliated Geophysical Institute, the university had never received anything like it. Brina Kessel, general supervisor of the project, finally relaxed. Twenty-five years later, she could still recall the day two university officials confronted her outside her office and made her aware of the administration's desire to land the big contract: "They didn't literally do so, but essentially, they grabbed me by the collar and said, Brina, don't let this research go through your fingers."

•   •   •

When Al Johnson put his mind to thinking about hiring field biologists for the big study at Cape Thompson, one of the first people he wrote was Les Viereck.

Leslie A. Viereck was reared just outside the old whaling port of New Bedford along the southern coast of Massachusetts. His father was a land surveyor who, whenever he could manage, fled the world of metes and bounds for the inconstancy of the sea: He liked to fish. Les grew up on the water, working on boats and crewing on yachts brought down by summer tourists. When he entered Dartmouth College in 1949, he took civil engineering courses, considering the possibility of a family business. But college, predictably, awakened other interests, which in turn suggested possibilities beyond his home town.

Viereck was drawn to the study of nature, in which Dartmouth offered opportunities. The Geography Department's David Nutt ran a schooner up the Labrador coast in summers. And Doug Wade, a naturalist, inspired Viereck with his knowledge of ecology, biology, and conservation. Wade was not required to teach formal classes; instead he led field trips nearly every weekend during the school year. Les Viereck was there, with long strides and wide eyes, his bird book and binoculars. He was there, too—not at Daytona Beach—when Wade took advantage of spring breaks to visit some interesting biotic community or other down in the Carolinas.

And, of course, Dartmouth had "Stef," Vilhjalmur Stefansson, the famous Arctic explorer and ethnologist who had lived for a year (1906–07) among the Eskimos, learning their language and culture and adopting their survival technologies. Later, he spent five consecutive years above the Arctic Circle and discovered the last four unknown islands of the Canadian archipelago. From 1947 until he died in 1962, he was at Dartmouth as Arctic consultant in the school's active Northern studies program. He brought with him his tremendous library of Northern materials and sat, more or less as a brood hen, incubating clutches of embryonic geographers, anthropologists, and naturalists. Seventy years old by Viereck's second year, the noted raconteur led a loose seminar, regaling budding explorers with tales of survival on the ice floes and the tundra. In the center of campus one winter's day, he taught Les Viereck how to build an igloo.

The summer following his second year at Dartmouth, Viereck was off to Alaska. He and a friend, hoping to find adventure and a summer job before starting their third year, drove a Model A Ford from New Hampshire to Fairbanks. The Alaska Highway was (and, until recently, remained) a narrow and winding track, at least 1,500 miles of which was unpaved. A potholed and dusty washboard while the long day's sun beat down, it could be a canal of mud in the rain. Between the isolated roadhouses where food, gas, and some repairs might be available, there could be 100 miles of road bounded by unpopulated wilderness.

His Model A survived, and Viereck landed work as a laborer in a coal mine at Healy near Mount McKinley National Park. He put in long, hard, dirty hours, but when the boss finally declared a day off one Sunday, Viereck headed for the park. Today a road links Healy and Denali National Park, as it is now called, but in the early 1950s there was no road. Just to have a look at the park, Viereck hiked the twenty-odd miles, through the Nenana River canyon to park headquarters and back, in one day. His Sunday stroll impressed the park superintendent enough that Viereck landed a job as a temporary ranger there in each of the next two summers. Whenever he had free time, he helped out the wildlife biologist and naturalist Adolph Murie. "This looks like the job for me," Viereck thought, "just wandering around the hills observing wildlife."

But he was wrong. He would return, by way of a meandering course, to botany, each detour adding to his kit of Arctic and subarctic experience. He completed a year of graduate work in wildlife biology at the University of Alaska in Fairbanks, then spent the summer camping out west of Mount McKinley, live-trapping marten and studying their food habits. He quit wildlife biology and was quickly drafted. Happily for Viereck, Uncle Sam decided to make use of his outdoor skills, assigning him to the Talkeetna Mountains, where he served as a military game warden, and flying him up to the Brooks Range to survey edible plants. He climbed 20,320-foot Mount McKinley. He spent a year in Labrador taking weather readings, studying the vegetation on an alpine summit, and getting married during a three-day leave in the spring. He earned a master's degree in botany at the University of Colorado; started a Ph.D. dissertation there on the revegetation of glacial moraines; explored glaciers around the Kenai Peninsula, the Alaska Range, Prince William Sound, and McKinley Park; and, before he was quite finished with his Ph.D., made the fateful decision to accept a job offer at the University of Alaska to work on Project Chariot.

Al Johnson had been a few years ahead of Viereck at Colorado and was now on the staff of the biology department at the University of Alaska. He wrote Viereck to say he was "about 85% confident that we can offer you a job" teaching and doing research. Johnson said he needed an assistant on studies to be sponsored by the U.S. Atomic Energy Commission at Cape Thompson in northwest Alaska. He himself was committed to several projects already and would only be able to spend about three weeks in the field during the summer of 1959. He could design the scope of the scientific investigations and assemble the equipment and camping gear, but he wanted Viereck to run the program in the field. Viereck was reluctant to abandon his dissertation before it was finished, but to be able to return to fieldwork in Alaska—and to be offered a job at the university—was more than he could resist.

As soon as school let out in Colorado, Les Viereck and his wife, Teri, who had just finished her Ph.D. in animal physiology, drove north. Within a day or two of arriving in Fairbanks, Les flew off to start botanical investigations at Ogotoruk Creek.

•   •   •

Thanks to Al Johnson's advance planning, University of Alaska biologists and field assistants were the first researchers to arrive at Ogotoruk Creek and were comfortably installed in their tents three weeks before construction of the AEC camp began. The party was favored by a rare two weeks of sunny, calm weather. Gerry Swartz's crew turned their attention to the seabird colonies inhabiting the Cape Thompson complex of cliffs. Nearly half a million puffins, murres, kittiwakes, and other cliff-nesting birds had already arrived from the south, swirling in gyres off the sheer rocky cliffs and pinnacles facing the sea. Viereck and his assistants headed up the valley to make a rough survey of the area. Seven miles up, they saw shorebirds nesting in the marshes around several ponds. Young longspurs were already trying their wings, their nests abandoned. Some of the inland ducks who had hatched their young trailed strings of ducklings over the black ponds. "The headwaters of the Ogotoruk was a great surprise to me," Viereck wrote in one of his notes. "From the aerial photos, I expected it to be all upland shale with the typical *Dryas* and other mat vegetation. Instead it is a very lush, *green* basin several square miles in extent . . . we saw two herds of 500 caribou each crossing the very headwaters of the Ogotoruk."

The serenity ended on July 5, at 9:00 A.M., when "the great confusion," as Viereck called the AEC's disorganized landing, commenced.

7   DROP US A CARD

> If your mountain is not in the right
> place, drop us a card.
>
> —Edward Teller

In 1959 and 1960, the University of Alaska conferred, or offered to confer, honorary doctorate degrees upon a remarkable gallery of personages.
J. Edgar Hoover, in the thirty-sixth year of his forty-eight-year reign as director of the FBI, was offered the distinction "in honor of [his] great service to our nation." Hoover, whom I. F. Stone called "the head of our
secret police," reviewed the bureau's files to check on President Patty and
to see if the University of Alaska might be soft on communism. "Nothing
derogatory in Bufiles [Bureau files] for Patty," Hoover was advised. But as
for the University of Alaska's past, there was one worrisome incident:

> Bufiles reflect that on 5-21-51 an effort was made to lower
> the American Flag and to raise the "red" flag over the Univer-
> sity of Alaska. Subsequent information reflected this was
> probably a student prank. . . . No investigation conducted by
> the FBI.

Perhaps the FBI did not mount a full-blown investigation of the red flag
incident, but Hoover did have his agents look into the matter, and a fat

sheaf of papers on the subject of "Communistic activities, University of Alaska" still resides in the FBI files today. Citing pressing official matters, Hoover wrote to decline the honor of the honorary doctorate.

The board of regents, led by Anchorage banker Elmer Rasmuson, Alaska's civilian liaison to the secretary of the army and reputedly Alaska's richest man, sent out more such invitations within the brotherhood of militant anticommunists and weaponeers. They wrote Wernher von Braun, the former Nazi rocket scientist who had helped to develop the V-2 rocket in an underground factory that brutally utilized slave labor, a clear-cut war crime under the Nuremberg laws; Hyman Rickover, "father of the nuclear submarine"; Wilber M. Bruckner, secretary of the army; and the biggest cold warrior of them all, father of the H-bomb, Edward Teller.

Teller was pleased to accept.

•   •   •

"I do not wish to speak to you about generalities," Edward Teller told the class of 1959. Instead, he preferred to discuss his work and how it might "have some bearing to the future development of the great state of Alaska." He spoke of nuclear power plants as a vital source of energy for Alaska's development, and mentioned the $6 million experimental reactor then under construction just 100 miles away at Fort Greely. He told the capacity crowd of his dream "to engage in the great art of what I want to call geographical engineering—to reshape the earth to your pleasure and indeed to break up the rocks and make them yield up their riches." Considering Project Chariot in particular, Teller found a sort of moral symmetry in the plan, quite apart from the "industry and progress" he usually boosted:

> Please God, that by making harbors here in Alaska, perhaps
> near coal deposits, by exporting this coal cheaper to Japan,
> the Japanese might become the first beneficiaries of atomic
> explosions as they have been the first victims.

The key to this bright future, where riches would abound and past wrongs would be righted, lay in overcoming "this hysteria about fallout." In a tape recording of his 1959 speech, Teller's words march in step, shouldering themselves forward in deep, *r*rolling accents, like great Hungarian warriors:

> Fallout, when it occurs in a concentrated form near a nuclear blast which has not been properly controlled, fallout can be very dangerous. But fallout, as it occurs dispersed over the world, carefully controlled by the work of many conscientious people, this fallout contributes to radiation less than the wristwatch I am wearing on my wrist. We know that radiation in great concentrations is harmful. The harmful nature of the exceedingly dilute form of radiation which we encounter in fallout has never been proved and in a strict scientific or statistical sense has never been noticed.

In a 1988 interview, environmental activist Barry Commoner refuted these assertions. And, in contrast to Teller's lugubrious intonation, Commoner's delivery puts one in mind of a street fighter from Brooklyn:

> Well, Teller, in this area, is a scientific charlatan. No two ways about it. I constantly recall the famous statement he made in connection with low level radiation effects. He said, *If low level radiation had an effect on genetic mutations, then it should be apparent among the Tibetans who live at very high altitude and are exposed to a great deal of cosmic radiation. No one has ever observed an increase of mutation rate among the Tibetans.* Well, of course, no one ever looked. There has never *been* a genetic study of the Tibetans. And that's typical of the way in which he handled data. The short statement that you read has at least three or four specific scientific errors.

"Total nonsense," Commoner says of the idea that fallout is "carefully controlled." "They set off the bomb and that's it. The weather takes it [fallout] wherever it's going to take it." And the wristwatch comparison, a Teller favorite, would only make sense if people were in the habit of eating their watches: "What you forget is that fallout radiation gets into your body. It's not on the outside." Strontium 90, for example, which is a constituent of fallout, emits beta rays that cannot penetrate skin. It would be relatively harmless on a luminous dial watch, but could be deadly if ingested. Because it is chemically similar to calcium, the body readily incorporates strontium 90 into bone. Once there it can irradiate bone, bone marrow, and other tissues adjacent to the bones, leading to bone cancer and leukemia.

Scientific charlatan or not, Edward Teller was awarded the honorary degree of Doctor of Science on May 18, 1959, by the University of Alaska "in recognition of his fearless endeavors to strengthen his adopted country against the menace of tyranny."

●   ●   ●

A month later, Teller was back in Alaska, this time accompanied by his wife, Mici. Again he gave press conferences and speeches, promoting Project Chariot from Juneau in southeast Alaska to Kotzebue in the far northwest. In Juneau, he tried to meet with Gov. Bill Egan. But because Egan was in poor health, a young biologist from the Alaska Department of Fish and Game named Jim Brooks was asked to meet with Teller in the governor's mansion. When Brooks pressed for details of Chariot's potential for environmental damage, Teller bristled, and mentioned his personal stake in the cold war with the Soviets. Brooks recalls their meeting:

> As I pursued this line of questioning, he became more irritable. And at one point he said that if Soviet Russia should surpass the United States in the development and use of nuclear energy . . . it would be one of the worst setbacks in his life, and he was very emotional in saying this. . . . I have a vivid recollection of that part of the conversation. [He] made it pretty evident that his motivation was not altogether concerned with domestic application of nuclear energy. If it were, I doubt that he would have invoked reference to the Soviet Union.

Brooks recollects that, despite the "emotion-charged atmosphere" of the conversation, he tried to keep the focus on public safety issues of importance to the state. "I tried to remain as calm as I could," he says, "but I also tried to phrase the questions that I knew would require answers, probably that he couldn't provide at the time, but that had to be asked anyway." Brooks was familiar with the scientific studies conducted after nuclear blasts in the Pacific and had concerns about radioisotopes moving up food chains and possibly concentrating in humans. Teller, he said, had no patience for these worries. "He interpreted my concern and probing questions . . . as being an expression of disapproval at the outset," said Brooks. "Of course, that wasn't at all my intention. I had no authority to approve or disapprove anything." Brooks felt that the purpose of the meeting was for Teller to brief the state government, and Brooks "wanted to have all the information he was willing to give in the area of environmental concern."

The sales campaign went better for Teller in Anchorage, where he spoke at Elmendorf Air Force Base. He had some fun with reporters, serving up trademark one-liners. Acknowledging that the harbor near Cape Thompson would have no economic value, he held out the prospect that other harbors in Alaska might follow if the experiment proved successful. When asked where such harbors might be built, he said, "That is like a little girl asking, what do I want for Christmas? It's up to you." Explaining his idea of "geographic engineering," Teller deadpanned, "If your mountain is not in the right place, just drop us a card."

At the University of Alaska a month earlier, and throughout his visits to the state, Teller recognized the rhetorical value in playing to Alaskans' collective self-image as intrepid, pioneering people known for their spirit of adventure—qualities that Americans had always thought to be associated with frontier life. Selling this image back to Alaskans, Teller was confident that these modern pioneers would seize the opportunity offered by nuclear excavation. "Anything new that is big needs big people to get [it] going," he was fond of saying, "and big people are found in big states."

After speaking at Elmendorf, Teller flew to Nome, where he was greeted enthusiastically by civic leaders at an impromptu luncheon. He was joined there by "Dr." Rodney Southwick, upon whom the *Nome Nugget* had apparently bestowed an honorary doctorate. Southwick was actually an AEC public relations officer who represented the Plowshare program. He liked to describe himself as a hard-bitten old newspaperman, but the *Nugget* called him one of "the world's foremost atomic scientists." Also in the group was Clifford Bacigalupi, an engineer in charge of construction operations for Livermore.

Southwick and Teller spoke at the district courtroom after lunch. In his remarks, Teller failed to note that Chariot had been redesigned, that it was now a cratering experiment, and that the AEC had disavowed any long-term commercial value. He continued to refer to the crater as a harbor that would make it possible to export known rich deposits of coal.

Bacigalupi also seems to have taken some liberties with the facts. According to the *Nugget*, he stated that "to date the studies indicate that early springtime will be the best time for blasting the harbor. For it's at this time the biological, weather and control conditions are the most favorable." But, as this was June 28, 1959, the environmental studies were only barely begun. The University of Alaska researchers had just set up their camp at Ogotoruk Creek a couple of weeks earlier, and the barges hauling the AEC camp were still en route. No studies in biology or weather had yet been done at Ogotoruk Creek in the spring months of March or April.

The AEC preferred a spring detonation because they needed at least six hours of daylight for flying before and after the shot, and for the photography that would help to determine the actual yield. The short Alaskan days between early November and early February did not offer sufficient daylight. But it was not true that as of June 1959, the commissioned biological or weather studies supported the AEC's preference for a spring shot. This was the first hint that the AEC might have predetermined the conclusions of the vaunted environmental studies.

After the presentation in Nome, the AEC party flew 200 miles north to Kotzebue. Teller addressed the townspeople and students at Kotzebue High School, then boarded a chartered light aircraft to view Ogotoruk Creek, 125 miles farther north. As he was considered to be a valuable national security asset, he flew in a twin-engine plane. And because the temporary strip at the Chariot site was not long enough for the twin to land, Teller simply looked things over from above. Below, field biologists paused from their sampling to shield their eyes and wonder about the circling Beechcraft. One can imagine the redoubtable physicist leaning into the window, looking down on the Ogotoruk Creek valley. Perhaps in his mind he saw a kind of reenactment of the third day of creation: his awesome explosion demarcating what would be land and what would be sea.

Then the plane banked a final time and disappeared into the southern sky.

●   ●   ●

From his camp at the mouth of Ogotoruk Creek, Les Viereck watched the first of two expected AEC supply barges heave into sight. The two barges, departing from Seattle and Anchorage, were supposed to have reached Ogotoruk Creek together on June 30. A temporary camp was to have been constructed before most of the labor force arrived, and then, once the men could be sheltered and fed, work on a more substantial, permanent camp would begin. But, as both barges were delayed, no shelter huts had been erected by the time men began gathering in Kotzebue, 120 miles south of the job site. This group included supervisory personnel from Holmes & Narver, the company regularly hired by AEC to run its field operations in the Pacific and in Nevada. They had flown up from the Nevada Test Site near Las Vegas and from H & N's headquarters in Los Angeles. Also gathering at Kotzebue were twenty-three local men hired for camp construction by B & R Tug and Barge Company of Kotzebue, which had been awarded the labor contract.

Several hours after the arrival of the first barge, which had picked up a load of fuel oil in Kotzebue, two aircraft landed at the site with an advance party of AEC personnel and supervisors from H & N. With the barge nosed into the beach, B & R bargemen worked steadily through the sunlit Arctic night unloading drums of oil. The next morning the barge from Seattle arrived, and with it came the kind of storm the Arctic coast of Alaska can serve up on short order. The second barge was only 25 percent unloaded when, as Viereck wrote in his notes, "all hell broke loose." High winds, a driving rain, and rough seas forced both barges to pull away from shore to keep from being pounded by breakers. They headed up to the shelter of Marryatt Inlet at Point Hope, and the critical unloading of provisions for the temporary camp stopped.

To add to the chaos, there seemed to be no coordination between H & N supervisors and the pilots. Two pilots flying low-winged bush planes for Safair out of Kotzebue dutifully met their schedules, dropping men at the job site despite the turbulent air. The low-winged planes were tougher to land in rough weather, as the rocking wings were that much closer to the gravel strip. The Alaskan researchers were amazed at the demonstration of flying skill and found out later that the pilots were two young Eskimo boys from Kotzebue.

Men were accumulating on the beach, and there was nowhere to shelter them. They ended up moving into the little tent camp set up by the University of Alaska researchers. In the race to unload the barges quickly, no care was taken to off-load essentials first. It would have been a near impossible job anyway, because the packing crates were not color-coded or otherwise marked in anticipation of an Arctic beach landing. As a result, wrote one Chariot contractee, "the beach was strewn with packing boxes, innumerous of which had to be opened to ascertain their contents."

Working straight through the storm, the crews managed to erect two sixteen-by-sixteen-foot Jamesway huts by the night of July 6. By July 7, the overtaxed camp cook was cooking for forty men on two two-burner camp stoves, one of these borrowed from the University of Alaska researchers.

Not until July 9, three days after the barges were forced to stand off, did the storm lift sufficiently to allow them to return to Ogotoruk Creek. Unloading was completed the next day; by then, eight Jamesway units were up, serving as temporary housing and galley facilities. It was still crowded, with eight men in each sixteen-by-sixteen-foot shelter, but work was gradually progressing on construction of the permanent camp and several investigations had begun.

•   •   •

By August of 1959, the Chariot camp was a bustling colony. The permanent camp included ten dormitories, a mess hall, a shower and laundry building with hot water available at all times, and a building housing water purification equipment and three gasoline generators. There were two warehouses, a communications building, a mechanic's shelter, wood-frame latrines, and sixty-six men at the mouth of Ogotoruk Creek.

Cargo trucks lumbered along roadways that the Holmes & Narver crews had bulldozed to connect the camp with the barge unloading area, the new airstrip, and the core-hole sites. Nine tanklike track rigs, known as "weasels," clattered noisily over the tundra, hauling scientists to distant test plots. Offshore, the *John Cobb* and the *Brown Bear* rolled in the swells as oceanographers and ichthyologists gathered samples and took measurements. Two sixteen-foot fiberglass skiffs with outboards carried biologists on short trips out to the sea cliffs and the resident bird colonies. Planes landed and took off, the generator shack throbbed with a constant *pop, pop, pop,* and the cooks served up three meals a day to a camp population that sometimes reached seventy.

On the whole, camp life was harmonious. Scientists and heavy-equipment mechanics could relax together, playing cribbage or watching films. Many of the Holmes & Narver hands had worked at remote sites in the Pacific and elsewhere and knew the drill. But once in a while there were words between the young men who were trained in the life sciences and the young men whose idea of recreation was target shooting at the wildlife.

A former Wien Airlines pilot tells a story that illustrates the dissimilarity in worldviews. Several of the Holmes & Narver personnel were deathly afraid of bears. When a big grizzly was spotted up the valley, a bulldozer operator named Hedrick took to sleeping with a .44 on the bedside table. One night the bear got scent of the camp and came down to take a look. He ripped a four-foot hole in one corner of the temporary hut Hedrick was sharing with the University of Alaska's Bill Pruitt. Hedrick leaped for his gun as Pruitt shouted, "Don't shoot him in the head! Don't shoot him in the head!" A biologist first and last, Pruitt was mainly concerned that the skull be kept intact for measuring.

Additional demographic diversity was provided by female researchers. Though it was not a common practice in those days, and though the AEC initially balked at the idea, integration was achieved when the University of Alaska's Brina Kessel insisted on sending women scientists to Ogotoruk Creek. The first two of these were each at a disadvantage physically. One woman had fallen off the back of her pickup truck and broken a leg, but the cast was not going to hold her back, says Kessel. "She had sewed up her jeans so that she had a zipper in the side of them, and several pairs of them." The other, who was pregnant, also found a sartorial solution to her predicament, "she took several different sizes of jeans up there, so that during the summer she was able to use the bigger and bigger size."

With Natives and Caucasians, men and women, construction workers and professors, the camp's population mix was not so different from any other town's. It was only the mission that was unusual.

Construction activity continued around the clock in a race against the coming freeze-up. Since the 800-foot runway along the little ridge next to the camp accommodated only small aircraft, H & N staked out a new one on the high ground across the creek. An outcrop of fractured shale provided the fill material, and two scrapers, four bulldozers, and a road grader were fired up to haul and spread the rock. Men and equipment worked two ten-hour shifts each day, building a 2,200-foot-long landing field that could accommodate Wien's twin-engine Beechcraft on its thrice-weekly run between Kotzebue and Point Hope.

Boyles Bros. Drilling Co. of Salt Lake City had less success. Working day and night with U.S. Geological Survey crews, they hoped to extract two or three 1,000-foot core samples for shipment to Livermore. But because of the frozen ground, they averaged only thirty-nine feet per day. Compounding the difficulties was the fact that drilling caused the surrounding permafrost to thaw, and the sides of the holes caved in each time the drill rod was removed. Frustrated, the drillers quit the first hole before reaching 600 feet.

Frozen ground caused more problems at the second drill site. The casing kept sliding down the hole, and cave-ins from thawing ground became so numerous that the recovery of core was abandoned. The technicians resorted to logging the drill cuttings that piled up around the hole.

By now, Boyles Bros. had a healthy respect for the difficulty of Alaskan fieldwork, but the point was cruelly reinforced at the season's end. On September 16, 1959, 1,500 feet of core samples, together with Boyles's drilling equipment, was loaded on B & R Barge No. 521, which departed Chariot camp under tow, bound for Seattle. Between Cape Saint Elias and Cape Spencer in the Gulf of Alaska, the tug and barge ran into three full gales. Heavy weather and rough seas pounded the vessels until, when they were almost across and into the shelter of the Inside Passage, seams in the barge hull opened. It took on water, listed to starboard, and capsized. Weather conditions at the time, according to the master's accident report, were full gale, rough sea, huge swells, and winds gusting to seventy miles per hour. All the core samples—the entire season's output—

and all the drilling equipment, except for one eight-foot length of pipe, were lost.

When Plowshare officials reported to superiors on their progress at Ogotoruk Creek, they conceded they had become "acutely aware" of the rigors of operating in the Arctic. But, if the AEC had learned this lesson with respect to the environment, they had yet to learn one concerning the necessity of liaison with the local people. No one from the AEC had visited Point Hope; no one had told the people about what was going on at Ogotoruk Creek, or why. And this oversight would prove a far greater impediment to the AEC's plans for Ogotoruk Creek than the weather would ever be.

8   POLARBASILLEN

We like him. He help a lot too. He help the people too, in springtime haul the meat, yeah.

—Kitty Kinneeveauk

He was my best friend and stayed close to us.

—Elijah Attungana

He lived closely with the people and learned with us.

—Teddy Frankson

At Dartmouth College, Don Foote was charismatic and handsome, lighting up a room with brains and energy and wit. Les Viereck was lank and craggy, soft-spoken and shy. Women were always falling for Foote's white-toothed smile and his attentive way of listening when they spoke. Viereck had the mild, intelligent eyes of a saint, though at times they could show wariness, even alarm. Although they were at Dartmouth at the same time, and had many mutual friends, they moved in different circles and were only casual acquaintances. No one who knew them both at Hanover would have guessed that in a few years, at the far end of the continent, they would become fast friends . . . or that in a few more years Les Viereck would unfold his long frame to stand up and deliver a simple, moving, Gettysburg-like eulogy over Foote, the *Wunderkind,* dead at thirty-seven.

●   ●   ●

Foote was born Don Carlos Barbito in New York City in 1931. His father, a Spaniard with a love of the racetrack, left the family when Don was two and his younger brother, José, a toddler. Eventually, the boys' mother Anglicized her sons' names and changed their surnames to Foote, her maiden name. The family lived for a few years in New Jersey, and briefly in Arizona and California, but from about the time when Don was eight, they lived in Bridgewater, New Hampshire, a place so rural that Joe called it a hermitage. Mrs. Foote wanted a better education for the boys than was available locally, and that meant a boarding school. Over Don's emphatic protests, the boys' new stepfather, with whom Don often had violent confrontations, prevailed in sending him to a military academy in Virginia.

Don hated it. After a disastrous first year, Joe Foote remembers his brother coming home and telling his parents, "Well, you've had your shot. Now *I'm* going to pick a school." His mother said that would be fine, but she would not be able to help him financially. Don selected the Proctor Academy in Andover, New Hampshire, landed a scholarship, and flourished there. He was elected class president, made captain of the baseball team, and named valedictorian. Delivering the valedictory address in 1949, Don spoke confidently of man's imminent destiny in space. Parents and teachers exchanged glances and quiet snickers, commiserating with each other in an amused tolerance of youth and illusion. But just eight years later, Sputnik whirled across the night sky, heralding the space age and validating one young scholar's prescience.

Foote entered Dartmouth with a full scholarship, but without definite career objectives. Then in 1951, after his second year, everything changed. He landed a summer job with the U.S. Weather Bureau and found himself aboard a research vessel bound for Greenland. There he was transfixed by the ice sheet, half a million square miles of it, averaging 1,000 feet in thickness. Once, he saw a polar bear. The clean, cold beauty of the Arctic world struck him with awe—he had been bitten by what the Norwegians call *polarbasillen,* the "polar bug." The twenty-year-old suddenly knew all at once and with complete conviction that he wanted to be an Arctic geographer. "From there it was a straight shot," says his brother Joe, "a single-minded pursuit of that objective."

The next summer, Foote went to Scandinavia and saw Norway, Sweden, and Iceland. Then, after graduating from Dartmouth in 1953, he worked on the Russian-Norwegian boundary study funded by the U.S. Office of Naval Research. He stayed on into the following year, studying at the University of Oslo, revisiting Sweden and adding Finland, Denmark, and the Spitsbergen Islands to his list of places seen, before the U.S. Army drafted him into service. Even during his stint in the army, Foote contrived to be posted to Norway, and by the time he was discharged, he spoke fluent Norwegian. In 1957, he began graduate studies at Montreal's McGill University, which, like Dartmouth, was known for its excellent Northern studies program.

Foote earned his master's degree in geography in 1958 and had begun his Ph.D. studies when restlessness overtook him. He thought McGill was too dominated by Cambridge dons who, though they were knowledgeable enough, had never really *done* anything in the Arctic. He wanted to be in the North again. He wanted to see Alaska. After securing a Carnegie Foundation travel grant, Foote began inquiring about summer employment in the newest American state. He soon learned about a research project that might fit his bill. A regional study under the auspices of the Atomic Energy Commission was to commence shortly in the Cape Thompson area of northwest Alaska near the village of Point Hope. Just what the purpose of the study was, Foote did not know. But he wrote to the AEC's Lawrence Radiation Laboratory to announce his availability for summer work. Writing in generalities, he mentioned his interest in "regional studies, planning and development in the northern fringe areas of the world."

He must have been bowled over when a Western Union telegram arrived shortly thereafter requesting that he draw up a budget. Still unaware of the AEC's plans for Cape Thompson, within twenty-four hours Foote sent off an outline for a program of studies that "involves a complete mapping of the study area, to include topography, soils (including permafrost), vegetation, drainage lines and run-off rates, existing settlements, communication and transportation links and industry" and more. Almost before he knew it, he had landed a $50,000 research project.

Crowding his luck, Foote made a handful of major life decisions all at once. He put graduate school on hold, invited his brother Joe to join him on the drive to Alaska, and arranged to visit his Norwegian sweetheart, Berit Arnstad, in Yellowknife, Northwest Territories. He wanted Berit to marry him, to head off with him to an Eskimo village in Arctic Alaska and who knew what adventures.

*   *   *

Don and Joe Foote left the East Coast for Alaska in a Volkswagen bus on the fifth of August, 1959. On the thirteenth, at Edmonton, they separated, Don catching a plane to Yellowknife, and Joe continuing north and west to Dawson Creek, B.C. At Yellowknife, on the shore of Great Slave Lake, Don met Berit, who was staying with friends and working at a fisheries job. The next day, he entered an uncharacteristically bland notation in his journal, "Berit and I, in a rather cool northwest wind, wrestled her fish cans about on the Wardair dock. In the afternoon we finally decided to be married."

They made arrangements with a somewhat overformal minister called Douglas, and the next day drove to the church with a couple who had agreed to be witnesses. The Reverend Mr. Douglas only grudgingly consented when Don turned off the fluorescent lights "to allow the setting sun to light the church."

> The important part of the ceremony was simple and impressive enough despite the good minister's dogmatic concepts of who should stand where, whose hand should be in whose hand, etc. Any atmosphere of meaning was exploded when Douglas started to preach a little sermon which had no place in the ceremony and certainly nothing to do with Berit and me.

At a small dinner party later, some of the artificiality of the affair was lifted when, as Don wrote, "old Jake Woolgar gave us one of the most sincere and impressive congratulations with a 'Goddamn it, God bless you both,' and a whiskey kiss for Berit."

On the first full day of his marriage, as might befit a bright, idealistic twenty-eight-year-old of the 1950s, Foote seems to have run through emotions ranging from the romantic to the existential:

> SUNDAY AUGUST 16, 1959: A magnificent day. Berit and I lingered over breakfast as perhaps only she and I can do. Some work for Fisheries, the sounding line, and short trips around town picture taking from the wildlife truck.
>
> Visits . . . took most of the afternoon and evening and it was not until midnight did we arrive back at Dick's hut. A near-full moon sawed through a high cloud layer, northern lights snaked low over our heads, Jake Woolgar opened another bottle of Scotch and an Indian beat out some slow monotonous rhythm on a sauce pan.
>
> I understand very little of what is happening here but there is no serious effect which would indicate fundamental violations of an inner self. Marriage remains as much a mystery to me as Berit

Two days later, on Berit's birthday, Don flew off by himself to rendezvous with brother Joe and the Volkswagen at Dawson Creek, the start of the Alaska Highway. Berit would shortly return to Norway, then fly back in early November to join Don in Alaska. The Volkswagen reached the end of the road, Fairbanks, on August 25.

Nine days later, Don Foote and his research assistant, Tom Stone, landed at Point Hope. When they had retrieved their gear from Wien's twin-engine Beechcraft at the Point Hope airstrip, they were directed to Browning Hall, an old army mess hall with a Quonset hut attached that served as both movie house and town hall. There they met David Frankson, an important *umialik* (whaling captain), who was also the village postmaster and had just succeeded Dan Lisbourne as president of the village council. At forty-nine, Frankson was thin, bespectacled, and a full head shorter than his wife Dinah. But this shrewd, tough man was one of Point Hope's strongest leaders in modern times.

Before he could brief Frankson on Project Chariot, Foote had to sit through a showing of the aptly titled film *Rebel in Town*. Then, in a three-hour session, he described the nature of his contract work for the AEC and told the council president that he would need the full cooperation of the Point Hope people in order to document their hunting economy and their use of the land.

The following night, nine-tenths of the village population attended a church party at Browning Hall. Tea and cake were served as the Reverend Keith Lawton showed slides. Afterwards, Foote took advantage of the assembly to explain Project Chariot. With maps and diagrams, he outlined the proposed nuclear blast and the scope of the environmental studies while David Frankson translated into Inupiaq. He said, "It is my personal opinion, and that of the AEC, that if Chariot proves harmful to the local people it should not be carried out." As a result of his up-front approach, Foote said later, he got 100 percent cooperation from the people of Point Hope. But that didn't mean they were signing off on Project Chariot.

The next day, to a great deal of excitement, the annual supply ship *Northstar* arrived. It sat at anchor about 800 yards off the beach; cargo was lightered ashore and practically the entire village turned out for the longshoring work. Straight-time wages paid by the *Northstar* varied according to the worker's age, ranging from seventy-five cents an hour for those over eighteen to fifty cents a day for six- to eight-year-olds. Foote noted that the women seemed to account for the bulk of the heavy work—and it was no small thing to haul fifty- and sixty-pound loads up the beach through sand and loose gravel. One woman carried even 100-pound sacks of flour by herself. The men seemed to concentrate on sorting the goods into piles according to owner. When the *Northstar* departed, fall began.

*   *   *

As a researcher, Foote had hit the ground running, logging in the village's hunting catch from his first day in residence. But between running down to the beach to count and weigh the animals whenever he saw a boat pull up and taking hunting trips with Dan Lisbourne, Antonio Weber, and the Reverend Mr. Lawton, he had to find a house and ready it for Berit's arrival and winter. Besides, he was getting tired of writing up his notes to the accompaniment of the sound tracks of the shoot-'em-up westerns be-ing shown in Browning Hall.

It took three weeks of searching and sending telegrams and letters to absentee owners. Finally, on September 28, he rented a small—perhaps fourteen-by-sixteen feet—two-story structure that he said was "without insulation, electricity, plumbing or water storage facilities. It lacks several windows, furniture, stove pipe and bed. Besides all this, the roof leaks." Foote might have been thankful it was of frame construction; some Point Hopers still lived in the traditional semisubterranean sod houses framed up with whalebone. In fact whalebone was still probably the dominant structural form at Tikigaq. When a Fairbanks artist named Claire Fejes visited the same year Foote arrived, she wrote about Point Hope, "Georgia O'Keeffe, the artist, would have loved the bone architecture; bone instead of wood for walls, clothesline supports, boat racks, grave markers, skin-drying racks, *umiak* racks, stakes for tying huskies, and for decoration. Some whale bones were twice as high as the dwellings next to them."

Foote set about repairing his wood-frame house "with those few materials available here in the village, i.e. sod and cardboard." He gathered popcorn cartons from the movie house and other boxes from the mission's used-clothes sale and used them to cover the interior walls. He convinced the AEC to send up some scrap lumber and insulation from the Chariot camp. To improve the exterior, he put plywood against the walls and stacked new sod up against them. He installed windows and scrounged parts for a crude wiring job so he could use a light bulb or two when the village generator ran between 8:00 P.M. and midnight. He plumbed an oil barrel to a small, gravity-fed furnace. Meanwhile the winds kicked up to nearly sixty miles per hour and temperatures dropped to below freezing. "House far too cold to work," he wrote one fall night, "I sit right next to the stove and my left side is much too uncomfortable." By October, he had the place in fair shape, though when it was windy, which was nearly always, the floor was quite cold. By November, he was writing the AEC again for "some type of sheeting to cover at least the north and east walls"—the walls facing the dominant wind. As he worked at his typewriter one night he noted that, with the furnace going full blast, a water bucket ten feet away and two feet off the ground was slowly freezing solid.

"He did everything to blend in," said one Eskimo woman who remembered Foote. "He wasn't like a teacher. Those were the three kinds of white people we saw: preachers, teachers and anthropologists. He ate the Eskimo food. Even his wife fit in." Foote ate the Eskimo food all right, but it's hard to tell if this was a matter of preference, necessity, or a sense of camaraderie and good sportsmanship. He wrote to the AEC in September that "both Tom [Stone] and I have suffered ill effects from the Native diet—while in the field and on local hunting trips—and perhaps the village water system." But a short time later, in a letter to a former classmate back at Dartmouth, Foote sounded no less an advocate of the native diet than Stefansson himself:

> There is little doubt that raw meat gives added warmth, and there is a continuous craving for fats. To date I have had *no* ill effects from the food which has included whale, oogruk (bearded seal), common seal, snowy owl, ptarmigan, caribou, walrus, tom cod, trout, grayling, salmon, squirrel, etc.

In the second week of November, Berit arrived, and like her husband she leaped into the Eskimo lifestyle with a vigor still talked about in Point Hope today. By the end of November, Don wrote a friend, "She has her new mukluks, her parkie is in the making, and she has started my 'fancy' mukluks for the Christmas feast." She "was just like an Eskimo woman," says one man from Point Hope who remembers her, "she learned how to do everything the Eskimo women do." Berit did learn to sew skins and butcher and cook sea mammals and caribou, but she was too adventurous to be limited to the woman's role in Eskimo society. "She started to dress the way Eskimo men—young men—would dress, rather than the way Eskimo women would dress," says Jackie Lawton, who lived at Point Hope with her husband Keith, the Episcopal minister. "She wore the tight mukluks that the young men would wear, not the type that the women would wear. . . . She wanted to be in the men's circle." For starters, she wanted to go hunting, and did manage to go on some caribou hunts, up into the hills by dog team. If Don was otherwise occupied, she sometimes teamed up with an Eskimo hunter on trips lasting several days—something native women did not do. And she very much wanted to be on a whaling crew. "They weren't too happy about that idea," remembers Jackie Lawton. "But finally when she, I guess, pushed herself a little bit more into the situation, they took her on. And . . . they treated her more like a boy . . . so she said they kept asking her to go and get things . . . the way they would use the young boys in camp. And this didn't sit too well with her, but she was glad to be part of the whaling scene anyway."

•   •   •

Those who knew Don Foote say he was a man of tremendous energy who could work for long stretches on very little sleep. And many of the hundreds of letters that issued from the drafty little shack on the Tikigaq spit indicate how he pushed himself through the long Arctic night, pounding out on the manual typewriter his reports, his letters and notes: "As I write the clock inches over 4:00 A.M. and an Arctic gale is into its 8th day." And another time, "I wrote Les in my thirteenth hour on this devil machine and now I try one to you in the fourteenth."

As the only investigator still in the area as the fall advanced, Foote was regularly beset with requests. The University of Alaska investigators wanted him to compile caribou data, and he tried to assist a disorganized marine-mammal study being run by other investigators. Then the AEC wrote him asking for his immediate assistance, to meet with the village council and "allay the anxiety of the natives." It turned out that Dan Lisbourne, the former village council president, had written Alaska's Sen. Bob Bartlett expressing his opposition to Chariot, and Bartlett had transmitted that concern to the AEC.

In Foote's reply, he made it clear that he would "be honest with the people within the limits of my own knowledge." The subtext here was that he had no intention of making unsupported guarantees of safety on behalf of the AEC. His letter went on to explain why the AEC might have expected the negative reaction from Point Hopers:

> most Point Hopers were angered by speeches and lectures in Fairbanks, Juneau, Nome and Kotzebue, hundreds of miles from Ogotoruk Creek, while not one AEC representative visited Pt. Hope, just 31.5 miles from ground zero. . . . Even after

Chariot operations were being completed for the summer [1959], most Point Hope residents were ignorant of AEC's intentions in Ogotoruk Creek.

Foote said that Lisbourne was "well read and as concerned about fallout etc., as any good New Englander," and that "Chariot should have been presented to the local people with the same philosophy as one would have used in any rural American area." The AEC had badly misjudged the Tikirarmiut. At a special meeting of the village council called at Don Foote's behest, the members declared: "The council discuss the possible dangers to people and animals and decide against this atomic explosion to be blasted at C. Thompson . . . the pres. asks Don Foote to ask the AEC that in future time some man come here and discuss with them more carefully and understand, and besides ask for films about this atomic explosion."

Rather than relying solely on Foote's intercession, the council two days later unanimously approved a petition to the AEC that put its position squarely on the record: "We the undersigned the Point Hope Village Council do not want to see the explosion at the near area of our village Point Hope for any reason and at any time."

Having discharged his duty to discuss the project with the council, Foote went hunting. From the beginning, he had taken advantage of every opportunity to get out into the country. He got the AEC's approval to spend a portion of the money he had budgeted for air surveys on dog-team rental instead. "In mid-October," he wrote to a friend back at Dartmouth, "I took a 115 mile sled trip inland along the Kukpuk River to check the caribou hunting and fish camps. Then, on October 28th I started for Kivalina by dog team, again traveling via the Kukpuk to visit the fishing cabins. Unfortunately, the worst storm of the season, so far, chose to keep us wind-bound for three and a half days." The meteorologist's log at the Chariot camp notes that it was blowing to seventy miles per hour the day before Foote arrived. "I won't be 'polar explorerish' and go into details," Foote continued, going on to sound polar explorerish, "but one does like the comfort of seeing the lead dog now and again, and whereas blowing snow can be tolerated, sand and small stones smart one's eyes." Foote had hired a Native dog driver, and may well have been riding in the sled as the stones pelted him. At Kivalina, Foote also hired a local man and his team. Still, it was an impressive gambit for someone only eight weeks in the country.

Eventually, Foote would request modifications of his AEC budget to eliminate funds for hired dog drivers and substitute the cost of some hickory boards from which he would build his own dogsled. Foote's intrepidity, his eagerness to do the things the local people did, and his nonchalance with respect to the cold—all of these were qualities respected in the North, and none of this was lost on the Eskimo people.

•   •   •

Earlier in the fall of 1959, the AEC hosted members of the press at Camp Icy Meadows—the name Holmes & Narver gave to the Chariot camp (though it never took). The *Fairbanks Daily News-Miner*'s Albro Gregory visited the project site on August 20, pleased that it had been "finally thrown open to the scrutiny of newsmen." In the five articles he produced on the subject over the next week, however, Gregory exhibited more sycophancy than scrutiny, which tendency was (and remains) typical of the Alaska press's approach to covering big dollar federal projects. Dutifully, he reproduced the AEC line, even when it contained obvious contradictions. On the twentieth, he wrote that after the shot, scientists would be able to inspect the irradiated crater lips "within three weeks." By the twenty-seventh, the period of restriction had shrunken and "scientists would be allowed to approach the big crater in a matter of minutes after the blast."

One of the three major purposes of the environmental study, wrote Gregory, copying accurately from the AEC press packet, was to determine the best time of the year for the detonation. But at the same time he was reporting that if the blast was approved, it "will be triggered in the spring of 1961." The AEC said both things simultaneously and, illogical or not, Gregory simply reported both. "The scientists now roaming over the countryside," he wrote, "appear to favor the March date for the blast because it is believed there will be less damage to the animal, bird, fish and plant populations at that time." It is not at all clear on what basis Gregory made this assertion. While this was the message put out by the AEC, it is unlikely that researchers—who had no data whatever on March conditions at Ogotoruk Creek—would have been inclined to make such speculations. It is hard to imagine why they would tell the press that such conclusions were so easily foreseen when they had just been arguing so determinedly that only scientific studies could give those answers. In fact, several of the biologists would shortly write a letter of protest to the AEC and mention the apparent predetermination of the spring firing time as scientifically insupportable.

Reporter Gregory even went so far as to claim that the experiment was one "which most scientists already believe to be feasible and in the public interest." He ignored the AEC's public admission that the downsized crater would not accommodate deepwater vessels or result in significant economic development, and wrote instead that the "harbor" would be "ready to accept the largest ocean-going ships and to get them loaded with the wealth of the area—coal, principally, and perhaps petroleum products from the Arctic slope."

It was typical *News-Miner* coverage of Native issues. The reporter repeatedly characterized the homeland of the Tikirarmiut as "this bleak outpost." Gregory declared that "these Natives, living largely a primitive life, would not be disturbed by the proposed action or suffer future ill effects," even though supporting data had not yet been gathered. Neglecting to mention that Point Hopers had utilized the Cape Thompson region seasonally as a hunting, trapping, or egg-gathering area for centuries, Gregory wrote that there were no "permanent residents" in the vicinity, that "only a few Natives pass by." This picture fit nicely with a classified summary produced by Livermore a year earlier:

> In the vicinity of the job site there are no known activities such as mining or trapping. . . . The only known population in the area is a community of Eskimos at Point Hope. These people are destitute and gain their existence from the sea, the Kukpuk River and the local reindeer herds.

In two of his articles, Gregory emphasized that the explosion "would apparently be the cleanest yet fired," and "the explosion will be the cleanest ever fired in the ground." Whether AEC spokesmen actually said this or whether Gregory missed the fine-print qualifiers, the fact is the AEC had already detonated several deeply buried shots in Nevada that produced no radioactivity at the surface. Because Gregory conceded Chariot would vent *some* radioactivity, it could not possibly be "the cleanest yet fired." With respect to controlling fallout, Gregory conceded that "it is a matter of [to] what degree this can be accomplished," but accepted on faith that it nonetheless "falls into the category of 'clean' use of the weapon." That same summer, at the "Undiscovered Earth" conference held in Birmingham, Alabama, Edward Teller outlined another contradictory version of the radiation picture. According to the Associated Press, Teller said Chariot could be fired without the venting of any radioactivity at all, promising the shot would not take place "until there is complete assurance that there would be no fallout." But, because a cratering shot such as the one proposed for Chariot would necessarily vent some radiation as exploding gases ejected the dirt, it *had* to create fallout; AEC publications on Plowshare clearly state as much:

> Any nuclear explosion designed for cratering will permit some radioactivity to enter the biosphere, the amount depending upon explosive yield, depth of burst, and fission/fusion ratio, as well as the chemical and physical properties of the medium.

●   ●   ●

As the fall of 1959 descended on Ogotoruk Creek, the camp population dropped sharply. Where seventy people crowded the mess hall at the beginning of August, only twenty-three sat down to eat by the sixth of September. Two weeks later, only the three Holmes & Narver caretakers remained. One of them, Harry Spencer, kept a daily journal through the coming fall and winter. On September 20, he wrote, the first snow fell at Ogotoruk Creek. On the twenty-third, the temperature dropped to fourteen degrees with roaring winds peaking at sixty miles per hour. Within a day, Ogotoruk Creek froze over, and two days later one of the men, finding nothing better to do, placed a case of dynamite on the creek and blew it up with rifle fire. Winter came anyway—and stayed a while.

On October 12, Spencer noted that all the ducks and birds were gone from the beach. On the twenty-seventh, a storm raged all night. Snow piled up high against the huts' walls and found its way through every chink, settling in little drifts on the floor inside.

OCTOBER 28: Wind worse today. Up to 50 knots, not drifting snow now as most of it is gone. Things shaking off the
shelves . . . buildings holding down OK.

OCTOBER 29: The 4th day of this. The wind got up to 70
mph several times today. . . . It was blowing so hard it was
picking up the sea water.

The next day, ice formed on the sea. Seals and *ugruk* appeared, and
Don Foote showed up with a Native dog driver. They were marooned
three days by the storm. On November 2, the Wien plane brought in gro-
ceries but forgot to bring the mail. "Everyone was disappointed as we
have had no mail since Oct. 23 and they have been here twice." On November 16, 17, 18, 19, and 20, Spencer's entries vary little: "Wind blew
hard all day, up to 60 mph."

NOVEMBER 22: Got up to 77 mph . . . all the stoves are operating on increased settings and the wind really takes the
heat out.

DECEMBER 7: Ed went out to Kotzebue today and was a
mighty happy guy to go. This is hard for some guys to take
with long nights and monotonous old day after day.

December nights grew longer still until, on the twelfth, the sun did not
rise. "Daytime" was defined by a few hours of twilight, and the yellow
glow in the southern sky grew dimmer each day. December brought the
darkness as well as the deep cold. On the thirtieth, Spencer's thermome-
ter stood at thirty-seven degrees below zero. The next day "a real howling
storm," with winds exceeding sixty miles per hour, blew out the year and
the decade.

JANUARY 3: Saw the sun for the first time this year. It was
visible for 35 minutes, approx. 1/3 of it was visible along the
horizon. It sure makes a lot of difference . . .

JANUARY 8: I had to use the rope to get back to the shack this evening as I couldn't see the radio shack from the weather house.

On the twentieth of January, along with "yowling, howling winds," Spencer reported that some visiting officials from Livermore were getting a good idea about what shot conditions would be like in January. "They all went outside for about 5 minutes," he said, which was about all a person could take.

February crawled by. "Good hard storms," with winds blowing more than fifty miles per hour and temperatures decades below zero, frequently kept everyone inside all day. Pruitt ventured out with his dog driver when the weather moderated to minus twenty and winds at about forty-five miles per hour.

On March 14, 1960, a clear cold day when temperatures never got above zero, Harry Spencer noted the arrival of another party of AEC officials. They stopped an hour for lunch, then flew off to Point Hope. The AEC party had been in Alaska since March 3 and was now in the midst of an informational tour of the villages along the northwest coast. Besides Juneau, Anchorage, and Fairbanks, they had already visited Nome, Wales, Shishmaref, Deering, Kotzebue, and Kivalina. After Point Hope they would fly on to Noatak, Kiana, Selawik, Noorvik, Point Lay, Wainwright, and Barrow. As Harry Spencer sat out an Arctic spring that included temperatures to twenty-eight below zero in mid-March and scientists suffering frostbite in April, the visiting AEC party was having its own chilling effect on the region. But at every village, a shy and no doubt utterly nonplussed and confounded population raised "no significant questions," as the AEC public relations officer noted in his trip report. At every village, that is, except one.

9   THE AEC MEETS THE ESKIMOS

I'm pretty sure you don't like to see
your home blasted by some other
people who don't live in your place
like we live in Point Hope.

—Kitty Kinneeveauk

If any market survey preceded the introduction of compact, mass-produced reel-to-reel tape recorders around 1960, chances are it overlooked one demographic group where the machines would sell like hot cakes: the Eskimo people of Arctic Alaska. Tape-recording became, as one newspaperman reported, "an Eskimo fad." Actually, the craze made sense for at least two reasons. First, the Eskimos' culture was based on oral history and traditions—they had no written language. Because the telephone had not yet appeared in Arctic Alaska, the tape recorder allowed Eskimos, for the first time, to communicate in their own language with distant friends and relatives. At Point Hope, more than half of the fifty households had tape recorders by 1962, and residents regularly sent off reels of tape to correspondents in Kivalina, Kotzebue, and Barrow. Secondly, the Eskimos were gifted tinkerers. Even the old people knew how to operate the gadgets. They spliced tape and argued over the best brands and the merits of stereo sound. If a machine broke down, they handled their own repair, ordering parts from catalogs.

No one who knew the Point Hope people would have been surprised to see Daniel Lisbourne and Keith Lawton one March afternoon in 1960 busily setting up tape recorders in the Point Hope parish hall in preparation for an important village council meeting. But the men from the U.S. Atomic Energy Commission, shortly to make a presentation there, *they* would be surprised.

●   ●   ●

On March 3, 1960, three men from the AEC met in Seattle and proceeded on to Juneau. They were Russell Ball, head of technical operations at the AEC's San Francisco office; Charles Weaver, safety coordinator at the AEC's Albuquerque office; and Rodney Southwick, the public relations man from AEC-San Francisco who had accompanied Teller on his earlier visit to Nome and Kotzebue. The stated purpose of the trip was to announce the AEC's 1960 program of continued environmental studies and to explain Project Chariot to villages along Alaska's northwest coast. Only incidentally did the AEC team mention that the engineering design of the project had been changed again. Now in its third configuration, Chariot's yield had been once more downscaled. The new plan called for a 280-kiloton explosion, down from 460 kilotons (which, itself, was down from 2.4 megatons). A single 200-kiloton bomb would excavate the turning basin, and four 20-kiloton bombs would cut the entrance channel. These details appeared in a March 4, 1960, press release coinciding with the AEC group's arrival in Alaska, and that primarily touted the agency's decision to continue field studies in the summer of 1960. The press release, however, gave no reasons for the change—in fact, it did not even note that the described design represented a change.

Internal documents show that Livermore had made the decision to scale back the Chariot blast in December of 1959. High explosive work in Nevada had demonstrated that reliable information on ditching required a minimum of four charges in a row, so the number of bombs to dig the entrance channel were increased from three to four. Because the channel bombs would give the desired data on row charges, there was no need to duplicate a row blast at the turning basin, and a single bomb there was judged sufficient.

Perhaps a more honest declaration of the AEC group's intentions on their tour of Alaska in March 1960 would have emphasized the objective of shaping public opinion to be favorably disposed toward Chariot. For Southwick, the PR man, that was his job. But Weaver too, the putative safety coordinator, on this trip at least seemed mainly interested in the *perception* of safety, rather than safety, per se. In his lengthy trip report, Weaver concentrated almost exclusively on public relations problems and analyzed the tour against the standard of a successful publicity campaign.

On arriving in Juneau, the group visited Governor Egan, and the next day they met with nine legislators from the northwest. Safety coordinator Weaver seemed to think the latter meeting went pretty well:

> Judging solely on my observations during these briefings, I would say they all seemed convinced that the Chariot detonations would be conducted safely. However, there is no concrete evidence to prove that once we leave they will not revert to their formerly biased opinions. Time alone will answer this problem unless a more informative and fact revealing public relations program is conducted.

The scientists at the Alaska Department of Fish and Game, whom the AEC men briefed next, also seemed inclined to harbor "biased opinions." Jim Brooks, William Smoker, and Allan Courtright objected to what they called the AEC's "propaganda" and pressed for the kind of information that would permit a bona fide scientific evaluation of the detonation's safety. Specifically, the Alaskans wanted to know the quantity of radiation that would be released. The AEC had limited their public statements to claims that only a small percentage of the total radiation would escape into the atmosphere. Weaver was telling people that less than 10 percent venting was a conservative estimate, and that "the laboratory was confident that less than five per cent would be released into the atmosphere." But Jim Brooks said that even 1 percent of an unknown quantity of radiation could be a tremendous amount, affecting people and food chains. Figures were not provided, however, and "this donnybrook," wrote Southwick in his trip report, "concluded with our hasty retreat to the airport to enplane for Anchorage."

The visit to Anchorage coincided with a March 9 and 10 meeting of the investigators working on the Project Chariot bioenvironmental studies. Southwick, Ball, and Weaver hit the radio and television stations as well as the newspapers, but, surprisingly, the most interesting public relations problem emerged from the AEC's own contract employees—the investigators conducting the environmental studies. Echoing the comments of the Department of Fish and Game, the scientists complained that Livermore physicists had been boosting Chariot by making statements in the area of radiation biology, which was outside their field of competence, and that the physicists had a stake in so doing because they wanted to see their experiments carried out. The environmental program investigators persisted in raising "certain questions as to the effects of radiation and the amount of radiation to be released," as Southwick noted in his report, as well as details of the AEC's test experience in the South Pacific.

Most vocal of the Chariot investigators was Don Foote, who "reported on unrest and apprehension building up at Point Hope among the Eskimos." Later, Southwick was able to engage Foote in a conversation:

> I discovered that he was concerned about "the honesty of the Commission" and he stated he had considered giving up his part in the Environmental Study program. This apparently was based on the belief that the AEC would proceed with Chariot regardless of the outcome of the environmental studies, that he was unable to answer Eskimo questions about radiation hazards satisfactorily, and that he was uncertain the AEC really was concerned with the safety of the Eskimos and protecting their means of livelihood. This attitude was reported to Dr. Wolfe. . .

As a result of the investigators' general dissatisfaction with the level of information the AEC was providing, a special session was arranged in which Wolfe and Seymour, along with Charles Weaver, presented to the assembled scientists more detailed information on the effects of nuclear explosives, the effects on fish and game in the South Pacific and Nevada, and estimates of the effects of deeply buried, versus atmospheric, shots.

Before leaving Anchorage, Ball, Southwick, and Weaver enlisted Robert L. Rausch of Anchorage to join the group on the village tour. Rausch worked for the Arctic Health Research Center in Anchorage, which was part of the U.S. Public Health Service. He was also on the Chariot bioenvironmental committee. But the real reason he was invited along, Rausch believes, had little to do with his area of scientific expertise. "I think I was taken because I was acquainted with the people at Point Hope," he said.

●   ●   ●

Rausch packed his Arctic gear. Even though it was March, the temperatures along the northwest coast would drop to near thirty below during the trip. The AEC men obtained parkas and flight pants, mitts and boots from Ladd Field, the army base at Fairbanks. After brief visits to the Fairbanks newspapers, the group flew by commercial airline to Nome and there picked up their chartered plane. Weaver had been sick for several days and felt too ill to go on to the villages. When Weaver flew home to New Mexico, Rausch found himself in the awkward position of inheriting some of the presentation dealing with safety matters. He had attended the Anchorage meetings where Wolfe, Seymour, and Weaver had described fish studies in the Pacific and sheep and cattle studies in Nevada. But Rausch knew he was not very well informed on the issue of radiation hazards.

At Nome, civic boosters insisted that the men from the AEC give short talks at a chamber of commerce luncheon before flying up the coast. This was easy duty for Southwick and Ball; the chamber was demonstrably gung ho on Chariot. The *Nome Nugget* captured the drift of the presentation in its story headlined, ATOM SCIENTISTS HERE TO REASSURE NATIVE POPULATION. The AEC team would "meet with the village councils and explain as well as educate the population on the importance of the project."

At the village of Wales, there were "no questions" raised by the forty people attending the meeting, according to the trip report. At Shishmaref and Deering, "there were no significant questions."

The only hint of anti-Chariot sentiment was picked up surreptitiously at Kotzebue. After discussing Chariot with those members of the community "who are considered influential in molding public opinion," Southwick managed to listen through the walls of his room at Rothman's Hotel to a conversation between "several men whose identity I was unable to learn." The gist of the conversation, which Southwick reported to his AEC superiors, was that the white polar bear hunters, who flew their clients into Point Hope, were against Chariot and might use whatever influence they had with the Natives to foster opposition.

At Kivalina, the last village they would visit before Point Hope, there were "no significant questions." At Noatak, Kiana, Selawik, Noorvik, Point Lay, and Barrow, villages visited after Point Hope, there were "no significant questions."

Though Project Chariot had been on the drawing board at Livermore since February 1957, by 1960 Point Hopers were still largely in the dark about the AEC's plans for Ogotoruk Creek. What information the villagers did have was generally "second-hand and warped," according to Don Foote. And until Senator Bartlett had responded to Dan Lisbourne's letter and successfully pressured the AEC to visit the villages, the AEC's efforts to inform the people were rather measly and ineffectual. A Kotzebue missionary, according to Foote, "presented a prepared and apparently incomprehensible document" at a time when most villagers were absent. At Noatak, a tape recording of the AEC presentation at Kotzebue was played "when 95% of the village was away," says Foote. "I found only two persons in that village who remember the recording but nothing of its contents."

Still, Foote said, the people picked up scraps of information, even if it was "a mixture of hearsay." Whenever AEC men spoke in Kotzebue, he said, news spread rapidly by word of mouth. And people traveling through Nome would clip newspaper articles reporting the reaction of the Fairbanks Chamber of Commerce to lectures on Chariot by prominent scientists. *Saturday Evening Post* articles on fallout were read by many. "Thus in several ways," said Foote, "knowledge of Chariot drifted north." Point Hopers with shortwave radios even picked up transmissions from the Chariot camp and from Kotzebue and so knew something of the comings and goings of the men from the AEC. That is how they knew the men were coming one cold Monday afternoon in March of 1960. They knew it was Tommy Richards—a part-Eskimo native of Kotzebue who was well liked along the coast—flying them up in a Cessna 180. They knew the plane left Kivalina late. They knew the group stopped for lunch at the Chariot camp. And the news passed by word of mouth.

●   ●   ●

At quarter to two on the afternoon of March 14, 1960, three men representing the AEC entered Browning Hall, a building so long and narrow that kids ran footraces there during the Christmas festivities. About 100

Eskimos sat, as they were accustomed to doing, on the floor, shoulder to shoulder against the two eighty-foot-long walls. As the government men made their way to the head table where two tape recorders sat whirring, it must have felt as if they were running a gauntlet.

David Frankson, the village council president, introduced the visitors in Inupiaq. They were Russell Ball, Rodney Southwick, and Robert L. Rausch.

Southwick took the floor. "Thank you, Mister Frankson. Ladies and gentlemen of Point Hope, we have come here as representatives of the United States Atomic Energy Commission. It is our purpose to tell you what the status is of the project at Ogotoruk Creek. . . . I want to emphasize and repeat that the Commission has not approved any explosion. The Commission's decision on whether there should be an experiment up here will be based solely on these studies—if such an experiment can be conducted safely, and only if it can be conducted safely. The Commission, by safely, means no one would be hurt, no one would be moved, your normal means of hunting and fishing would not be interfered with."

It would be difficult to imagine a more effective way to get the attention of an Eskimo audience than to raise the possibility of risk to the animals upon which their economy depended, or the possibility of moving the people from their homeland. In 1960, the Inupiat people at Point Hope were both staunchly protective of their traditional lifeways and opportunistic when it came to incorporating those artifacts of American culture that appealed to them. The annual Christmas party featured both traditional Eskimo dancing and singing accompanied by the banged-out rhythms of the drum, and the "Tikiraq Playboys" sporting Western shirts and scarves strumming guitars and crooning "Tom Dooley." They were deeply interested in Elvis. They still hunted sea mammals in skin boats, but the boats were powered by outboard motors, and they shot the animals with rifles. They still ran dog teams, but often tucked away in the sled bag was a tape-recorded "letter" bound for friends in the next village. The people assimilated the white man's language as well; many could not only speak, but could also read and write English.

Notwithstanding that the Inupiat of Point Hope lived in a hybrid culture, most of the people had difficulty with the AEC presentation: It was in English—and jargon-laden English at that; it dealt with technical matters such as the physics of cratering and the biological effects of radiation, subjects that would have presented a problem for most listeners at any

American town-hall meeting; and the people simply had no affinity whatever for the technology proffered by the AEC.

After a brief power failure, Russell Ball was able to introduce the film *Industrial Applications of Nuclear Explosives*, which Livermore had produced and shown to the Second International Conference on Peaceful Uses of Atomic Energy in Geneva. "We now have available to show you a brief moving picture which describes several of the ways in which we think it will be possible to develop peaceful uses of these nuclear explosives. The first of these that will be in the movie has to do with the Cape Thompson project although it is not mentioned by that name. . . . You will find that it talks about the creation of a harbor. And I would first point out that the film was made at a time when thought was being given to the creation of a much larger harbor than is now considered. Because of this, the explosive devices which the film mentions are much larger than the ones we now propose to use."

Council president David Frankson said a few words in Inupiaq, someone turned out the lights, and the film began. The first thing the Eskimos saw after the title faded was a white flash followed by a churning, red fireball lifting off the earth. Over the quavering strains of the music track, they heard the smooth, euphonic tones of a narrator: "Now that man has learned to control a vastly greater force than chemical explosives, his thinking has turned to methods which will make effective use of this power within industry. The potential applications? Consider just one. The development of a harbor in the hard rock of a remote coastal area."

The still-growing mushroom cloud dissolved, and in its place appeared, in animation, the Ogotoruk Creek valley. The audience recognized the green, rolling hills where they hunted caribou, and the slanting limestone cliffs along the coast where they gathered crowbill eggs. Graphic overlays dissolved in to show the keyhole-shaped outline of a harbor and the location and depth of the nuclear explosives. "The procedure: bury an in-line series of four equally spaced one-hundred-kiloton shots at a depth of thirty meters, and a terminal shot of one megaton, fifty meters deep. All shots of minimum fission yield. Depending on the nature of the surrounding medium, surround the devices with sufficient borated compounds to reduce neutron activation by several orders of magnitude. Detonate."

The green valley floor instantly heaved into a massive dome and blew apart in streamers of dark rocks and boiling dust. The audience cried

"Yeeeee! Ahhhh!" Under the sound of the wailing, and as the screen showed the blue sea rushing in to fill the channel and crater, the narrator continued reassuringly: "Activity in the region of the crater would be washed out into the ocean rapidly and essentially removed from the biosphere. The activity on land near the crater would rapidly decay. . . . In remote areas, where large yields can be safely used, and economic factors are favorable, such explosions are demonstrably feasible."

After the film, Russell Ball took questions. Keith Lawton, the Episcopal priest at Point Hope, stood up to speak. He had been cutting meat before the meeting and was still wearing bloody clothes and carrying a knife with a thirteen-and-a-half-inch blade. Lawton asked several questions dealing with the suitability of the site's geology, the seismic effect of the blast, and the time of year of the detonation. He also wanted to know how long it would take for the various radionuclides in the fallout to decay to harmless elements. "Perhaps Dr. Rausch would answer that," said Ball.

Rausch was a slight, bespectacled and pedantic-looking scientist with the handshake of a blacksmith. He had a peculiarity of speech wherein he pronounced each word—each syllable—distinctly: "in-for-may-shun," he would say. Rausch forbore contractions. The whole effect was one of great precision, as if he took in the world in data bits, and processed them with robotic exactitude.

Rausch was not an employee of the AEC, but a parasitologist who had lived in Alaska since 1949. In connection with his work, he had traveled widely in northern Alaska, including to Point Hope, and was well known to the Eskimo people. That he was well liked is clear from an account of one trip Rausch took with four colleagues to Anaktuvuk Pass, a village in the central Brooks Range. The village correspondent filed a brief item with the Fairbanks paper noting the visit and describing the scientific party as including "four white guys and Bob Rausch."

When he traveled with people new to the Arctic, Rausch drew secret delight in supplying his companions with prosaic descriptions of the raw side of Northern life. With studied blandness, he might detail the pathology of some ghastly parasitic disease endemic to rural Alaska. Or he might casually recount the habit of Eskimo hunters who, after downing a caribou, would rush to pull back the hide and expose hundreds of encysted warble fly larvae. The hunters would greedily pick off the thumb-sized grubs and pop them into their mouths. If his listener's jaw dropped and eyes widened, Rausch seemed not to notice, merely observing that "they are quite delicious."

Bob Rausch had no particular problem with the Project Chariot experiment, though he did think some of the Livermore scientists were "like little boys" showing an almost "pathological glee" at setting off these explosions. Mainly he considered Chariot "a very unusual opportunity to obtain some very good information." In any event, he had received the AEC's briefing from Allyn Seymour and Charles Weaver in Anchorage, and now found himself on the village tour explaining the Chariot environmental studies and the effects of radiation.

"According to the information that I have," Rausch began, in answer to Lawton's question about how long it would take the fallout to decay to harmless elements, "the amount that will escape, to begin with, will be a quite small part of the total yield. And the half life of the radioactive elements that are produced will be so short that some will be gone in a matter of hours, others will take longer." Here Rausch was speaking quite outside his area of training, and the impression he was leaving with the people was erroneous. Almost certainly Russell Ball knew that radiation with a half-life of many years would be released. But Ball said nothing. Nor did he correct Rausch when Rausch told the people another thing he understood from AEC scientists: that "the amount of radiation that is likely to be released, according to the information that is available, will probably be so little [as] to make it impossible to detect the actual increase over the present amount of radiation that is there." Rausch had understood this to be true from the statements of the AEC's John Wolfe at the Anchorage meeting.

Returning to this topic later, Keith Lawton wanted to know what the AEC would do "if anything *did* happen . . . that some pocket of radioactivity due to a wind shift landed on a particular portion of Point Hope." Russell Ball replied, "At this distance the amount of airborne radioactivity which could reach here would be, could not *possibly* be enough to cause any injury to the people or the animals. There's just no chance of that." And at the crater site itself, Ball said that "after a short time—in terms of months, at least—it will be possible to remove all restrictions on access."

The effect of radiation on the animals was a matter of grave concern to the Point Hopers. When Lawton raised the question of radiation and fish that might swim into the crater, Ball tossed it to Rausch, saying, "Why don't you mention the experience from Eniwetok [the AEC's nuclear test facility in the Marshall Islands] with regard to fish."

Rausch had not been involved with the fish investigations at Eniwetok, and he qualified his information as coming from others, namely Allyn Seymour. "But," he said, "they found no evidence that fishes were destroyed or that there was any significant amount of radiation in them."

Ball added that "the amount of radioactive material released into the sea there was a thousand times greater than that which would be released here. And even there it was safe to eat the fish which were caught, even in that water."

At another point, again relying on information provided by the AEC's Seymour and Charles Weaver, Rausch added: "According to the tests that have been carried out in other places, there appears to be very little reason to believe that any of the animal life is likely to be harmed. In the Bikini tests that were carried on above ground, the amount of radioactive material in each was many times greater than will be the case here. And even then, there was no effect on the fishes in the sea. None were made radioactive and therefore really impossible for them to be eaten." Rausch said that the studies on fish after the detonations in the Pacific "did not show anything that was considered of any danger to anyone if those fishes were utilized."

Daniel Lisbourne, the previous village council president, raised the question of the project's interference with caribou hunting, and Lawton wondered if steps would be taken to keep the caribou away from Ogotoruk Creek. Rausch, again at Ball's prompting, mentioned the AEC's experience with a herd of cattle the government maintained on the Nevada Test Site. "Yes," he said, "they have been keeping cattle in the Nevada area, in the immediate vicinity where the shots have been made. That herd is maintained there and the animals are killed periodically and examined and tested in various detailed ways. And they have yet to find any evidence of any damage."

Ball added, tellingly, that the herd was brought in "to provide evidence to the farmers that it was safe."

Keith Lawton was beginning to see that the villagers' questions as to radiation safety were perhaps being asked of the wrong people. "Who is the radiation biologist who works with the AEC—who is the radiation biologist who is working on this program?"

Southwick perked up. "Dr. Allyn Seymour is a member of the committee."

"Dr. Allyn Seymour," repeated Lawton. "And he wasn't available for tours around the villages at all?"

"He was with us at the three-day meeting which the committee held at Anchorage," replied Ball. "But we didn't at that time consider it important enough to bring him up here."

"We didn't think it was necessary to pull him off the job he's on right now," said Southwick, hoping to put a slightly more polite spin on Ball's comment. Then, probably because he didn't want the session to become the sort of "donnybrook" he'd encountered in Juneau, Southwick moved to wrap up the meeting: "Are there any questions? If not, I would like to thank you very much Mr. Frankson, and the people who came here—"

He was interrupted by a woman's voice. Dinah Frankson, David's wife, who had great influence in the village because she was an *umialik*'s wife, spoke in Inupiaq while Daniel Lisbourne translated. "Ah, the woman here mentioned, all of these people here, all these people, most of them are just silent right now and they have great fear in, in this detonation and the effects, and how the effects of it will be."

Southwick, who could be a little slow on the uptake, asked, "Internationally?"

"No, here," said Lisbourne.

"What?" asked Ball. "I don't quite get what her question was."

"The effect of the blast," explained Tommy Richards, the pilot.

"The effect of the blast to the people," echoed Lisbourne.

Dawn was breaking in Ball's mind. "On your own Eskimo people?" The audience looked back at him without speaking. "Oh, well, ah, I believe we've covered that already. Ah, I think we'll have to . . ." Ball's voice trailed off as he turned to Richards to discuss their return flight to Kotzebue.

"I don't think that's true," said Richards. "You haven't."

Many people then spoke at once, in Inupiaq. Daniel Lisbourne spoke for the people. "Ah, I think, Mr. Ball, that this, the majority of us here, right now, have no understand what you have said previously, having not know enough about the English language." Lisbourne said Dinah Frankson wanted to remind the AEC that she was a citizen of the United States and that she feared for the men who hunted in the Ogotoruk Creek area, especially the seal hunters who, in April, went out on the sea ice near the planned ground zero.

"All the way down there?" asked Ball.

"Yeah. All the way down there," said Lisbourne.

"And collect eggs," said Keith Lawton. The murre, known locally as the crowbill, laid a large, blue-green egg that was a favorite of the people. Crowbill eggs had a pronounced taper at one end so that they could roll only in a little arc, not waddle off the ledge. The people of Point Hope gathered them by the hundreds of dozens; they boiled them, peeled them, and preserved them in pokes of seal oil.

"When they do get them, they get them by boatloads from Cape Thompson," said Lisbourne.

"The eggs?" asked Ball.

"The eggs," said Lawton.

Dan Lisbourne said that last year the Point Hopers gathered 733 dozen, and that that was a low year.

"Ah," said Ball. He paused to think. "All I can say is, briefly, is that the studies we are conducting of your people, of your hunting, of your fishing, of your catching seals and so on, all of this will guide us in finding the right time of year for such an experiment. Only if these studies show that we can pick a time that will not be of harm to your people, only then would we do the experiment. We have your welfare at heart; we cannot afford to do you harm. This program is of great importance to us. This is one of many experiments we would like to do. If we do this carelessly, so as to bring harm to your village, the reaction would be to stop our whole program! We, we can't afford this! We can't afford to be careful—careless. We must be very careful not to injure your people or your way of life. So these studies we are conducting will tell us if there is a time when this can be done safely. If there is such a time, then we'll choose to do it and it will not harm you. If we cannot find a time to do it safely, we won't do it."

"I hope you don't!" said Kitty Kinneeveauk to the laughter of the audience. "Once I read some news from magazines about Indians where you work on this too, blasting their town, and none of these atomic people help them. It didn't help them. Injure their food, their game and their water and their homes. And none of the atomic people help them, or turn back and see what's going on out there."

"We, ah, so . . ." began Ball.

But Kitty Kinneeveauk wasn't through. Kitty, all five feet of her, was well known in the village as a woman who would stand up to injustice wherever she found it. To the frequent alarm of her more reticent husband, Kitty did not hesitate to challenge white men at public meetings. At such a meeting in Kotzebue once, according to Kitty, "I told my husband, 'You are a whale captain! You better talk!' [He said] 'I just take care of your boat.' So I raise my hand. I go down there. They were from Washington, D.C. . . . My husband said he want to hide all right, but he couldn't hide."

Another time there was a public hearing dealing with a controversial fish and game regulation that would restrict the bag limits allowed the Eskimo hunters. Kitty turned the tables on the game manager: "I told that man, 'Well, if you go back to States, one chicken every one month! Not more than that,' I told him. 'And don't kill no pig! If you want to kill a pig, just only one, once a year.'"

And Kitty had already had a run-in with the Chariot men at Ogotoruk Creek. When the construction workers put in the first airstrip, they bulldozed the sod house Kitty and her family used as a seasonal trapping base. On discovering the situation in the late fall, Kitty confronted the caretakers, as she remembers in a 1989 interview: "'How come you ripped our house off here? You didn't even ask! Didn't ask me, just working.' And my husband got scared, 'Dear, you are scared and we better go.'" But Kitty said, "I want to talk to them some more." She mentioned that the men now had built many houses, and pointing to them, she said, "Maybe you can replace it [with] one of these ones." When the men "didn't say nothing," Kitty told them, "You're just bad men." Though her claim for damages was ignored, Kitty had tried "to get action," she said, "because I pity my kids. And besides, we work hard."

Perhaps it was because of the bulldozing incident, or perhaps it was that Kitty's husband was finally resigned to her ways. In any event, when Kitty leaned over to him at the AEC meeting and said, "You talk," Mark Kinneeveauk wearily encouraged his outspoken wife, "Dear, you never be scared. You talk."

Interrupting Russell Ball, Kitty continued: "So, I've been thinking about we really don't want to see the Cape Thompson blasted because it our homeland. I'm pretty sure you don't like to see your home blasted by some other people who didn't live in your place like we live in Point Hope."

"Oh, I understand that thoroughly," said Ball. "We know this. . . . What I'm saying is we will do the work only if it will *not* injure your home. If we cannot find a way to do it without injury to you, we will *not do it.* . . . Now, in answer to your first question, ah, the testing we have done so far has had *no* effect on the Indian people anyplace. There are no Indian people within many, many miles of where we test. There are many, many, er, Americans, white people."

"I've read it on a book," insisted Kitty. "It happened while they blasted their homes."

"No, not by us," replied Ball. "I think you must have misunderstood the book because we have never done that to any people. Indian or otherwise, we have never done that."

"In the book they said it was done by white people," said Kitty.

"We have only tested weapons in two places," replied Ball. "In Nevada and *way out* in the Pacific."

"At Eniwetok," said Southwick.

"At Eniwetok," repeated Ball. "Now, er, at, ah," Ball turned to Southwick and quietly asked, "I wonder if she could have in mind the natives at Eniwetok?" Then, to the audience, "Perhaps you have in mind the native peoples of Eniwetok?"

Kitty was uncertain, but Daniel Lisbourne remembered a detonation in the Pacific. Wasn't there a problem with a fishing boat?

"That's true, there was a fishing boat," said Ball.

"Seventy-five miles away, yeah," pressed Lisbourne.

"That's right and because they were where they were told not to be. . . . They were within the danger—we had publicized around the world an area which we asked all shipping to stay out. They ignored that and were within the danger area."

Daniel Lisbourne turned up the heat on the AEC delegation by raising the issue of compensation. He said that Point Hope hunters brought in 175,000 pounds of meat as a result of hunting "every day of the week."

What would the AEC do, he wondered, if the men were kept from hunting for several days. Would there be cash compensation?

"This is a question we'll have to take up," said Southwick.

"We don't really know the answer to that," said Ball.

"There's a method to do it," added Southwick, "by suit, *if* anything should go wrong."

"Yeah, *if,*" said Lisbourne.

"What I mean is you can always sue, but that takes five years and it has to go through the courts and it is expensive—"

"I was not thinking about the sue at all," replied Lisbourne, "I was just asking if we should get laid up for a few days, that would cripple us, you know."

"We do not have an answer for that," said Southwick, "other than to say you can always sue. But we don't consider that satisfactory."

Don Foote, who had been silent throughout the long meeting, attempted to explain the Eskimos' position. "Mr. Ball, I think Daniel's question stems from the fact that there are many boys right here who don't have dog feed today and they have to get it. And *one day,* if kept in the village, can be serious and they want to know what—"

"Yeah, we know that," interrupted Southwick.

". . . and what—" began Foote again.

"Yeah, we understand this thing," said Ball, cutting off Foote again.

". . . and—" continued Foote.

"All I can give in the way of answer is . . ." said Ball.

". . . the question is now under consideration for Plowshare," finished Southwick.

"But has not been acknowledged as such and—" said Foote, trying a fourth time to finish his sentence.

"We understand that, Don. What I was going to say was that, er . . ." said Ball.

"You're turning up a lot of these things," Southwick complained to Foote.

"This is obviously a question of great importance to your people," said Ball, "a matter which must be very carefully considered by us . . . it seems to me we would have to lay plans to assist you, perhaps lay in food ahead of time . . . we *must* work out a solution which protects your interests."

David Frankson had heard enough. "We council at the Point Hope that sent the protest letter to Atomic Energy Commission stating that we don't want to see the blast down there. And *when we say it, we mean it!* For any reason." Frankson was referring to the unanimous petition the Point Hope Village Council had sent to the AEC on November 30, 1959.

Then Frankson went on to mention another letter, this one sent to the AEC by the business interests in Nome and Kotzebue, who together comprise the Arctic Circle Chamber of Commerce. He said the chamber's letter had denounced the Point Hope petition and supported Chariot. "Now, which of these letters are most important to the commission? Because these people that write against our protest, they be the people that are planning to make some money out of this program. And we are not!"

Frankson was on target—more than he realized. The January 18, 1960, letter from the Arctic Circle Chamber of Commerce claimed that the Point Hope protest represented a "change in attitudes" of the Point Hope Village Council. It went on to note that in spite of the protest, the chamber of commerce, "by unanimous vote," was "reaffirming our confidence" in the AEC and in Project Chariot. This bullish review of harbors from nuclear bomb craters was signed by the chamber's secretary—and owner of B & R Tug and Barge Co. in Kotzebue—Mrs. Edith Bullock. In the two field seasons prior to writing her letter, Mrs. Bullock's company had landed AEC contracts for barging and labor worth more than $200,000.

"Well," began Ball in response to Frankson's question about which letter was more important, "they are all important, obviously."

"He has a very good point there," said the pilot, Tommy Richards.

"Yes, he does, indeed," replied Ball.

"He was asking which of these letters had the most weight," pressed Richards.

"Well, I think—" began Ball.

But Southwick, the public relations man, once again interceded quickly, "Yours."

"Because," said Frankson, "we feel we are going to be the people who stays under the dust that's blowed up and take more harm than you will do to the other people."

"Well, obviously your attitude has more weight than people far away, if that answers your question," said Ball. He began assuring the audience that AEC personnel would be on site before, during, and after the detonation. But then a voice broke in, claiming that fifteen years after the blast children would be deformed. "Oh, no," said Ball. He turned to Rausch. "Would you like to answer that?"

Rausch, perhaps not liking the direction in which the meeting had turned, declined to address the question of genetic hazard. "Ah, let me indicate, of course we have . . ." Ball paused and thought for a moment. "I guess the best answer I can give there is to summarize for you the studies which have been made in Japan. You remember at the end of the war we bombed two cities in Japan. Destroyed them with atomic bombs. And there many people were exposed to very great exposures to radiation. Many were killed by it. Many survived. And of those who survived, a number had very great exposure to radiation. Thousands, many thousand times what could possibly result from this. Ever since that time, even to the present day, there have been very careful studies among those people. And we have found no evidence for *any* effect upon the children of those people from the radiation dose they received. No evidence of any indication at all."

Ball picked up the point again when Alice Weber, who had been trained as the village health aide, mentioned her late husband, Chester Downey. He had been in the army, she said, and participated in the cleanup of Nagasaki. When he came home he was sick and sterile. "Well," said Ball, "these problems which he experienced were in no way connected with his visit to the Japanese cities. . . . The real answer to your question lies in the studies of the Japanese people who were very seriously exposed. And once those who survived recovered from the illness, there was no other result."

A woman spoke to Richards in Inupiaq, and he translated to the group, "The more you talk, the more scared she gets." Suspecting irony, the AEC representatives laughed. But the Eskimos, who heard only the intended seriousness, did not. Richards said the woman wanted to know who would take care of her kids if her bad heart failed when the bomb went off.

Again not taking the question straight, Ball laughed. "You've got me there. I sure don't know. . . . I don't think the noise would do her any harm. It would not be a very loud noise." Richards asked about the seismic effect, because oral tradition at Point Hope held that an earthquake had once hit the area followed by a tidal wave, which partially submerged the spit. "Well," said Ball, "you won't feel any seismic effect at all. Very sensitive instruments might detect some, but you'd never feel anything."

Dan Lisbourne wanted to know if the local people would be invited to observe the detonation. Southwick wisecracked, "It might interfere with their hunting."

Ball said they would be welcome to watch from any vantage open to other observers, then desperately sought to end the meeting by thanking the audience for attending. "As I say again, we are taking very great pains to assure you that we will protect your interests completely. We can't afford to do this any other way." He turned to the pilot, Richards, and said, "Tom, what time do we have to leave to get back?"

"Oh, we can give them a little—" began Richards.

To the audience, Ball said, "We have a meeting in Kotzebue. We have to get back before dark."

Keith Lawton thanked the delegation and expressed the hope of more visits and more information. A white polar bear hunter then staying in the village asked if any part of the Chariot program was restricted from public knowledge. Ball said: "The only thing about this entire program which is in any sense restricted is the design of the explosive itself. Outside of that, *absolutely everything* will be public knowledge and will be published." It was nearly 5:00 P.M. when the three men filed out of the parish hall, into the cold, and made their way to the airstrip. As the little plane headed south over the frozen Chukchi Sea, the Point Hope Village Council took a vote. Again, unanimously, they declared their opposition to Project Chariot.

•   •   •

Robert Rausch had spoken what he understood to be the truth. But, unfortunately, he had had no formal training in the biomedical effects of radiation. As a busy Public Health Service scientist, spending lots of time in the field and working long hours on his own research, he had not read substantially in the field of radiobiology. He had not visited the Pacific Proving Grounds, where the United States conducted nuclear tests from 1946 to 1958 on the Bikini and Eniwetok atolls in the Marshall Islands. Nor had he worked on the animal studies associated with those shots or with the ones in Nevada. He was relying on the briefings supplied by the AEC at the Anchorage meeting. According to Don Foote, who also attended that briefing, Rausch's presentation at Point Hope seemed to consist of "the verbatim statements of Allyn Seymour" of the AEC-funded Laboratory of Radiation Biology at the University of Washington, as Rausch himself had more or less acknowledged to the group.

Ball, for his part, had made an attempt to speak plainly and to avoid technical jargon. And Southwick, whose performance Rausch considered "rather aggressive," was, after all, a public relations specialist doing the job assigned.

Notwithstanding these rationalizations, three representatives of the U.S. Atomic Energy Commission told the people of Point Hope explicitly that the nuclear tests at the Pacific Proving Grounds had not contaminated fish with radiation such that the fish were unfit to eat; that radioactive fallout from the Chariot blast would be so little that it would probably not be measurable with radiation detection equipment; that the harmful constituents of fallout would for the most part be gone from Ogotoruk Creek in a matter of hours; that people at Point Hope would not feel the seismic shock of the Chariot detonation thirty-one miles away; that a study of cattle in the Nevada desert offered evidence as to harmlessness of fallout moving through an Arctic ecosystem; that American nuclear testing had not harmed "Indian" people anywhere; and that once Japanese survivors who received "very great exposures" recovered from radiation sickness, they suffered no further effects.

And the Eskimos got it all down on tape.

10 BIKINI, NEVADA, TIKIGAQ

A Trail of Empty Words

> Hooray! I'm a radioactive Eskimo
> With a radioactive mother,
> A radioactive sister,
> And a radioactive brother.
> My wife can't suckle our babies,
> Our milk must come from cans,
> My wife's too radioactive,
> Say, we're real atom fans.
>
> —From the song
> RADIOACTIVE ESKIMO,
> by Peter La Farge, 1965

In 1815, Otto von Kotzebue sailed around the world for the Russian navy and in the process became both the first European to contact the northern Alaska Eskimo and the "discoverer" of the Marshall Islands in the central Pacific. Shortly after Kotzebue's voyage, American ships sailed around the Horn and up through the Pacific islands to the whaling grounds of Arctic Alaska. One hundred years later, the U.S. Atomic Energy Commission charted a roughly similar course, offering both groups of indigenous people—the Marshallese and the Eskimos—not simply inclusion on

Western maps, but a bit part in the drama of the century. And if contact with the white man in the nineteenth century had brought the risk of flu and smallpox epidemics, the nuclear age offered its own plagues: cancer, leukemia, and genetic mutations.

Though the AEC representatives at Point Hope blithely characterized the experience of the Marshallese as benign, history shows it to have been anything but. In 1946, with less than one month's advance notice, the Bikini Islanders were compelled to leave their ancestral homeland to accommodate the U.S. nuclear weapons testing program. The military governor of the Marshall Islands, speaking at the conclusion of church services, told the Bikinians they were like the children of Israel whom the Lord "led into the Promised Land." He told them the American scientists needed to develop nuclear bombs "for the good of all mankind and to end all world wars." And he said the U.S. military had searched the world for a test site and found Bikini to be the best. When the trusting Bikinians had deliberated, according to the governor's account, their chief replied:

> If the United States government and the scientists of the world want to use our island and atoll for furthering development, which with God's blessing will result in kindness and benefit to all mankind, my people will be pleased to go elsewhere.

Over the next few years, the Bikinians were shunted around the Marshall Islands where their condition alternated between a condition of "near starvation" and the dubious advantages of the U.S. welfare system.

Meanwhile, Operation Crossroads, the name for the first detonations at Bikini, commenced, and even test officials described the series as "poorly conceived and inexpertly executed." Furthermore, contrary to the statements made at the March 1960 meeting at Point Hope, these blasts certainly did irradiate fish. After a 1946 shot code-named Baker on Bikini Atoll, small reef fish that feed on coral and algae picked up considerable quantities of radiation and selectively concentrated it in various body organs. Dr. David Bradley, a navy radiological monitor at Bikini, reported that the absorption of radioactivity by fish was documented rather dramatically by X-rays. Technicians cut the small reef fish in half down their lengths, then laid the halves, cut side down, on a photographic plate. The fish were so radioactive that they left images of their bodies on the film. Bradley called these pictures "radioautographs."

> We have several of them. They demonstrate clearly the pattern of selective absorption of fission products by the various tissues. One can see the gills darkly outlined, the long coiled intestine, the large burst of radioactivity associated with the liver, and the less dark areas of the gonads. The rest of the fish, muscle, bone, scales shows only a trace of the material. This is hard on the fish but a most fortunate thing for man.

It was fortunate for man because the edible portions of the fish showed less radiation, and several of the fish had been eaten already. But to anyone concerned about the health of the fish stocks, the situation was not so fortunate. As Bradley noted:

> What is true of the reef fish now will become increasingly true of the larger migratory fish—the tuna, the jacks, the sharks, and so on—as the larger, the predatory fish eat more and more of the smaller fish who are sick with the disease of radioactivity. We know that this process is going on. Almost all the seagoing fish recently caught around the atoll of Bikini have been radioactive.

While Bradley reported no sightings of large numbers of dead fish floating in the water or washed up on shore, neither did he expect them. Once a fish's reflexes were slowed by radiation poisoning, he speculated, the fish quickly fell victim to a larger predator and disappeared, the radiation continuing up the food chain. Bradley recognized the need for more precise measurements of radioactivity than those obtained by the "gross and spectacular" radioautographs. And subsequently his laboratory undertook some additional measurements by periodically collecting and testing fish samples. This time they found that larger fish taken by spear gun on the ocean side of the reef showed no radioactivity, while those taken within Bikini lagoon "all showed considerable radioactivity." Bradley wrote, "I believe that there is enough radioactivity at the bottom of this lagoon to kill fish either by total radiation to the body, or by the destruction of vital organs by absorbed radioactivity."

Obviously, the AEC men at Point Hope had presented a very different picture to the Eskimo people. Perhaps Russell Ball felt justified in claiming that "it was safe to eat the fish" because the radiation tended to concentrate in body organs that are not typically eaten. By this reasoning, a fish might be dying of radiation poisoning, its liver and intestines intensely irradiated, but if its flesh was below the AEC-set standard, it could still be pronounced wholesome. It is harder to conjure up a reasonable explanation for the AEC representatives' contradiction of Dr. Bradley's direct observations of "considerable radioactivity" in fish. The AEC men repeatedly told Point Hopers there was "no effect" on the fish, that "none were made radioactive," and that there was no evidence "that there was any significant amount of radiation in them."

•   •   •

In 1954, the AEC tested the prototype of the country's first aircraft-deliverable H-bomb at Bikini. Like the Crossroads series in 1946, the Bravo shot also irradiated fish. In the days after the spectacular detonation, U.S. servicemen reported seeing "dead sea life all over, floating around by the millions." In the months following the shot, nearly a million pounds of irradiated tuna were condemned by the Japanese government and destroyed. AEC chairman Lewis Strauss denied the conclusions of the Japanese health authorities and claimed "the only contaminated fish discovered were those in the open hold of the Japanese trawler." The trawler in question, the *Lucky Dragon*, was fishing northeast of Bikini when pulverized and irradiated coral began to fall like snow and cover the deck. By the time the crew returned to their home port two weeks later, the boat was quarantined and the twenty-three-man crew hospitalized with acute radiation sickness. Six months later one fisherman, aged forty, died.

Chairman Strauss denied that the crew had suffered from radiation poisoning at all. He said the skin lesions and burns were "due to the chemical activity of the converted material in the coral, rather than to radioactivity." The statement outraged Japanese medical professionals, who since the bombing of Hiroshima and Nagasaki had become only too expert in diagnosing radiation pathologies.

Strauss further claimed the fishing boat was "well within the danger zone." Strauss's claim can fairly be called a lie, because all the evidence available to him was to the contrary, according to the AEC's own official history. The *Lucky Dragon* was apparently eighty to ninety miles from Bikini, well outside the exclusion zone and well beyond the thirty-mile perimeter established by the AEC for the protection of U.S. Navy ships. Actually, the boat's exact distance away—thirty miles or ninety miles— amounts to an irrelevant quibble. The AEC had expected and prepared for a six-megaton shot, but Bravo turned out to be two and a half times more powerful—at least fifteen megatons. And a widely read book published by the AEC in 1957 acknowledges that the detonation spewed lethal levels of radioactivity over a vastly greater area than shot planners had secured:

> The resulting fallout . . . seriously contaminated an elon-
> gated, cigar-shaped area extending approximately 220
> (statute) miles downwind and varying in width to over 40
> miles. In addition, there was a severely contaminated region
> upwind extending some 20 miles from the point of detona-
> tion. A total area of over 7,000 square miles was contami-
> nated to such an extent that survival might have depended
> upon evacuation of the area or taking protective measures.

Subsequent editions of this book, *The Effects of Nuclear Weapons,* extend the area to 330 miles by 60 miles.

Six years after the Bravo incident, Russell Ball insisted at Point Hope that the fishermen were injured "because they were where they were told not to be." But the truth was the crew could have been *hundreds of miles* beyond the established exclusion zone and still have been in peril due to the AEC's mistake. It would be interesting to know how Ball would have explained the heavy doses that fell on the atolls east of Bikini, which were all outside the exclusion zone. Presumably, he would have had to concede that the islands were where they were supposed to be.

Even Edward Teller allowed that the Bravo fallout debacle was due to a "tragic error." The radioactive cloud from such a large detonation as Bravo can climb twenty miles into the air and spread over hundreds of square miles. Winds at various altitudes may be moving in different directions. The task of predicting which way the fallout will be blown is, as some Bravo test officials admitted later, "at best an educated guess."

Several hours after the shot, readings of radiation detection equipment on Rongerik Atoll, 180 miles from Bikini, exceeded the highest levels the instruments could measure. Seeing the gauges pegged out, the twenty-eight air force weathermen on the island washed themselves down, put on protective clothing, and took refuge inside a weather-tight building. They were evacuated by plane a day and a half later. But it was nearly a day after that before eighty-two Marshallese on Rongelap Atoll were picked up—by boat. And Rongelap is sixty miles closer to Bikini and was directly in the path of the fallout. It was a good thing the Rongelapese all happened to be on the south end of the island. There they took radiation doses of about 175 roentgens before they were evacuated. Had any of them been ten miles up island, chances are half of them would have died. And had they been on the north end of the island, just thirty miles away and where they often fished, the 3,300 roentgens floating in on the tropical breezes would have left every one of them dead. The 157 people on nearby Utirik Atoll, also exposed to fallout, were not evacuated until more than three days after the blast.

Besides being evacuated sooner, American servicemen and test personnel exposed to fallout were better equipped to minimize the effects. As one survivor wrote: "We had meters, film badges, we understood what we were doing and for the most part what was happening; we knew how and what precautions to take." The people of Rongelap did not have these advantages. Though radiation showered down on them for hours, they were given no instructions whatever on how to minimize their dose. They suffered from itching and burning of the skin, eyes, and mouth, and from nausea, vomiting, and diarrhea. They were not even told that they could have washed the radioactive dust off their bodies. Nor were they told they could also have washed their food and thereby avoided ingesting the large doses of radiation that accumulated in their thyroids and other internal organs.

In the days and weeks after their evacuation, the Marshallese showed the symptoms of acute radiation poisoning. As Dr. Bradley writes, they were "severely burned by beta-gamma radiation on heads, hands, feet, eyelids, necks, shoulders, beltlines—anywhere the dust could catch and stick to sweaty skin." Feet injuries were especially severe; those who had gone barefoot had weeping, open sores. The AEC issued a press release ten days after the Bravo shot, at a time when the AEC knew of the suffering of the Rongelap people. The statement characterized the test as "routine." It mentioned that "individuals were unexpectedly exposed to some radioactivity," but added: "There were no burns. All were reported well. After the completion of the atomic test, the natives will be returned to their homes." Even a month after the injuries, AEC chairman Lewis Strauss dishonestly told reporters that the affected natives were essentially uninjured and untroubled:

> None of the 28 weather personnel have burns. The 236 natives also appear to me to be well and happy. The exceptions were two sick cases among them, one an aged man in advanced states of diabetes, the other a very old woman with crippling arthritis. Neither of these cases have any connection with the tests. Today, a full month after the event, the medical staff on Kwajalein have advised us they anticipate no illness, barring of course disease which might be hereafter contracted.

Of course, any cancers and leukemias that might develop as a result of the exposure would not be expected to show up for years. The statement that no such illnesses had appeared in "a full month" had propaganda value, but from the standpoint of medical science—or from the standpoint of the honest administration of public affairs—it was tantamount to a lie.

It was three years before the islanders returned to Rongelap, and nine years before the first cases of thyroid cancer were noticed, especially in the children. Of the young Rongelapese who were under twelve when Bravo exploded, nineteen of twenty-two developed thyroid tumors. Many had to have their diseased thyroids removed. Eventually government studies disclosed that of Marshallese children who received in excess of 1,000-rad thyroid doses from U.S. hydrogen bomb tests, eighteen of nineteen died.

As a result of the tragedies associated with the 1954 Bravo test, the public-health hazard of radioactive fallout became, overnight, an issue of international concern. The uproar was covered widely in the popular press, including *Life* magazine. Several Point Hopers subscribed to *Life,* including Kitty Kinneeveauk, the woman who had insisted she'd "read some news from magazines about Indians where you work on this too, blasting their town, and none of these atomic people help them."

Many people considered Bikini Atoll, with its coconut palms, lovely lagoon, and excellent fishing, the crown jewel of South Sea atolls. Though the people had agreed to leave, they always thought they'd be able to return to their islands at the end of the testing. About a hundred Bikinians did go home in the late 1960s, when the AEC said the atoll was safe. They found the topsoil had been bulldozed into the sea and that the important food plants were missing. In 1975, plutonium began showing up in the people's urine. The AEC told them not to worry, the levels were safe. Then unexpected levels of strontium and cesium were detected and the Bikinians were moved again. They left the main island of Bikini for a smaller one, Enyu, at the entrance to the atoll. When contamination was found there, the people were told to stay off Enyu for at least twenty more years.

The United States detonated sixty-six nuclear bombs at Bikini and Eniwetok before consolidating its testing in Nevada. But the U.S. military didn't leave the Marshall Islands. Today the U.S. tests missile defense weapons systems at Kwajalein Atoll. And missiles fired from Vandenberg Air Force Base in California cross several thousand miles of Pacific Ocean to zero in on one or another of the islands at Kwajalein. The Marshallese and the U.S. government continue to negotiate over what type of association should exist between them. But the U.S. brings one nonnegotiable demand to the bargaining table: the long-term use of Kwajalein as a missile range.

Meanwhile, the Bikinians have been displaced for about fifty years. Kitty Kinneeveauk had it about right.

●   ●   ●

After the first Soviet atomic test in 1949, the National Security Council wanted a more secure, continental nuclear test site. The next year, the AEC and the Department of Defense sought permission from President Truman to conduct an underground nuclear test at Amchitka Island in Alaska's Aleutian Chain. Truman granted permission, but the location was not very convenient for the weapons designers at Los Alamos, and the

AEC continued the search for a site in the States. (Much later, in 1965, 1969, and 1971, the AEC and DOD did conduct underground nuclear tests at Amchitka, code-named Long Shot, Milrow, and Cannikin, respectively.) On December 18, 1950, Truman established the Nevada Test Site (NTS) within the Las Vegas-Tonopah Bombing and Gunnery Range. The very next month, the air force began dropping nuclear bombs—one of them larger than the Hiroshima bomb—over Frenchman Flat, about fifty miles from the outskirts of Las Vegas. In all, the AEC conducted 100 aboveground detonations at NTS prior to the moratorium on testing agreed to by the United States and the Soviet Union in 1958. The moratorium ended in August 1961 when the Soviets violated the agreement, and the United States resumed testing almost immediately. Thereafter, the AEC fired only underground detonations at NTS, though most of the explosions still vented radiation into the atmosphere. Nuclear blasts in the atmosphere continued in the South Pacific. Finally, on August 5, 1963, the United States and the Soviet Union, together with the United Kingdom, signed the Limited Test Ban Treaty, which prohibited nuclear testing in the atmosphere, under the ocean, in outer space, or "in any other environment if such explosion causes radioactive debris to be present outside the territorial limits of the state."

When, in 1979, the U.S. Congress finally looked into the health effects of the years of atmospheric testing in Nevada, Rep. Bob Eckhardt's Subcommittee on Oversight and Investigations (of the Committee on Interstate and Foreign Commerce) concluded that the government had been negligent. Internal AEC memoranda brought out at the hearings showed that the agency was well aware of the likely health risks to downwind residents, but had subordinated its responsibility to ensure public safety in order to advance its mission to build nuclear weapons.

The subcommittee said the AEC failed to inform the nearby residents of evidence in its possession that suggested that radiation was causing them harm. In fact, "all evidence suggesting that radiation was having harmful effects, be it on sheep or on the people, was not only disregarded but actually suppressed." The evidence also showed, according to the committee's report, that before the tests even started, the AEC had "sufficient information available" to anticipate that "people living nearby needed protection." Nevertheless, the AEC "totally failed" to give adequate advance notice of impending tests, failed to monitor properly the population's exposure to radiation, and falsely reported the radiation ex-posure data in order to derive more favorable dose assessments.

Additionally, the Eckhardt committee found that it was more likely than not that various health problems cropping up among the "Down-winders," as they called themselves, were attributable to fallout. By 1979, the year of the hearings, a growing body of evidence showed that cancers occurred among the people downwind of the Nevada Test Site at a rate higher than would normally be expected. Sheep injuries and deaths were also the result of radiation exposure, the committee said, declaring that the federal government made a "concerted effort to disregard and dis-count all evidence of a causal relationship between exposure of the sheep to radioactive fallout and their deaths."

By 1984, even a court decision judged that the Nevada Test Site oper-ators "negligently and wrongfully breached their legal duty" to protect the off-site residents. Writing on the case, one political scientist called the court's opinion "a factual portrait of agency carelessness that, legally, rose to the level of a willful violation of the rights of the downwind residents and others who involuntarily suffered the consequences of the testing program."

Perhaps Russell Ball and Rodney Southwick were not involved in the AEC's disinformation program. Perhaps they were not even aware of the agency's distortion and suppression of the evidence that Nevada testing was injuring people. But these deceptive practices were standard proce-dure at the AEC long before they told the Point Hopers that testing in Nevada had never harmed the people nearby.

•   •   •

When experts in the Eckhardt committee's employ examined one of the AEC's animal studies, they discovered experimental irregularities that seem properly labeled scientific fraud. Dr. Harold Knapp, working for the committee, undertook research into the AEC's sheep studies. Knapp, for-merly with the AEC's Fallout Studies Branch, found "critical omissions, distortions, and deceptions concerning experimental data on the effects of ingested radioactivity on sheep."

One example of government deception Dr. Knapp found concerned an AEC study of lambs born to ewes fed radioactive iodine. In the first lambing season, according to the AEC, the lambs were "normal in size." In subsequent seasons, the lambs showed "a significant reduction in birth weight." These results might be of some slight concern to sheepherders, but would not register alarm with the general public. But Knapp located other scientific accounts of the same experiment that revealed that the first season's lambs, the ones "normal in size," had shown symptoms of thyroid anomalies shortly after birth. This suggested that the thyroids were damaged before birth when the embryos had received doses of radioactive iodine. The thyroid observations were omitted from the report, while the lambs' weights were duly logged in.

Secondly, with respect to the later generations of lambs, those that showed "a significant reduction in birth weight," Knapp discovered that the lambs were all either born dead or so ill that they died in a matter of days. In a study purporting to assess whether radiation might cause health problems in sheep, a neonatal death rate of 100 percent would seem a rather pertinent detail. But the government scientists declined to note it. Instead they apparently carried dead lambs to the scales, weighed them, and wrote up the results as if they believed that birth weight, rather than fatality, was the notable characteristic.

The Nevada cattle study mentioned by Russell Ball at Point Hope fits the same pattern of science in service of public relations. The study was formally known as the Offsite Animal Investigation Project, according to an AEC press release issued just weeks after Ball's mention of the study at Point Hope. For this study, the government put a herd of cattle out to graze on the Nevada Test Site or on the surrounding bombing range. The term *offsite* was used to show that the findings were expected to apply to the animals on ranches beyond the test site. The idea was that the herd would graze on "old ground zero areas and other close-in pastures which have had maximum fallout." Then, according to the press release, if the animals showed no "injury or bad effects of any kind, it becomes obvious that other animals grazing on pastures that received much less fallout will not be affected in any significant way." Over the course of two years, nineteen of the cattle were slaughtered so that tissues could be tested. And, sure enough, laboratory analysis determined there was "no noticeable effect on these animals."

The experimental design of the Offsite Animal Investigation Project would not have passed muster as a term project for a first-year graduate student. For one thing, the study could not draw conclusions about such health effects as cancers or blood disease because the duration of the test was too short to permit these pathologies to develop. The herd began grazing in the area in late 1957, and the AEC was announcing its "findings and conclusions" two years later. The latency period, i.e., the time between exposure and the onset of disease, is likely to be three to five years in the case of leukemia, and ten years in the case of cancer.

Furthermore, if exposure to radiation in fallout concentrations caused increases in cancer rates by, say, 1 in 500 per year, that might be considered a significant health issue. But the AEC's experimental design did not allow such an effect to manifest itself. At its zenith, the study involved forty-seven animals. In such a small sample size, no increase in cancer would be expected to show up in two years.

In addition, it is not reasonable to presume that the potential for biological hazard from fallout is always a strict function of distance from ground zero. Different fallout radioisotopes tend to be found at different distances from the blast and at different times after the shot. Most of the radioisotopes that fall to earth quickly, and relatively near to ground zero, have short half-lives. Isotopes of the gases krypton and xenon, which change into strontium 90 and cesium 137, can move more easily through the churning dirt, be carried aloft in the fireball, and be caught by the winds. The very fine particles in the radioactive cloud sometimes take weeks, months, or years to fall out. They can travel great distances and may be richer in the long-lived radioisotopes, particularly strontium 90 and cesium 137. And these radioisotopes constitute two of the greatest biological hazards associated with fallout because they may be ingested with food, metabolized in the body, and there irradiate adjacent tissues for years. It was not unusual for AEC's radiological monitors to encounter downwind radioactive "hot spots"—where range animals were killed—well outside the boundaries of the test site and bombing range.

Finally, because the arid test site offered little in the way of "pastures," feed and water were regularly hauled to the herd. Again, these are not test conditions from which meaningful scientific results can be obtained. In order for the study to be a legitimate indicator of the cancer and leukemia hazards of fallout to grazing animals, it would need to involve many more animals; they would need to eat irradiated grass, not imported feed; and they would need to do so for many years.

Aside from the flawed experimental design, it was improper and misleading of the AEC team at Point Hope to suggest that studies of cattle in the desert offered insight into the case of caribou in the Arctic. That an Arctic ecosystem might differ in important respects from a desert one should have occurred to scientists representing themselves as authorities on radiation and ecology.

•   •   •

As the tape recorder reels turned at the March 1960 meeting at Point Hope, Russell Ball told the Eskimo people that studies at Hiroshima and Nagasaki showed that among "people who were very seriously exposed, once those who survived recovered from the illness, there was no other result." Ball apparently ruled out such health effects as cancer and leukemia. But even Edward Teller, who was inclined always to put the best possible face on fallout, acknowledged the connection. In a book published two years before Ball's statement, Teller wrote, "Large doses of radiation increase the likelihood that an individual's life will be shortened by leukemia and possibly other cancers." And with respect to the Japanese survivors in particular, the AEC's definitive work by Samuel Glasstone, *The Effects of Nuclear Weapons,* contradicts Ball:

> The first definite evidence of an increase of the incidence of leukemia cases among the inhabitants of Hiroshima and Nagasaki was obtained in 1947. At least 2 years elapsed, therefore, between exposure and the development of the symptoms. The number of new cases reported has increased fairly regularly in succeeding years.

Glasstone's 1957 edition of the same work shows that, years before Ball told the Point Hopers there were no such risks, it was common knowledge among people conversant with nuclear issues that exposure to radiation could produce "late effects":

> There are a number of consequences of nuclear radiation which may not appear for some years after exposure. Among them, apart from genetic effects, are the formation of

cataracts, leukemia, and retarded development of children *in utero* at the time of the exposure.

In response to a question asked at Point Hope about possible birth defects as a result of exposure to radiation, Ball pointed to the studies of Hiroshima and Nagasaki survivors. He said: "Ever since that time, even to the present day, there have been very careful studies among those people and we have found no evidence for *any* effect upon the children of those people from the radiation dose they received. No evidence of any indication at all." But again Ball's account does not square with the facts as published by the AEC itself. Among the Japanese women who were pregnant at the time of the attack, and who suffered large doses of radiation, Glasstone noted "a marked increase over normal in the number of still-births and in the deaths of newly born and infant children." When a study of the surviving children was done a few years later, it found an increase in the frequency of mental retardation.

Reviewing the scientific literature dealing with the Japanese survivors, Nobel laureate Linus Pauling wrote two years before the Point Hope meeting: "There is no doubt that the increased incidence [in cases of leukemia] is to be attributed to the exposure to radiation." Pauling also noted a significant increase in leukemia deaths among U.S. radiologists who, in the course of their work, were inevitably exposed to repeated doses of low-level X-radiation. "I conclude," he wrote, "that it is highly probable that small doses as well as large doses of radiation cause leukemia in human beings." Nearly every scientist who expressed an opinion at the hearings before the Special Congressional Subcommittee on Radiation in 1957 agreed with Pauling. They expressed the belief that even small doses of radiation could produce leukemia, bone cancer, and other diseases, and caused a shortening of life expectancy.

Even if Ball had not stayed current in his reading, evidence of long-term health problems from radiation dated back to the last century. Within a year of discovering X-rays in 1895, early experimenters reported burns on their hands, forearms, and faces that led to skin cancer. The tragic case of the radium-dial women in the mid 1920s had been widely reported. Women factory workers who painted watch dials with a radium-based luminous paint had the habit of using their lips to put a point on the tips of their brushes, and in so doing they ingested minute quantities of radium daily over many years. Once in the body, radium, which is chemically similar to calcium, takes up residence in the bones. There it continues its radioactive decay, irradiating adjacent bone, tissues, and the blood-producing marrow. The women developed assorted bone diseases, bone cancers, and anemias.

Also common knowledge for years—especially to anyone in the atomic energy field—was the plight of the uranium miners. It was known that air in the mines was contaminated with radon gas, a product of uranium decay. The gas, in turn, decayed to isotopes of polonium, bismuth, and lead, the so-called radon daughters. As the miners breathed the air, radioactive particles lodged in their lungs. Dr. Pauling reported in 1958 that one-half of the uranium miners who had died by 1939 had developed lung cancer.

Hiroshima and Nagasaki survivors, as well as radiologists and some patients treated with X-ray therapy, had shown aplastic anemia, a disease of the bone marrow that inhibits the production of red blood cells. And among physicists who operated cyclotrons (particle accelerators), patients exposed to X-rays, and many of the Japanese survivors, many cases of radiation-induced cataracts were noted.

Ball had also essentially denied the effect of radiation on human heredity when he dismissed the long-term health effects of exposure to large amounts of radiation. Radiation causes genetic mutations, and genetic mutations are nearly always harmful. On this issue, Ball seemed to present to the Point Hope villagers a view that was more extreme than that argued by even the AEC's most tendentious spokesmen. When Edward Teller visited Alaska to promote Project Chariot, he brought University of Alaska president Ernest Patty a signed copy of his 1958 book, *Our Nuclear Future*. In its pages, Teller granted that radiation does cause mutations, and that "mutations can be caused by any small amount of radiation. The less radiation, the less the chance. But the chance will always be there." Teller felt that the degree of increase caused by atomic testing was not serious and argued his point with the canard about Tibet that so amused the environmental activist Barry Commoner:

> The Inca empire existed for many generations in the high
> country of Peru. The people of Tibet have been exposed for
> generation after generation to the greater cosmic ray intensity

which bombards them through a thinner layer of atmosphere. These people have been exposed to much greater additional radiation than anything which is caused by atomic tests. Yet genetic differences have not been noticed in the human race or for that matter in any other living species in Peru or Tibet.

Logically, if Teller had taken the trouble to assure himself that genetic effects had not been noticed in scientific studies of the Tibetans or Peruvians, then he must have found out during his research that no such study was ever done among those groups. As Barry Commoner has pointed out, such medical statistics simply did not exist in those countries. But, if the data *were* compiled, Linus Pauling expected they would show "an estimated 15-percent increase in the incidence of seriously defective children in Tibet."

The fact that radiation caused genes to mutate was known since Professor H. J. Müller discovered the phenomenon in 1927 by exposing plants and animals to X-rays. As early as 1954, editorials in *The Bulletin of the Atomic Scientists* had warned of the "potentially fateful dangers of a long range damage to the hereditary endowment of the human race" from the low levels of radiation found in fallout.

Writing in 1955, Dr. Müller, by then the foremost American authority on radiogenetics and winner of the Nobel Prize for his 1927 discovery, said of radiation that "no exposure is so tiny that it does not carry its corresponding mutational risk." Müller believed that many Hiroshima survivors must have had radiation doses sufficient to cause mutations, and that the inability to prove it from inconclusive statistics should cast no doubt on this certainty. While AEC physicists sometimes disputed these claims, particularly the linear relationship between the dose and the chance of genetic damage, no trained geneticist doubted them. In his widely read book of 1958, *No More War,* Linus Pauling cited the conclusions of the Committee on General Effects of Atomic Radiation of the U.S. National Research Council-National Academy of Sciences: "Any radiation is genetically undesirable, since any radiation induces harmful mutations . . . the basic fact is—and no competent persons doubt this— that radiation produces mutations, and that mutations are in general harmful."

All of this information was reported in scientific journals, the popular press, the testimony of Congressional hearings, as well as in many widely read books. And all, save the Eckhardt committee findings and the documentation of thyroid abnormalities in the Marshallese, appeared years before the March 14, 1960, meeting at Point Hope. There was universal scientific agreement among those with competence in the field that long after the immediate symptoms of radiation sickness subsided (such as skin lesions, nausea, and hair loss), those "very seriously exposed" to radiation had an increased risk of developing various cancers, leukemia, anemia, and cataracts. The health effects of *low* doses of radiation were, in 1960, a hotly debated issue. But as of that date, the history of the "permissible" dose standard for workers in nuclear industries was not reassuring. The permissible doses were repeatedly and radically reduced. The only constant was the inveterate confidence that each new level was the safe one and should be trusted:

1928–1936	100 rads per year
1936–1947	35 rads per year
1948–1957	15 rads per year
1957–	5 rads per year

*   *   *

Reading a transcript of the Point Hope meeting twenty-eight years later, Robert Rausch said: "Looking back on those statements from present knowledge, I would say they sounded quite stupid." Of course, a reasonable person familiar with atomic testing could reach the same conclusion on the basis of what was known *then,* in 1960.

Rausch had said that the Chariot blast would probably contribute so little radiation that it "would be impossible to detect." This is what he had understood from his AEC briefing, but the statement was inaccurate. Radiation detection equipment then in use was sensitive enough to measure fallout down to very low concentrations. There would be no difficulty detecting the additional radiation that the Chariot experiment would add to the local environment, even if it was, as the AEC predicted, very small. Russell Ball should have known this, but he did not correct Rausch.

Actually, the amount of vented radiation was not planned to be small at all, though the AEC liked to say that radiation effects "would be negligible, undetectable, or possibly nonexistent" beyond the immediate crater area. A tremendous quantity of fission products would have vented, according to documents from Livermore declassified in the course of this research. High levels of fallout would be scattered across hundreds of square miles of northwest Alaska. A comparison is useful: The nuclear power plant accident at Three Mile Island near Harrisburg, Pennsylvania, in 1979 is said to have released about fifteen curies of radioactivity into the atmosphere. When the top blew off the reactor at the Soviet Union's Chernobyl nuclear power plant in 1986, the explosion released 40 million to 86 million curies, according to one estimate. But the Project Chariot detonation, at its smallest configuration of 280 kilotons, would have vented fission products into the atmosphere totaling 1.5 *billion* curies, according to the Livermore document. Per standard practice, this figure represents the radiation as measured at one hour after detonation. That much radiation would equal the amount vented by seventeen to thirty-eight Chernobyl accidents, according to the range quoted above for Chernobyl's venting.

Another government estimate, this one prepared by the Weather Bureau's Special Projects Section in Las Vegas and titled "Surface Radiation Estimate for Project Chariot," predicted a different figure. This document questioned the Livermore physicists' assumptions with respect to the percent of the fallout that would vent to the atmosphere. The Livermore physicists believed that the fallout fraction for an underground burst would depend on the ratio of the depth of burst to the depth of the crater produced. They further believed that the fallout fraction would decrease logarithmically with increasing depth ratios. Livermore considered these assumptions to be "based on the best available estimate." But the scientists of the Weather Bureau noted that "other scaling methods which have been suggested result in lower depth ratios and therefore higher fallout fractions for the Chariot detonations." The Weather Service declined to use Livermore's optimistic logarithmic formula, which resulted in fallout fractions on the order of 10 percent. Instead, while recognizing the uncertainties as to which method might be the better predictor, they applied the linear model and from it obtained "a fallout fraction of about 30 percent for the 200 kiloton device and about 40 percent for the four 20 kiloton bursts. Using 300 megacuries per kiloton, this amounts to a total of 27,000 megacuries of fallout at one hour after detonation." Twenty-seven thousand megacuries is equal to 27 billion curies. If this prediction was accurate, the vented radiation of the Chariot explosion would be equivalent to between 314 to 675 Chernobyl disasters.

The magnitude of the potential health effects to the Eskimos is suggested by the fact that within a five-mile radius of the Three Mile Island power plant, schools were closed, and children and pregnant women were required to evacuate the area. Within a ten-mile radius, people were advised to stay indoors. But in the case of Project Chariot, the AEC planned to vent a quantity of radiation greater than the Three Mile Island accident by *100 million times* (using the lower, Livermore estimate), and to do it thirty-one miles from a village of 300 people. If the wind cooperated with the shot planners' predictions, perhaps no Chariot fallout would land on Point Hope. But if the weather forecasters were wrong, or the wind shifted as happened during the Bravo shot at Bikini, Point Hope could be plastered with fallout concentrations of 100,000 curies per square mile. It is difficult to reconcile these figures with Russell Ball's statement at the Point Hope meeting: "At this distance, the amount of airborne radioactivity which could reach here could not *possibly* be enough to cause any injury to the people or the animals. There's just no chance of that."

Ball had also stated that radiation exposure from the Hiroshima or Nagasaki blasts was "thousands, many thousand times what could possibly result from this." By this Ball may have meant that radiation doses to people in the Japanese cities were greater by thousands of times than would be possible at Point Hope. But many people may have interpreted the remark to mean that the Chariot explosion could not possibly *release* as much radiation as either of the Japanese detonations. And this was not true. As Ball had said in his discussion of the percent of vented radiation expected from Chariot, vented radioactivity could be on the order of 10 percent of the total yield. If, as the AEC had said, fission bombs were to be used, rather than fusion bombs, then the total yield would be attributable to fission, and 10 percent of 280 kilotons would vent. "Let's round that up to three hundred," said Ball, "a nice round number to talk about. If only ten percent of that is to get to the surface, we would then have on the surface an amount of fission activity roughly equivalent to that which would have been produced by a surface explosion of thirty kilotons." But the Japanese explosions were only about twenty kilotons each, according to AEC publications of the period. In other words, the Nagasaki bomb vented twenty kilotons of fission products, while the Chariot shot, by Ball's own calculation, might vent as many as *thirty* kilotons of fission products. Far from being "many thousand times" less than Nagasaki's, Chariot's vented radiation could well be more.

Russell Ball also erred when he said that Point Hopers would not be able to feel any tremor from a 280-kiloton detonation thirty miles away. The AEC's own predictions, published a few months after the Point Hope meeting, were that the seismic shock would be felt for eighty miles.

Point Hopers were seriously misled regarding the longevity of the harmful radioisotopes to be produced by the blast. Rausch had said "some will be gone in a matter of hours, others will take longer." What he did not say, and possibly did not know, is that two of the fission products most likely to harm man, strontium 90 and cesium 137, have half-lives of approximately thirty years. In other words, thirty years after the detonation, half of all the strontium and cesium deposited on the Eskimos' hunting grounds would still be radioactive. Sixty years later, a quarter of the original amount would still be giving off radiation. With strontium 90 and cesium 137, therefore, a significant amount of radiation would persist for several centuries. Anyone familiar with nuclear testing knew this. But neither Ball nor Southwick offered any clarification, and the people of Point Hope were left with a grossly inaccurate understanding of the persistence in the environment of harmful radionuclides.

Finally, Russell Ball had said, "The only thing about this entire program which is in any sense restricted is the design of the explosive itself. Outside of that, absolutely everything will be public knowledge and will be published." Edward Teller liked to make the same claim. In the same month, March 1960, he wrote in *Popular Mechanics,* "the whole Plowshare subject is now declassified with the exception of the black box that contains the nuclear explosive itself." But an internal, "official use only" AEC document explains that the agency had decided to operate under a system of "classification restraints," which prevented federal officials from giving the Alaska state government information that it had requested about Chariot's fallout. At this writing, Chariot documents unrelated to device characteristics are only released following great perseverance, usually through repeated requests and appeals under the Freedom of Information Act.

●   ●   ●

When interviewed in connection with this research, Robert Rausch was in his seventies. Of all the scientists who worked on the AEC's Bioenvironmental Committee, and of all the key people who were generally on the AEC side of the Chariot controversy, it seems that Rausch alone is inclined to look back on the episode with regret and self-criticism:

> I doubt if I would ever do anything like that again. I enjoyed the association with my colleagues . . . and I think that the scientific information that was eventually published was very valuable. But it's not the kind of situation I especially care to be in, dealing with either industry or something of this kind. They have too many of their own objectives and it's difficult for them to look at anything else.
>
> I think I learned about how the government works in situations of this kind. And I think that gradually led to my being a little more concerned about what is going on in the country. I tended originally to be very apolitical and not much concerned with other than my own interests. And I think I soon matured somewhat and realized we have to watch what is going on and be informed.

Perhaps that is why Rausch is willing to talk about the day when a "quite high-up AEC official" offered the Project Chariot Bioenvironmental Committee "all the money you need for research" if the committee would just sign off on Project Chariot.

11 THE COMMITTEE AND THE GANG OF FOUR

If mankind is going to advance, you
have to take certain risks and chances.
To think anything else is ignorant.

—Norman Wilimovsky

The majesty of technology is not sufficient unto itself. It is also encumbered
with the responsibility of asking itself
before its accomplishments, what else
these monumental achievements do
to man's environment.

—John N. Wolfe

The Committee on Environmental Studies for Project Chariot had been organized by John Wolfe two years before the meeting at Point Hope. As
Wolfe had begun to think about organizing a broad series of scientific investigations in the Arctic, the first thing he did was call his old friend Max Britton. Britton and Wolfe had been pals for twenty years, since they were in
graduate school together at Ohio State. At the time of Wolfe's call, 1958,
Britton was working in Washington, D.C., heading up the navy's arctic research program. Among his duties was the administration of the Naval Arctic Research Laboratory at Point Barrow, the northernmost tip of Alaska.

In a 1988 interview, Britton remembered how his friend was "filled with excitement" over the potential of the environmental studies. Wolfe had never been in the Arctic, but he knew that Britton had been there and was acquainted with the scientists who worked in the North. "He wanted to know who knows what about the area . . . he wanted to organize an independent research group" to run the program. And Wolfe wanted Max Britton to be a member of that group.

Wolfe, says Britton, recognized that an extraordinary opportunity existed for ecologists to make an enormous contribution to a large government project. Not only could environmental studies launch a project from a better foundation of knowledge, but, where it was warranted, the studies might determine that a proposed project was not feasible. "The point was," says Max Britton, "that the Committee, if he [Wolfe] was going to run it, was going to be independent in almost every degree from other sections of the Atomic Energy Commission."

Britton recommended for service on the committee Norman J. Wilimovsky, a cigar-chomping fisheries biologist then working for the U.S. Fish and Wildlife Service in Juneau, who would soon move to the University of British Columbia. Wilimovsky was "a fighter, a slugger, hard worker . . . a bulldog," says Britton, ". . . a highly vocal participant in our committee meetings. . . . There were no shy mouths in this group. We had knock-down, drag-out fights at every one of our frequent meetings." Actually, there were a couple of "shy mouths" on the committee, but the truculent Wilimovsky was certainly not one of them. He relished the meetings. "There were no holds barred," Wilimovsky says, "and some people frequently got their ears burned." "I'm too blunt, too honest," says Wilimovsky charitably of himself, "and I don't like the diplomatic niceties of saying no." Despite his pugnacious temperament, Wilimovsky developed "a kind of respect that borders on love for your colleague—meant in a masculine sense."

Another of Britton's selections was Robert L. Rausch, the parasitologist who accompanied the AEC's Russell Ball and Rodney Southwick to the March 1960 meeting at Point Hope. Though Britton was nearly a generation older than Rausch and a noted Arctic scientist himself, Britton says Rausch was "an old man in Arctic science before I ever saw the Arctic." Rausch had worked for the Arctic Health Research Center as a parasitologist since finishing his education in the late forties. And he happened to be one of the first men Britton had met on his initial visit to Barrow:

> I thought that the guy was absolutely crazy because he was talking about once kicking a brown bear in the face, about eating bott flies out of caribou, about eating half-hatched eggs with the Eskimos. I got an introduction to Arctic folklore that night that I will quite honestly say . . . I thought was as phony as a three dollar bill, that nobody had ever done all those things.

Britton eventually learned that the half-legendary Rausch, who was the only Alaskan selected for the committee, had done all those things and more. He had even been known to crawl into dens to take the rectal temperatures of hibernating bears. He was learning the Inupiaq language. He was the rare sort of scientist whose work earned him the highest regard of his colleagues, but whose respect for the wisdom of the indigenous culture earned him the friendship of the local people. Decades later, when the Inupiat of Alaska's North Slope established an Arctic Science Prize, they selected Robert Rausch to be its first recipient.

"The last of the ones of my selection is maybe the brainiest, the most thoughtful, certainly the most analytic of all the scientists on the committee," says Britton. That was Art Lachenbruch, a geophysicist from the U.S. Geological Survey whose specialty was permafrost. Lachenbruch was a driven researcher who loathed interruptions to his work. Max Britton remembers:

> He came on to the committee with great reluctance . . . I implored him to do so, begged him to do so, and appealed to all his better instincts as to what the committee needed and how he was probably one of the very few people who could provide it. . . . A gem of a scholar who offered more thought to our tempestuous meetings than any of the rest of us could.

Rounding out the committee were two men from AEC-funded laboratories. One was Kermit Larson, who worked at UCLA's Laboratories of

Nuclear Medicine and Radiation Biology and whom Britton remembers as "a very quiet chap . . . [who] really did not participate very much in committee discussions unless directly asked to speak." His background was in radiochemistry and fallout phenomenology.

The other was Allyn Seymour of the AEC-funded Laboratory of Radiation Biology at the University of Washington, the man who would brief Rausch before the 1960 meeting at Point Hope. Britton also described Seymour as "a very quiet, unassuming, distinguished, highly knowledgeable scientist." Seymour had worked in the laboratory since 1945 when the work was funded by the Manhattan Project. He conducted classified research dealing with the effects of radiation on fish, particularly the salmonides in the Columbia River adjacent to the AEC's Hanford nuclear plant in eastern Washington. For two years, from 1956 to 1958, Seymour worked directly for the AEC in Washington.

Finally, to keep the committee's work organized, the Atomic Energy Commission assigned a nonvoting secretary. He was Ernest D. Campbell, who worked out of the AEC San Francisco office and also served as "field coordinator" for the environmental program. Max Britton remembers him as "an eager beaver on transacting business," the one person at the meetings who could supply the paper trail on everything. "Ernie would typically arrive at the meetings with more briefcases than any three men ought to carry," says Britton, ". . . multiple, fat, robustly stuffed briefcases—let's say three or four—which at a meeting he would have clustered around his knees." Campbell would constantly anticipate from the context of the meeting what subject would come up next. "And whatever it was, he would be whipping out the papers in advance. My admiration for the guy had no bounds," recalled Britton, ". . . he could correspond interminably."

So the Committee on Environmental Studies for Project Chariot consisted of seven voting members and a secretary. Five of the members were biologists. And, though AEC press releases liked to make the point that "only the chairman, Dr. John N. Wolfe, is on the AEC staff," four were affiliated with the AEC: Wolfe (chairman), Seymour (vice chairman), Campbell (nonvoting secretary), and Larson. All eight relied on federal money for their paychecks.

But even if the committee seemed to lack representation from some obvious quarters—such as the University of Alaska or the Alaska state government—those members who are still alive insist that the character of the committee was such that agency pressure had no influence on their deliberations. They were a "hard-headed" bunch, says Britton. They were willing to accept AEC money for environmental studies, then use those studies to "turn right around and criticize [the AEC], which we did with impunity if we found reason to." Art Lachenbruch agrees: "If people tried to get me to say things, or do things, or knuckle under to things that I didn't think were right, well, of course, I didn't do it." Seymour says, "I can guarantee you this, there wasn't one member of that committee that would rubber stamp anything from Livermore. We, many of us had arguments with the Livermore people." And Wilimovsky leaves no doubt how the arguments were resolved: "I can remember once somebody suggested—from the PR office at Livermore—to do something that didn't have to do with the firing of the shot, but to make some statement to placate somebody . . . and in direct words, we told him to go to hell."

The insistence with which the committee members assert their ability to stand up to AEC pressure suggests that there were times when such assertions of independence were necessary. Yet when pressed on this point, most of the members waffle. Seymour proclaims that "many of us had arguments with the Livermore people," but when he is asked what occasioned the arguments he hems: "No pressure was put on us"; and haws: "Well, I guess I'm thinking more about the, our laboratory in the Pacific." Wilimovsky boasts of telling LRL officials to "go to hell," but when he is asked to talk about specific episodes, suddenly the agency pressure slides into the realm of the hypothetical: "If it happened, it must have happened very early in the game." Art Lachenbruch is clear about his reaction if anyone suggested he knuckle under. But when asked about occasions when he might have made such outspoken assertions, he allows that "we might have had hassles like that," but no incidents come to mind. Max Britton talks about the committee having criticized the AEC "with impunity" when it became necessary, but when asked about AEC pressure that might have prompted the criticism, he can hardly be more categorical in denial: "I do not recall a single event in all history in which we were asked to go along with anything."

Even though it seems that the former committee members might be shocked at the suggestion, it is hard to imagine that the AEC did not press its agenda. It would be much more remarkable if the Atomic Energy

Commission—an agency of considerable institutional inertia, unaccustomed to abdicating power, that operated largely in secret during the height of the cold war—did not use its appreciable power to lean on the committee when it saw fit to do so. Besides, the AEC was footing the bill for the committee's research to the tune of well over half a million dollars a year.

Documents in the archival record support the notion that agency pressure did exist. According to one account, Robert Rausch told Don Foote that the committee was sharply divided between those favoring all AEC proposals, on the one hand, and those favoring a more objective approach, on the other. Rausch is said to have complained that an AEC official approached the committee with a prepared statement at the beginning of a meeting and asked all members to sign. According to this account, Rausch and several others objected, but finally agreed to sign a watered-down version that Rausch considered "misleading" to the public and "meaningless" to the scientific community. "I absolutely remember nothing like that ever having occurred," says Max Britton, "and I would love to know where any such comment as that came from, and I would love to know what Robert Rausch would respond to that [laughs]."

When asked about the nearly thirty-year-old incident, Robert Rausch said:

> I remember the day, but I don't remember the details very well. But it is my recollection that we were asked to provide some positive reaction to their proposal prior to the time we felt was appropriate, and I think there was general agreement that this was the case. . . . I suppose we came up with some more or less useless statement.

Rausch says the committee did not change its views to accommodate the AEC, but "nonetheless we were aware of a certain amount of impatience and pressure." And, unlike his colleagues, Rausch was willing to elaborate:

> RAUSCH: It is my recollection—I don't know if this has been mentioned before or not, and I'm somewhat reluctant to go into some of these less pleasing details—but it is my recollec-

tion that the committee was more or less offered ample continuing support for further work if we would go ahead and give approval for some fairly immediate action. I don't know whether that has been mentioned by anyone else or not, but that is my recollection.

QUESTION: That agency support for future research in your respective fields back at your universities—

RAUSCH: No, no. Continuation of the Project at Cape Thompson.

QUESTION: Oh, I see. So, recognizing that you folks had an interest in seeing this basic science done, the kind of quid pro quo was: maybe we can keep the project going longer and get more data provided you—

RAUSCH:—don't make an obstacle. I, that's my recollection.

QUESTION: And what was the committee's reaction to that?

RAUSCH: Quite adverse. No, we were not influenced by that kind of thing.

Later, Rausch added to the account:

RAUSCH: I recall quite distinctly a quite high-up AEC official coming to us and saying if you go along with this you will have all the money you need for research.

QUESTION: Let me be clear on that. He was saying, if the committee would conclude that the biological risk of the project was low, that the AEC would continue to fund the Cape Thompson studies after the blast?

RAUSCH: Yes, that's right.

QUESTION: So it was kind of a bribe, not as far as offering you financial gain, but offering you the chance to do more good science?

RAUSCH: Yes. And that's how we thought of it, as a bribe and, of course, we declined. We were not going to be influenced by that.

While direct financial gain may not have been a factor in any attempt by the AEC to influence committee findings, the fact remains that three of the committee members, through their respective agencies, were working under contract with the AEC on the environmental studies at Cape Thompson. Had the committee gone for the "deal," it may have meant additional federal contracts for some of its members.

Rausch is the only member of the committee to acknowledge what seems implicit in the others' statements, even in their denials: that pressure was exerted on the committee to endorse AEC positions. It was his perception that the committee was indeed somewhat divided, that the AEC-affiliated members seemed to have allegiances that he, Lachenbruch, Britton, and Wilimovsky did not.

●   ●   ●

Once the entire committee was onboard, they set about writing a charter to define their role in the operation. From the very beginning, before the committee was fully organized, John Wolfe had been proclaiming that the shot would not go off if the studies showed that it was too dangerous from a biological standpoint. And AEC press releases and public statements frequently reminded Alaskans that the experiment would be conducted only if the studies showed that the project would be safe.

The committee's charter, however, stopped well short of any authority to cancel Chariot. According to the charter, the committee would decide what studies should be done so that the biological cost of the experiment could be determined. It would recommend who should do those studies. It would analyze and summarize the investigator's reports for the AEC, and pass on "the Committee's recommendations for further action, if any." What is *not* clear from the charter is that the committee had any license to do anything other than plan the program of studies and transmit the scientific results to the AEC. It is not clear that the document conveys any mandate to interpret the studies so as to determine Chariot's safety.

Internal documents show that the AEC did not consider the environmental studies to be important indicators of safety anyway. Gen. Alfred Starbird, head of the AEC's Division of Military Applications, which oversaw the Plowshare program, told the AEC commissioners (with John Wolfe present):

> The studies are not inherent to the safety of the proposed project, but the bioenvironmental knowledge gained would

be extremely valuable to science, and would be helpful from
a public relations standpoint.

A month later, the AEC would nonetheless claim in a press release that
the project could go forward if the studies should "*prove* the safety and
practicality of the experiment." (Emphasis added.)

But some of the committee members also doubted that they were responsible for determining the safety of the nuclear explosion. Allyn Seymour wrote Wolfe that he wanted the committee's responsibilities to be
better defined. "For instance," he wrote, "is it up to us to say whether
or not the detonation is safe? . . . our responsibility on this point is not
clear to me." Seymour was still unclear on the issue twenty-nine years
later:

> You'd have to kind of predict what the impact was going to
> be, wouldn't you, in order to say whether you should or
> should not go ahead? And I don't think that was really [the]
> objective here. The objective was to provide a baseline, and
> that doesn't tell you whether to go ahead.

Norman Wilimovsky appeared to agree. "Whether the shot would go
ahead or not was not this committee's responsibility," he said.

But even assuming that the committee was charged with proving the
safety of the explosion, as the AEC press releases continually suggested,
how were the studies going to permit such an evaluation? What were the
questions that, if answered by the studies, could allow the committee to
predict the biological cost of the detonation? While the character of the
physical environment—the geology and climate, for example—were examined, the committee focused more on studies of the biota, which included humans. But a large part of these biological investigations
comprised little more than inventories of the life-forms present in the region. These baseline data were to be compared with postshot measurements. Those studies that did address the dynamic aspects of biological
assemblages—such as food chains—stopped short of a design that
would permit definitive conclusions about the potential for harm to
man. Though AEC public statements stressed that the environmental
program was designed to "assure that the public health and safety will

be protected," the studies did not always seem designed to provide a scientific basis for such assurances.

For example, over the first three field seasons, studies were designed to measure existing radiation levels in the air, water, soil, vegetation, fish, and animals, but not in people. Not until the summer of 1962, at a time when Livermore had lost interest in Project Chariot and recommended to the AEC that it be canceled, were radiological studies extended to include the human inhabitants of the Cape Thompson region. And it wasn't until that fall, when Chariot's demise was actually announced by the AEC, that tracer experiments were conducted at the Chariot site. Here scientists sprinkled actual fallout mixed with dirt onto the tundra to track how it would move via groundwater and surface runoff. This latter study wasn't even part of the Chariot investigations, but was proposed by the U.S. Geological Survey and apparently funded outside the AEC's Plowshare program.

Other seemingly obvious pathways by which radiation might move through the environment to man were likewise neglected by the battery of studies. Livermore scientists frequently produced drawings showing the expected contours of fallout distribution. But if the radioactive particles were landing on snow in an area notorious for its winds, how valid would be the plotted fallout densities after the snow and its radiation had been blown away and concentrated in drifts? None of the Project Chariot program of studies focused on this phenomenon, but the above-mentioned tracer study noted the potential for an "Ogotoruk wind" to "within a single day, redistribute fission products from a large part of the area and carry them far downwind mingled with snow." And snowdrifts, the report said, might create "notable 'hot spots'":

> Assume a snowdrift 30 feet deep, such as occurs locally in the lee of minor ridges athwart the dominant wind from the north. . . . Assume further that the drift resulted from wind erosion of outlying snow surfaces to a depth of 1 inch and that all fission products from the eroded areas were removed with the snow. The drift, therefore, would contain fission products from an area 360 times greater than its own extent.

In effect, John Wolfe's committee organized studies to assess whether conducting Project Chariot would involve biological cost at the ecological level—that is, might the blast render species extinct, break food chains, disrupt animal migrations, and so forth. But even if the conclusion was that the chance of damage at this level was remote, it would say very little about the potential health hazard to Eskimos. It would not mean that radiation from the Chariot shot might not move via runoff and groundwater transport, thus contaminating local water supplies. It would not mean that, owing to various unique features of the Arctic ecosystem, radiation might not move freely up the food chain to man. And it would not assure that fallout mixed with snow might not even be blown directly into the village of Point Hope on the frequent gale-force winds. It would merely signify that Chariot was not expected to cause the sort of catastrophic biological disruption that the AEC had caused on Bikini Island, for example, where most coconut palms and other important plants had been eliminated and replaced by a dense layer of scrub.

●   ●   ●

When the five commissioners of the U.S. Atomic Energy Commission originally voted to approve money for Project Chariot on June 12, 1958, they did so with the proviso that the necessary port facilities should be built by private industry. But after the Alaskan businessmen had all but laughed the ice-locked-harbor promoters out of the territory, chagrined Chariot planners went back to the commission in early 1959 admitting that "outside sponsorship . . . has not been forthcoming." They pleaded for continued funding to support surveys and planning for a smaller, less expensive harbor. The commission agreed with the plan, but again added a proviso. Before committing to any major expenditures, it directed that the AEC's general manager should review the upcoming summer's environmental surveys and, based on that information, "by January 1, 1960, recommend whether the project should proceed or be indefinitely suspended."

It may have been a milestone in public policy that a decision to proceed with a large federal project should be predicated upon the results of an environmental study. But from a pragmatic standpoint, the backers of Project Chariot needed by year's end some sort of positive expression from the Bioenvironmental Committee that the summer of 1959 fieldwork showed that the excavation could be safely done. Consequently,

Russell Ball of AEC San Francisco sent all of the contractors working on environmental studies a letter asking them to submit by December 1 "Preliminary Evaluation Reports," which would "be very important to Chariot. They will be used as a basis for forming the recommendations as to whether or not to continue the project into the next year." All the investigators worked long hours, day and night, organizing the summer's data in order to meet the deadline. Don Foote's report was "hammered out on the typewriter in less than a week." The same week saw Berit's arrival at Point Hope and the last-minute completion of two important surveys Foote had been conducting under extreme weather conditions. It was nearing five in the morning on November 23, 1959, when he finally tied up the last details and got the manuscript ready to send to his brother Joe for editing, reproduction, and delivery to the AEC.

On December 10, the Bioenvironmental Committee met to review these reports and begin to draft a statement. The members finally signed off on what would be a bombshell—a document entitled "Statement of Committee on Environmental Studies for Project Chariot" on January 7, 1960. It began:

> It is the unanimous opinion of the Committee, based on the data known to it as of December 10, 1959, that Project Chariot may be carried out under the following conditions:
> 1. Time—The preferred time of the year for the detonation is Spring, i.e., March or April.
> 2. Debris—The preferred disposition of the debris, especially that of a radioactive nature, is to sea; placing the debris to land is also considered acceptable.
> 3. Qualifications—In no case should radiation be delivered to humans in excess of that specified as acceptable for the general public; nor should the detonation cause significant damage to the food sources of the indigenous human population.

The statement went on to list its reasons for recommending springtime as the best season for the detonation:

> 1. Few birds are in the area.
> 2. Most small animals are under snow cover.

3. Most plants are under snow cover and their metabolism is low.
4. Local hunting activity is low.
5. The sea and inland waters are under ice; snow is on the ground. It is expected that radioactive debris will be flushed from the frozen landscape by the spring runoff of rapidly melting snow.
6. Weather is generally good and daylight is increasing which will facilitate project studies.

With this document in hand, the AEC's general manager reported to the commissioners that "based on the results of these studies, the Bioenvironmental Committee is of the unanimous opinion that Project Chariot can be conducted safely under certain conditions." In response, the commission authorized the continuation of Project Chariot on March 2, 1960.

The very next day, the AEC's Ball, Southwick, and Weaver arrived in Alaska to give the news to state officials and to commence the village tour that would culminate in the difficult meeting at Point Hope. The Bioenvironmental Committee also wanted to show the statement to the Chariot investigators about to convene in Anchorage. The reception was not good. Scientists from both groups, the state of Alaska and the Chariot investigators, challenged the accuracy of the statement and registered their disapproval of it. Notwithstanding these objections, the AEC released the statement to the local and national press.

The trouble with the statement was just about everything. It professed to enumerate "conditions" under which the shot could be safely fired, then offered two preferences (time of year and direction of wind) instead. The third condition, the limitation on radiation doses delivered to humans, was proscriptive, certainly, but it seemed to have more to do with anticipating a result of the explosion than with establishing the preconditions that might ensure that result. Its import, therefore, was that of a meaningless truism: The shot can be conducted safely provided it is conducted safely.

Regarding the analysis of the advantages of a springtime detonation, the statement purported to be "based on data known to it [the committee]," but was not based on the research data at all. With the first studies begun in June and the reports due in to the AEC by November, there had been no opportunity to gather data for March and April. Of course, the AEC's preference for a spring shot was well known to Alaskans. For months, news articles had been quoting officials as saying that they favored that time of year, though, ostensibly, all options were being weighed as scientific data were reported. What wasn't known was that, internally, AEC planners considered the spring to be the only possible time of year in which the experiment could be conducted. Considering the need for sufficient daylight and good weather, both immediately prior to the shot and for postshot work, the AEC had concluded in documents classified "Secret" that "the Chariot detonation can only be conducted in the spring and any delay means a minimum of a 1-year delay." It was in this context that John Wolfe led the Bioenvironmental Committee through its elaborate justification of a March or April blast.

Overall, the committee's statement of January 7 struck University of Alaska researchers as so destitute of scientific rigor that, when the committee failed to issue a correction after the Anchorage meetings, four of them took the unusual step of writing their funders in protest. On June 8, 1960, Albert Johnson, William Pruitt, L. Gerard Swartz, and Leslie Viereck mailed the committee a letter "to formally register our disapproval" of the committee's statement. They said the document lacked scientific objectivity, contained misinformation, and incorrectly implied that the professors themselves had furnished the data to support it.

"We take particular exception to the statements concerning snow cover," wrote the university professors, noting that the committee's assumptions "reflect the popular idea that the Arctic is completely snow-covered in the winter." But, they said, because the Ogotoruk Creek valley is so windy, "the snow cover is very sparse." And this was not simply a matter of opinion. The experimental results noted by the scientists in their reports had shown it to be so. Some of the botanists' study plots had been "70% bare of snow." And on the basis of their late winter visits to the valley, they believed that it was likely that "most of the plants are exposed" at that time.

The Gang of Four, as the dissenting professors sometimes referred to themselves years later, also took the committee to task for its assumptions concerning the nature of sea ice. The committee had said that fallout could be "deposited on the ice and would decay appreciably before enter-ing biotic cycles through the sea after breakup." The Alaskan biologists had found "no evidence" to support such a notion and suggested that the committee's scenario was based on "another popular misconception of the Arctic":

> The sea ice is not solid and immovable. It is constantly in mo-tion, freezing, refreezing, churning, grinding, occasionally solid for a while but soon resuming its motion in answer to currents and winds. If the ice were solid, few air-breathing marine mammals could live there.

Finally, while the committee had declared that hunting activity was low during March and April, the professors noted that the opposite was more likely to be true: "There is considerable evidence to indicate that hunting activity on land is high during late winter and early spring, and may even be concentrated in the Ogotoruk Creek Valley."

Concluding their letter, the university researchers addressed the cen-tral issue in language no scientist could fail to understand:

> The scientific investigations will provide many answers to some of those questions, but in the meantime, it is unfortu-nate that the Committee felt it necessary to offer opinions for which it had little or no special information.

Jim Brooks of the Alaska Department of Fish and Game also reacted strongly to the committee's statement. Writing to the governor of Alaska, Brooks went point by point through the committee's assumptions, disput-ing them and offering his own observations and experience as counter-point. Migrant birds began arriving in April, he said, and were abundant in May. Plant metabolism might be low in March and April, but shortly it would be rising rapidly so that "a fall or winter shot would be preferable by far if one wished to consider plant metabolic rates." Regarding the po-tential of the spring runoff to flush the landscape of radiation, Brooks said that "anyone who has been in the Arctic during spring knows that the snow melts gradually, bare places appear in increasing number and size until the tundra is bare." Much of the melting snow filters down into the plant cover and is lost by evaporation or plant respiration.

Besides rejecting the specific findings of the Bioenvironmental Committee's statement, Brooks also doubted that the research program was of sufficient scope to draw the anticipated conclusions:

> Even now the investigations are credited with the objective of establishing whether or not the blast is safe. In reality, the bulk of the studies at present are slanted towards cataloguing the kinds, number and distribution of plants and animals and the background levels of radiation. These have value mainly in the post-blast studies.
>
> The uptake of various types of isotopes by different species of plants on a whole range of soil types is not being adequately studied. The course of isotopes through food chains is likewise receiving scant attention. If the studies continue long enough it might be possible to fairly estimate some of the hazards. At this time no one knows that the blast will be safe, though that is the impression that the layman receives from the AEC traveling trouble shooters. . . .
>
> The Nome Chamber of Commerce may be satisfied, but the scientists of this department are not.

●   ●   ●

When Don Foote received a copy of the committee's January 7 statement, he too was upset. At the Anchorage meeting, he stood to protest that his report—the only one that dealt with the hunting patterns of the Point Hope people—had *not* shown that hunting activity on land was low during March and April. Foote had written in his November 1959 report that "a decline in marine hunting during February through March was compensated for by *increased hunting on the land.*" (Emphasis added.) He also reported that, during the whaling season, if a south wind closed the lead, Point Hope hunters would "take advantage of these interruptions for day or overnight trips to hunt caribou near Cape Thompson." His subsequent research, he said, "proved, without doubt, the validity" of these preliminary conclusions.

Later, he was astonished to learn that, "despite my protest, the Committee's statement was released to the United States public and the world through such news media as *The New York Times.*" When neither the AEC nor the committee offered to correct the publicized misinformation, Foote laid his cards on the table in a pointed letter to John Wolfe, the Bioenvironmental Committee's chairman:

> I will not allow the results of my research to be misused. If the Committee continues to feel springtime is a desired season for Chariot and justifies this decision on the grounds of minimum human activity on the land or on the sea, then . . . I cannot, in good faith, continue to remain employed by or identified with the Environmental Program for Chariot.

Wolfe wrote back in July both to Foote and to the University of Alaska professors hoping to close ranks a bit. He told the latter group that the snow-cover and sea-ice comments made by the committee were, in part, the result of "regular, thrice-daily observations of weather at the site." But, as Les Viereck pointed out, Wolfe was playing a bit fast and loose with the facts: "It is clear from the date of the original Committee news release (7 January 1960) that *no weather information was available from the site for the March and April period.* By the time Dr. Wolfe wrote his reply in July, of course, it was." (Emphasis in original.)

Writing to Foote, Wolfe implied, but did not plainly say, that the committee hadn't received his report at the time of the December 10 meeting. But Foote had traced the parcel through the mail to its arrival in San Francisco four days before the meeting, and he had had comments from AEC personnel that the report "was received by the committee with a great deal of satisfaction." Wolfe went on to say—and he told the professors as well—that the committee had prepared a new statement that now brought its perspectives into line with the researchers' data. Wolfe promised that the new statement would be forwarded along, and the dissenting scientists took no further action while they waited to see what sort of correction would come from the committee. But no such revision was ever received by the scientists, nor was any ever given to the press, as an examination of the press releases shows.

John Wolfe did, however, give a news conference a short while later. And what he told the press so angered the University of Alaska scientists that one of them publicly denounced the AEC's management of the studies as biased and formally resigned in protest.

WOLFE, MEAT, AND WOOD

> There are no biological objections to
> the shooting on the basis of our in-
> vestigations.
>
> —John N. Wolfe

It is the rare dog team that waits calmly, like Sergeant Preston's dutiful huskies, for its driver to finish lashing down a load and step on the runners. Alaskan sled dogs are "crazy to go," as the dog drivers say. And the din and chaos of eight or ten screaming dogs—lunging into their harnesses, leaping into the air, and near to fighting with frenzy—can be utterly unhinging. When the driver finally yanks loose the slipknot on his snub line, there is no need for a command to "go." There is only the rush of the runners over the snow and the racket of the sled going airborne between every hump in the trail and landing with a banging of the stowed gear.

After freeze-up, the hunters of Point Hope will load their lightweight basket sleds, hitch up their yapping dogs, and head upriver on the frozen Kukpuk far into the hills behind Ogotoruk Creek to look for caribou. They will travel light, planning to be gone three to five days, though they might be out twice as long. They will carry a tent or at least a canvas tarp. Each man will have caribou hides to sleep on and a sleeping bag. There will be a two-burner Coleman stove, a Coleman lantern, a can of gas, rifles, boxes of ammunition, binoculars, and extra clothes. The grub box will hold coffee, tea, salt, sugar, rice or noodles, oatmeal, crackers, sourdough biscuits, whale or seal oil, and meat. Dogs and men will share the same meat supply.

They will spend the first night at their fish camps on the Kukpuk, relaxing and picking up information from other hunters about the location of the herd. The next day they head upriver, sticking close to established trails and with an eye out for weak ice or open water. They might cut off the wider bends by crossing portages, saving a few miles by bouncing over tundra hummocks, careening down steep-cut banks and over the rocks on the river bars. They will stop often to boil tea and glass the hills with binoculars. At the end of the day, the men will camp in a place with hard-packed, drifted snow. Here they will cut and remove snow blocks, creating a sort of quarry pit. They will stack up the blocks to form a short wall around it. Over the top of these walls goes the tent or canvas tarp.

If the hunters spot caribou, they must maneuver downwind and get in close. They must not only keep the caribou from seeing the team, but they must also keep the team from detecting the caribou. The dogs' inborn desire to chase and attack can quickly turn an otherwise disciplined team into a berserk and uncontrollable pack. After staking out the dogs, the men approach on foot until they are in a position to shoot. Each man hopes for a sled load of caribou, preferring the cows and younger animals to the large bulls. It takes an Inupiaq hunter only half to three-quarters of an hour to skin, gut, and quarter a caribou. In the process he may eat—raw—the heart, liver, kidneys, and windpipe, and drink the blood. The dogs will get chunks of bloody meat and a slithery pile of intestines. When the work is done, sleds, each loaded with hundreds of pounds of caribou meat, will slowly wend back down the frozen Kukpuk River to Point Hope.

In this way, in one year, Point Hope hunters can bring into the village more than 100,000 pounds of caribou meat. In 1959 and 1960, when Don Foote was keeping track of every caribou killed by the Eskimo hunters at Point Hope, every single caribou taken was shot within twenty-five miles of Ogotoruk Creek.

•   •   •

As they began the 1960 summer field season, Don Foote and the University of Alaska's Gang of Four were confronted with the sort of ethical quandary that most scientists would prefer to avoid at all costs. On the one hand, they believed fervently in the proposition articulated by John Wolfe that environmental scientists "*are,* and need to be" part of such large technical projects as Chariot. In addition, they were just plain grateful—overjoyed, even—to be doing fieldwork in the Arctic, drawing pay for adding to the body of knowledge in their area of specialty. For outdoor people like Pruitt, Viereck, and Foote, this was the payoff for the years of rigorous training in college and graduate schools.

But as the scientists learned more about Project Chariot and its potential to do environmental damage, as they witnessed the AEC's tendency to blur science and public relations, and saw firsthand the dependence of the Native people on a subsistence way of life, they began to question whether it was ethical to continue to participate in the AEC's program. For some of the scientists, the issues came into clear focus after a press conference at the Chariot camp at Ogotoruk Creek.

John Wolfe, Allyn Seymour, and Norman Wilimovsky were waiting at the Chariot camp on August 15, 1960, when reporters representing papers ranging from the *Nome Nugget* to *The New York Times* flew up from Kotzebue. That day, the twin-engine Beechcraft couldn't land on the long runway at Ogotoruk Creek because of fierce crosswinds. The shorter runway alongside the camp could not accommodate the speedy twin, so the Beechcraft flew on to the Point Hope strip. An icy wind was slicing across the spit when the pilot transferred the newsmen into a smaller plane for a second try at Ogotoruk Creek. The little plane bounced along above the whitecaps into a wind gusting to eighty miles per hour, then dropped onto the short strip on the little ridge above the beach. When the pilot dropped the flaps, "the plane appeared to hover, almost like a helicopter in the teeth of the howling wind," according to one of the reporters. Men from the camp rushed to hold the plane down long enough for the passengers to scramble out. When they let go, the plane rose nearly straight up and was off.

When the news conference began, the message was succinct and emphatic, as the *Times* science writer Lawrence Davies wrote in his lead:

> Point Hope, Alaska, Aug. 15—A proposed nuclear blast on the Arctic coast thirty-five miles below this Eskimo village was pronounced safe today from a biological standpoint.
>
> Dr. John N. Wolfe, an Atomic Energy Commission scientist, said a fifteen-month field study costing $2,000,000 had

produced no evidence that the detonation would damage the
Eskimos' relationship to their environment and livelihood.

Davies went on to quote Wolfe directly, "I would say that there are no
biological objections to the shooting on the basis of our investigations."
"Pronounced safe." Yet John Wolfe knew—and in fact he had been
trying to get the AEC to accept—that there were too many uncertainties
to be able to pronounce the shot safe. The meteorology of the area was
not well known; it was not clear whether the fallout should be directed
over land or sea; and the fallout characteristics of bombs not yet invented,
but which might become available before the shot took place, were unknown. Wolfe argued this position privately with the AEC and Livermore,
both of which wanted the committee to issue "firm conclusions." Publicly, however, Wolfe gave the press to understand that a definitive evaluation had been made.

"[The study] had produced no evidence that the detonation would
damage the Eskimos' relationship to their environment and livelihood."
Two months before John Wolfe made this statement to newspaper reporters at Ogotoruk Creek, Don Foote had sent him a report on the hunting patterns of the Point Hope Eskimos, details of which are incorporated
into the account of caribou hunting above. Foote's meticulous fieldwork
scientifically documented the dependence of Point Hope people upon
caribou taken in the immediate vicinity of Ogotoruk Creek. At the same
time, fallout contours drawn by Livermore physicists showed that, if the
fallout was directed inland, the heaviest contamination would move directly into the heart of Point Hopers' caribou hunting grounds. Clearly, if
the Chariot detonation took place, the hunting of caribou would need to
be restricted for some indefinite period.

Wolfe's dissembling before the press prompted Foote to calculate the
net effect to Point Hope caribou hunters if the Chariot blast had been
scheduled for April 4, 1960, and if the AEC's plan was to direct the fallout
over the land. That day and month had often been used in Chariot prediction models, and a landward disposition of fallout had always been mentioned as an acceptable plan. Foote showed that in such a case, Point
Hopers would probably be restricted from scheduling hunting trips into
the downwind area after March 26 so that all hunting parties would have
returned before the shot date. But there was another consideration. The

winds on April 4, 1960, blew to sea. If it had been planned to direct fall-out over the land, travel restrictions would have been extended as the fir-ing team waited for favorable winds. In April of 1960, suitable conditions for the detonation did not exist until the twenty-first. By that time, the hunters would have been idle for nearly four weeks. After the shot, travel in the fallout zone would have continued to be banned for some addi-tional period while the shorter-lived isotopes decayed to less dangerous levels. Foote speculated that the men would be free to enter the caribou hunting area about May 6. By then, only two weeks remained in the sea-son during which conditions would be suitable for sled travel. Foote con-cluded that based on the actual 1960 caribou harvest statistics, if the Point Hope hunters had been restricted from entering their hunting area from late March to early May, it "would have meant the loss of about 9,000 pounds of caribou meat."

In addition, he said, "If residence on the sea ice, south of Point Hope, was restricted from April 3 to April 22, the loss of marine-mammal meat would have been about 25,500 pounds." This figure, of course, included one bowhead whale.

Wolfe, Seymour, and Wilimovsky were men with a worldview shaped by their training in science. For them, accuracy in factual matters held near religious importance and scientists were the high priests of truth. (Wilimovsky once despaired "that we're ever going to get the social side of our society up to the scientific society.") Such men might have argued that alternate hunting grounds, while perhaps not as productive as the fa-vored areas, might provide some hunting success and reduce the calcu-lated hardship. They might have pointed out that Foote's data did not amount to conclusive proof of harm to the Eskimos, or that AEC-supplied "store food" might tide the village over, or that in any case the harm was offset by Chariot's value to science. And perhaps at other times, on other days, they did advance these arguments. But on August 15, 1960, they took part in a news conference at Ogotoruk Creek where Wolfe conveyed to the world the message that the extensive studies had produced "no ev-idence" that the detonation would "damage the Eskimos' relationship to their environment and livelihood."

Foote's data *did* exist. They *were* produced by the environmental stud-ies, and they *were* evidence that the detonation could interfere with the way the Eskimos had been making their living. Perhaps the data would

have struck the scientists as less intangible if the food lost to the village—
the four and a half tons of raw caribou meat—could have been gathered
in one spot. It is tempting to imagine a convoy of grocery trucks pulling
up to one of the scientists' suburban homes and, with curtains parting
along the block, grocery boys heaping up on the lawn a mountain of glis-
tening, purple caribou parts. As the stupefied neighbors—still holding
their newspapers or drying their hands on dishtowels—emerged to investigate, a tractor-trailer might pull up to offload a whale.

*   *   *

Albro Gregory was a good old-time Alaskan and a natural-born newspaperman. He had worked as a reporter all over the lower states, from New
York to Seattle, and in Alaska he wrote for papers from Ketchikan to
Nome, a stretch of geography nearly as vast. In an interview shortly before his death, Gregory let rip with the sort of feisty banter that characterized his approach to life, and to his craft:

> I'm just proud as hell to be an Alaskan. I can't stand to hear
> people say nasty things about Alaska, and I tell them off, usu-
> ally in pretty plain language. People usually say what they
> think here, whereas down in the states they pussyfoot about
> everything. . . .
>
> I love the newspaper business, especially where I can
> write the editorials and say what I think. I know I'm loved
> and hated all over the place.

Covering the Ogotoruk Creek news conference for Fairbanks's number
two paper, *Jessen's Weekly,* Gregory acknowledged the existence of some
criticism of Project Chariot. "But this, admittedly, has come from misinformed people," he told his readers. "To the critics of the plan there are
good, sound answers," he said. He implied that there was little at risk in
"this bleak spot on the snow-covered outlands" where "an occasional Eskimo or caribou walks." John Wolfe, "an eminent scientist," had said the
shot could go off without danger to human or animal life. "And he should
know," Gregory said. As for the blast harming whales and the like, "it's
silly to talk about it," according to "another eminent scientist," Allyn Sey-

mour. "A caribou might be hit on the head by a rock" was Gregory's way of summarizing the biological hazard. William Wood, the new president of the University of Alaska, put the issue in proper perspective, Gregory felt, when he said: "If the United States government decides that the project is a safe one, there is no reason for concern."

Albro Gregory did not give voice to the Native people's concerns about Chariot, though he reported that they were "feeling better about things." Winding up the piece, he suggested that the project in fact offered benefits to the opportunistic Eskimos:

> This last winter, for the first time in many years, Natives hunted in this Cape Thompson area. Why? The Jamesway huts which lodge the scientists in summer are abandoned in winter and make comfortable temporary homes while on the trail.

This last comment so puzzled Don Foote that he wrote Gregory a letter. Foote politely asked for the reporter's source for the statement that the Natives had not hunted at Ogotoruk for many years and pointed out that the assertion "contradicts the facts presented in my studies." Gregory wrote back, huffily, and declined to reveal his source: "I feel that I do not necessarily have to reveal the sources of any statement I make." And, if it wasn't clear already, Gregory offered his personal view on the matter of Project Chariot: "I take the position that if the AEC's committee decides the projected detonations of nuclear devices is necessary and practical, then that is the thing to do. Who knows more about it than these people?"

Foote wasn't inclined to let the matter rest. The next time he traveled in to Fairbanks, he stopped by the *Jessen's Weekly* offices, which a traveling writer once described as a "windowless hut smell[ing] of printer's ink and damp fur." Gregory, a thick-built man with a bristly mustache and a bristly crew cut, was more congenial in person. He allowed that he knew the AEC men personally and had been out drinking with them when they came through town. When Foote brought up the matter of the Eskimos not hunting at Ogotoruk, Gregory volunteered: "Now that you mention it, I must tell you that it was Dr. John Wolfe himself who said that." In fact, Gregory said that he had shown Foote's letter to Wolfe on his last trip to Fairbanks and that Wolfe again stated the Eskimos had not been hunting there before.

Not only had Foote's reports on hunting patterns shown the Eskimos' use of the Cape Thompson area, but earlier researchers had also reported that use as well. Froelich Rainey, who lived at Point Hope in the late thirties and early forties, noted that the region was used for reindeer herding, trapping, and caribou hunting. James Vanstone lived in the village in the mid-1950s and corroborated that intensive caribou hunting occurred in the Cape Thompson area and in the hills along the Kukpuk River just northeast of the headwaters of Ogotoruk Creek. Both scholars believed that the Cape Thompson area had been a hunting ground of the Point Hope people for at least 2,000 years.

John Wolfe knew that his statements to the press contradicted the AEC's only data on the topic, namely Foote's report. Bill Pruitt wrote Foote: "Wolfe had read your June report before he talked to the newsmen. We had a session that night and he said he had." Wolfe had no additional scientific information on the local people's use of the area. He was apparently relying on the journal kept by the Chariot camp caretaker and meteorologist, Harry Spencer. Spencer's entries over the winter of 1959–60 show that he felt the Natives were taking advantage of the camp's hospitality:

> A native from Point Hope and two from Kivalina stopped in today and will spend the night. This is getting to be too much of a good thing so will stop it right away as this is the second bunch in a week.

A few weeks later, Spencer again found reason to grouse about traveling Natives, whom he regarded as intruders:

> Two Eskimos staying all night from Kivalina tonight. I think the word has got around we serve good meals and this is the end. I told them to pass the word.

The reporters had flown off by the time Pruitt and Viereck trudged in from the day's fieldwork. "The AEC arranged it very nicely," wrote Viereck in a letter, "so that the reporters were not allowed to talk with any of the scientists—except Wilimovsky, Seymour and Wolfe during their visit." In the evenings, however, John Wolfe and the University of Alaska researchers had an opportunity to exchange grievances. Viereck wrote in a letter that Wolfe was "very unhappy about our attitude. . . . One evening he came in with a very aggressive attitude . . . Pruitt matched him point for point without loosing [*sic*] his temper and Wolfe finally went away in a rather apologetic mood." But shortly after that, Wolfe let it be known that he was "going to straighten things out with Dr. Wood" at the University of Alaska.

•   •   •

The Dr. Wood that Wolfe was going to see was William Ransom Wood. A month before the press conference at Ogotoruk Creek, Wood had become the University of Alaska's new president, taking over from Ernest Patty, the mining professor who had taught the local prospectors mineral identification through "rock poker." The new president who would play an important role in subsequent events had come from the University of Nevada and was familiar with the Atomic Energy Commission's testing activities.

Wood had been a public school teacher in the lower states and, after graduate training, returned to the public school system as an administrator. His career as a university executive did not begin until 1954, when another former public school official, Minard Stout, brought him to the University of Nevada at Reno. As academic vice president at Nevada, Wood was the number two man in an administration whose concept of a university seemed to spring from the ethos of the old school disciplinarian. President Stout imposed a "chain-of-command" system of governance, eliminating the traditional participation of faculty in policy making. His notions of hierarchical organization even extended to the belief that faculty members should not concern themselves with issues outside their area of specialization. The disparity in philosophies between administration and staff came to a head when a biology professor passed on to colleagues an article that took a critical view of some practices in public schools and in schools of education. Stout censured the man, telling him: "You are hired to teach biology and not to be a buttinsky all over the campus."

But many of the faculty supported what seemed a modest measure of academic freedom: the collegial exchange of scholarly writing and opinions. When the dispute failed to die down, Stout summarily fired the professor together with several others on the faculty whom he considered members of "a small, dissatisfied minority group." Rather than putting the dispute to rest, the firings precipitated a statewide controversy that was only resolved when a panel of noted educators was appointed by the state legislature to investigate. That panel concluded that Stout's "quasi-military, almost authoritarian theory of administration" was "quite inapplicable to a college or university community." The board of regents was expanded until Stout's conservative backers on the board were outnumbered and he was forced to resign.

William R. Wood took over as acting president at UN as the search for a permanent replacement began. In response to the period of repression under Stout, a democratically selected faculty committee was set up to preside over the nominating process. But the carefully worked out procedure was nearly subverted when a few of the "Stout" regents attempted to install an apparently acquiescent Wood as the new president even though he had not been recommended by the faculty committee. Fortunately for the faculty, who thought that Wood was too closely associated with the Stout administration, the academic coup d'état failed.

Passed over at the University of Nevada, Wood still sought the presidency of a state college somewhere in the West and corresponded with half a dozen such institutions concerning openings. He had already bolstered his résumé by writing to the publishers of *Who's Who in the West* and *Who's Who In America,* nominating himself for inclusion in the compendia and promising to purchase a volume. When the publisher assented, Wood included the listing under "Honors" in his résumé. But the Stout connection followed him, and it did little to improve his prospects as a university chief executive. In fact it accounted for his rejection at one of the schools, at least. "I think I should say that the position was not offered to you," wrote an official from Idaho State, "because of complications and entanglements in connection with the situation at Nevada."

But to the regents at the University of Alaska, Wood looked like presidential timber. The word on Wood was that he had been the major factor in Nevada's doubled campus enrollment and doubled campus plant. In addition, he had helped to establish the Desert Research Institute and a nuclear engineering program. He told University of Alaska officials when he applied for the job that he thought he had done "more in the promotion of scientific studies and scientific research than any of our good people in the scientific fields." And among the University of Nevada's most important benefactors in the area of sponsored scientific research was the Atomic Energy Commission, which accounted for millions of dollars in grants during the 1950s and 1960s.

•   •   •

When the AEC's John Wolfe stopped in to "straighten things out" on the matter of the protesting biologists from the University of Alaska, he found a strong ally in William R. Wood. Earlier in his career, Wood had had a professional interest in the civil defense role of schools in the aftermath of a nuclear war. He had attended seminars on the potential for nuclear technologies. And, in connection with work for the Defense and State departments, he had been given security clearances to the level of "Top Secret." Wood believed, as he wrote years later, that the launch of the Soviet satellite Sputnik in 1957 represented "a threat to all our tomorrows" that could only be met when "university leaders, industrialists and heads of government agencies put aside doubts of one another . . . to work together in pursuit of the same objective, establishing and maintaining a top, world-scale, leadership position in science and industrial technology for America." Finally, and perhaps above all else, Wood was a natural ally because of his consummate interest in securing government-sponsored research at the University of Alaska and the growth that such funding would nurture.

Brina Kessel, the young head of the Department of Biological Sciences who was responsible for managing the university's contract with the AEC, wanted badly to know how the president's meeting with Wolfe had gone. She tried for three weeks to get an appointment with Dr. Wood, but his schedule was a busy one. Kessel did get word from Wood that the AEC was pleased with the scientific work being done by the university. But she wanted the president's counsel, and to enlist his support on behalf of her staff, which was up in arms—Pruitt and Viereck, especially. They were, as she put it, "completely down on the AEC." Viereck told friends privately that he was "ready to do anything to counteract Wolfe's statements," and thought that "some public statements of our own" were in order. "I am ready to quit the whole project." he said. And Kessel herself felt that the agency's statements to the press were "extremely misleading and are probably meant to be."

On October 26, 1960, Kessel finally got in to see President Wood. The day before, she had assembled all the department personnel working on Chariot in order to get a current "list of 'gripes' and get some more specific examples to illustrate our problem to Dr. Wood." She had also sat down and reread her collection of newspaper clippings dealing with Chariot, and on the basis of all of this, felt she had a good outline of material for her presentation to Wood. It included what she called the "'ethics' of the AEC, including material on news releases, their apparent lack of attention to scientific data . . . etc." But the meeting did not go as she had expected:

> [Wood] has his own definite opinions on the matter. He, of course, has been close to the Nevada test site, and knows much more about the AEC than Patty did. . . . Wood's philosophy is definitely that the use of atomic energy is important to the welfare of mankind . . . and that man is going to have to learn how to use it, and is going to have to take 'calculated risks' in its use.

The notions that nuclear technologies represented a reality that people "had to learn to live with" and that "progress" involved acceptance of certain "calculated risks" were standard fare in Edward Teller speeches; Wood had attended several of them. Still, Brina Kessel managed to give Wood a fairly complete picture of the faculty's reasons for displeasure with the AEC. And she made sure to present "some of the more radical views of others of the staff" in addition to her own "more tempered" ones. Wood believed that Kessel's staff had reason to be dissatisfied on certain points, but that there was nothing very serious to worry about. He tried to soften the biologists' objections to AEC statements to the press by blaming inaccuracies on newspaper journalism. But Kessel pointed out that in some instances, she herself had seen the original press releases issued by the AEC and that she felt they were misleading.

Nonetheless, Wood felt strongly that the university should continue with the contract. One of his principal arguments, said Kessel, concerned the potential to develop an Institute of Arctic Biology. A few weeks earlier, the National Science Foundation had approached the University of Alaska with the idea of funding such a research center—an opportunity that one UA official called "the most silverish of platters." According to Kessel, Wood stressed that the university "shouldn't do anything to upset our relations with [the AEC]" because ". . . if we get a bad reputation with one governmental organization, it will give us a black eye with other governmental organizations" that might be likely to fund the biology institute.

Kessel felt she was "enough in agreement with Dr. Wood's thinking" that she was willing to continue working with the AEC despite her misgivings. But she was not prepared for the direction in which Wood next turned the conversation:

> The unexpected slant was [Wood's] apparent feeling concerning the staff members that were engaging actively in what is being called "obstructionist" activities. He was particularly concerned about Pruitt's "unethical?" activities. Dr. Wolfe had complained to Dr. Wood last summer that Bill was intentionally biasing the Eskimos with what Wolfe called misinformation.

Here, then, was the first suggestion of what John Wolfe had in mind when he set about to "straighten things out with Dr. Wood." And, according to Brina Kessel's contemporaneous written account, President Wood dismissed the concerns of his faculty and came down squarely on the side of the AEC. Wood felt that "the AEC has bought the material and it is theirs to use," according to Kessel. She said the president's view was that so long as the AEC did not "come out with an exact misquote of a particular scientist's work, that particular scientist doesn't have much to say about how the material is used." Wood apparently saw no need for a public correction of AEC misrepresentations on the basis of the scientist's loyalty to his discipline, to say nothing of his moral obligation, generally:

> He said that if some of the scientists believed that they were being compromised, they had one easy out—not to sign up for this kind of contract research.

If Brina Kessel was taken aback by Wood's academic Machiavellianism, she didn't have much time to register the shock before the new president delivered another jolt. Kessel wrote:

> And the surprise came when he actually suggested that maybe we (as the university) shouldn't rehire members of the staff that felt so strongly against the activities of the agency that was providing money for their research (meaning Pruitt and possibly Les). His comment was to the effect that with the impending expansion in the biological sciences, wouldn't it be possible to get along without many of the specific individuals presently in the department, i.e., couldn't we replace them and be just as far along five years from now whether these particular individuals were here or not?

Kessel had gone to Wood to seek his support for her Chariot scientists who believed that the AEC was violating the accepted ethical ground rules for contract research. Instead of coming to the defense of his faculty, Wood suggested the scientists be fired. There was no tenure at the University of Alaska in 1960; faculty served at the pleasure of the president.

As Brina Kessel wrote it down, Wood seemed to deny the relevance of morality when it came to preserving good relations with important funders:

> He pointed out that already the AEC was somewhat biased in the use of [Pruitt's] work because of personality factors—whether this bias on the part of AEC is justified or not, the fact that it is there is a discredit to Pruitt and thence to the university.

In other words, if the AEC was disinclined to hire Pruitt—even if the rationale was improper—then Pruitt had brought dishonor to the university perforce. His removal could be justified. Clearly, Wood's logic derives from a view of the university that is—to put it charitably—circumscribed. Here the university is seen as a collection of buildings and programs and institutes, and not as an idea sustained by such values as probity, integrity, and honor. If to most thinkers a university represented a sort of island of intellectual anarchy within the social order where widely disparate ideas might be pursued on the basis of critical thinking and scientific evidence, to President Wood it was an enterprise akin to a corporation, staffed by "team players" who developed public credibility by the homogeneity, rather than the diversity, of their speculations.

Brina Kessel thought Wood's tactics "seemed pretty cold blooded." She said she "didn't take it lying down," and pointed out "how valuable Les and Bill were because of their knowledge and liking of the North." But Wood had an answer for that, too. Maybe, after the department had been expanded, it could better encompass this type of person. "My first reaction was one of objection," she wrote, "in thinking it over, however, I have to admit there is a lot in what Dr. Wood has to say."

●   ●   ●

After Brina Kessel's meeting with Wood, she called together the Project Chariot personnel and, though she knew they would not be happy about it, she gave them "the word" pretty much as Wood had given it to her. Except she left out any mention of firing Pruitt and Viereck.

Two days later, John Wolfe was on campus, accompanied by environmental committee members Seymour and Britton. He proposed to talk to the Chariot biologists. Wood was in and out of the meeting, which, predictably, proved to be a very subdued encounter. Kessel and Wood were disinclined to make waves with the AEC, and the committee members were senior scientists, skillful at smoothing over past false steps. Perhaps there was little the comparatively younger biologists felt they could say. As Les Viereck wrote a few weeks later, "it seemed of little use to go through it all again because we did discuss it at great length last summer with no satisfaction on anyone's part." Still, he was a little disappointed in himself and his colleagues when their grievances—as presented by Kessel—were "passed off, as before, as someone else's fault."

One thing of importance Viereck took from the meeting was that Wolfe pledged that no statement concerning the committee's recommendations would be made public until all the committee members had approved it. But the very next day Wolfe seemed to renege on his promise in a speech before the state chamber of commerce. Celia Hunter, secretary of the newly formed Alaska Conservation Society, was in the audience taking notes:

> Toward the conclusion of his talk, Dr. Wolfe made the follow-
> ing statement—and this I took down verbatim: "The commit-
> tee will recommend in a very short time that the blast need
> not be called off on biological grounds. . . ." Following this
> statement, Dr. Wolfe read very rapidly a series of stipulations.

The stipulations concerned the time of year in which the shot could be
fired, the direction of the fallout, etc. In other words, it was just the sort
of public statement that Wolfe had promised to let the committee review
first. To Les Viereck, it seemed that Wolfe was declaring the committee's
conclusions in advance of their deliberations. And according to Viereck's
account, Robert Rausch of the AEC's environmental committee was also
angered by Wolfe's public statement:

> Rausch claims that he submitted his resignation from the
> committee at the last meeting but was then talked out of it by
> Wolfe and the rest of the committee after they had promised
> to mend their ways, be careful of publicity, make no recom-
> mendations yet, etc. Rausch was therefore furious when we
> told him about Wolfe's talk to the C of C. . . . He is extremely
> unhappy with Wolfe and the way the AEC is handling the
> whole thing. He indicated that Britton feels much the same
> way.

By November 1960, Les Viereck had had it with the AEC. He wrote Al
Johnson, the man who had originally brought him to the University of
Alaska, and who was away on sabbatical:

> I just can't take any more of it and I told Brina that I definitely
> won't work for the AEC contract next year. Brina informed
> me in turn that this was the only possibility for me working
> with the department so it appears that I will be looking for
> another job. . . . I do hate to stop working for the university
> because it is just what I would like to do.

Kessel also wrote to Al Johnson, but put a slightly different spin on
Viereck's willingness to sever his ties with the university:

Les . . . wants to cut his relationship with the AEC. I told
him, and he obviously had come to the same conclusion previously, that if this was the case that there would probably
not be enough money to hire him in the department next
year. This didn't seem to bother him particularly. . . . I told
him that if we got the expansion we were looking forward to
that I hoped that we could hire him back within a year or
two. And that is where the situation now stands.

Kessel said that she "admired Les for 'sticking to his guns,'" but that
"for department planning purposes it is best that things be left in this severed state for now." The "severed state," along with its promise of a possible rehire in the future, came to pass just about one week after President
Wood had suggested exactly that scenario. On December 29, Viereck for-
malized his resignation from the Project Chariot work with a letter to Dr.
Wood. Working from a tape recording that Don Foote had prepared,
Viereck presented a brief history of Project Chariot and a résumé of the
agency's duplicity as he saw it. He said that "the situation has now
reached the point where I can no longer maintain my personal and scientific integrity and work for the AEC project." He also took a stand on the
issue of who owns the results of contract research—a stand in direct opposition to the president's views, as relayed to the biologists by Brina
Kessel. Viereck wrote:

It has often been stated that a scientist working under a government contract should not worry about the interpretation
of the data he is hired to collect. I strongly disagree with this
attitude and feel that it is the duty of every scientist to protect
his data and to be sure that it is interpreted correctly. A scientist's allegiance is first to truth and personal integrity and only
secondarily to an organized group such as a university, a company, or a government.

Viereck felt ethically bound to resign from the AEC contract, but he
didn't want to give up his teaching post. He knew that his appointment
was temporary; the letter offering him the job doing part-time research
and part-time teaching said he "could continue on the same basis for as

long as the project runs." And Kessel said the two were inextricably linked. Her rationale went like this: Before Chariot, the biology professors were all teaching full time with very little funding available for research. When a research opportunity came along, they all wanted to lighten their teaching load and add some research. So Viereck and Pruitt were hired both to help with the AEC contract and take over some of the courses. Because, according to Kessel, few opportunities existed outside of Chariot for sponsored research in biology, if Viereck were allowed to drop the research component of his appointment, it might mean he would be teaching only, while someone else would have to be hired to do research only. And Kessel felt that everyone in the department ought to do both.

Of course, all this was a bit hypothetical and arbitrary. No one knew, for example, whether Viereck might not land research funding from a source other than the AEC. Still, though Kessel insisted that his resignation from Chariot work necessitated his dismissal from the university, she did want to rehire Viereck sometime, in some capacity. Her business letters show that she was "really quite pleased with Les as a teacher, worker and departmental member," and that a position at the university's herbarium might go to him. But President Wood signed off on all hiring, and he had other ideas.

A few days after he'd turned in his letter resigning from Project Chariot sometime between January 30 and February 2, 1961, Viereck spent about an hour in President Wood's office. Wood steered the conversation to such issues as worldwide disarmament, then drew Viereck out as to his ideas for future expansion of the biological sciences program. But he stayed clear of the specific charges Viereck had leveled at the AEC. Viereck found Wood "a delightful person to talk with." As for keeping Viereck on staff the next academic year, Wood said that he would be glad to do so except that there wasn't enough money.

Actually, the money did become available. "We have plenty of money," Kessel wrote in May 1961, "but I haven't been able to get the proper talent lined up." "Ordinarily Les would be an obvious choice to step in and fill our gap in the teaching area and in the herbarium," she said, "but Dr. Wood has asked that he get his Ph.D. before we consider him for another position." Wood's new grounds for refusal to consider Viereck stirred Kessel's sympathy, and she left little doubt as to what she considered the real reason for the president's unyielding position: "I surely wish [Viereck] had used a little better judgment in some of his activities this year," she wrote, "because I am sure that this is having an effect now."

Having the doctorate degree might have seemed a reasonable criterion to Viereck if it weren't for the fact that 64 percent of the University of Alaska faculty in 1961 did not have one. Besides, as he pressed for retention, Viereck had offered several proposals to the university that would have given him the month or two he needed to finish his Ph.D. thesis and still earn some income through part-time teaching or working at the university's herbarium. The proposals would have benefited both the university and Viereck: Both would profit from his finishing the degree, and the university still needed someone to teach his courses. All of Viereck's proposals were rejected.

When it came time to assign instructors for the coming school year, Kessel offered the general ecology course, which Viereck had taught for two years, to a graduate student who, as Viereck wrote, "is less far along toward the Ph.D. than I am . . . who has other part time research and teaching work in the department, and who will, in addition, have a baby during the middle of the semester." The general botany course, which Viereck had also been teaching, was offered to a botanist from Juneau who would have to be away from home for five months to teach the class. And finally, Viereck was passed over for the herbarium position as well.

Les Viereck was out. From the circumstances of his separation from the university, it is easy to see why he left "with the bitter thought that I am being relieved of my job because of my stand on Project Chariot."

13   GOING PUBLIC

An imprudent young fellow named Pruitt
found poison in caribou suet.
"If I tell Dr. Wood, it's not any good,
I'll rue it," said Pruitt, "but screw it!"

—Anonymous psychology
professor, University of
Alaska, 1963

During World War II, Ginny Wood (no relation to William R. Wood) and
Celia Hunter, young women in their twenties from Washington state,
served as Women's Air Force Service Pilots (WASPs). At the time, the
army air force was ferrying thousands of fighters and bombers to Alaska as
part of the Lend-Lease program. Russian flyers picked up the planes in
Fairbanks and flew them west across Bering Strait and Siberia, halfway
around the world to the Russian front. "All the guys we flew with would
always tell us about these places they went [Fort Nelson, Watson Lake,
Whitehorse, Northway, Fairbanks] and turning these planes over to the
Russians," says Ginny Wood, ". . . but we couldn't bring them to Fairbanks because they didn't have any—they said—'facilities for women.'"
The WASPs resented the restriction because they'd flown countless long,
cross-country trips, often through the rough weather around the Great
Lakes. And none of the other air bases they used around the country had
any special "facilities for women."

200

When the war ended, however, Wood and Hunter got their chance to see Alaska. They heard of a Fairbanks flying service looking for a couple of pilots to bring up two war-surplus planes. One was a four-place Gull-wing Stinson, the other a two-seat Stinson L-5 spotter plane with "no heater and plenty of ventilation," says Hunter. This was a different proposition from the wartime mission, which "would have been a snap," according to Ginny Wood, "because they had a B-25 that had a radio, you just had to fly in formation with this B-25 with a radio giving you the signals." Now it was December. And launching into the dim skies of the subarctic at the darkest and coldest time of year, just the two of them, might bear some rethinking. Only a couple of minutes' worth, as it turned out.

They took off. And they got no better breaks than they could have expected. Temperatures dropped to sixty below zero and colder, grounding the aircraft. Snowstorms stopped them, too. Out of twenty-seven days in all, they could fly on only four. The planes did have radios—or rather each had half a radio. Luckily, the defects were complementary: The spotter plane's radio couldn't receive, and the Gull-wing's couldn't transmit. But together they were whole—so long as they could use sign language. Ginny would ask for a weather report and Celia would pick up the message. She'd fly over to Ginny and flash her a "thumbs up" or "thumbs down." In this way they pressed north, the two radial engines rattling noisily in the cold air above the Alaska Highway, and crossed into Alaska territory on the last day of 1946.

At Fairbanks, temperatures stayed between fifty-five and sixty-five below for three weeks. Commercial airline service was suspended. They couldn't fly home. Then, as Ginny Wood recalls, one thing led to another:

> By the time it warmed up a bit, the guy asked us if we wanted to take some cargo out to Kotzebue. And we got weathered in another 18 days in Galena and a week in Kotzebue. And by the time we got back it was the end of February and it was getting warm and we decided maybe we better stay and see what the summer was like. And then we never went home.

Alaska summer days of twenty-four-hour sunlight tend to be action-packed, especially for flyers. In a country dominated by mountains and glaciers, mile-wide rivers and boggy muskeg flats, a gravel airstrip could be built much more easily than a road. Bush planes were the clear choice for access into hundreds of thousands of square miles of roadless wilderness. In the years that followed, Wood and Hunter settled in Fairbanks, did some more flying, and built and operated a wilderness lodge near Mount McKinley. Then, "somewhere along in the fifties," says Celia Hunter, they became involved in conservation:

> We got acquainted with [the naturalists and authors] Mardy and Olaus Murie who were working to get the Arctic Wildlife Refuge set aside, and we helped prepare testimony for that hearing by Senator Bartlett. And ultimately we founded the Alaska Conservation Society of the people who had prepared testimony for the Range.

That was in 1960, and the Alaska Conservation Society (ACS) was the first such organization in Alaska. After the successful campaign to establish what is today called the Arctic National Wildlife Refuge (ANWR), the handful of Fairbanksans who comprised ACS's active membership turned their attention to Project Chariot. Hunter and Wood, together with a few of the biologists from the university (Viereck and Pruitt, Bob and Judy Weeden, and Fred Dean), were the nucleus. But ACS boasted an Alaskan membership of 205, and 219 nonvoting, out-of-state dues payers. "It's when environmental groups operated out of living room floors with a mimeograph machine and a typewriter," says Ginny Wood. Actually, in the case of the ACS, business was transacted from the floor of a log cabin (where the mimeograph was a primitive handcrank affair) and environmentalism had not yet succeeded conservationism.

●   ●   ●

As the spring of 1961 approached, it seemed to the members of the Alaska Conservation Society that the Project Chariot detonation was imminent. "We all feel," wrote Viereck to a concerned citizen on the East Coast, "that there is a great deal of urgency about acting quickly." He thought the shot was probably scheduled for the fall of 1961 or the spring of 1962 and that "something fairly drastic has to be done." In February, the ACS sent out a press release noting that the organization's board of directors had endorsed a resolution introduced in the Alaska state legislature by Rep. Jacob Stalker, an Inupiaq from Kotzebue. It called for halting all preparations for the Chariot blast until the AEC gave adequate assurance of the safety of the people and wildlife in the area. The ACS called for an "independent study of all raw data," noting that "so far, assurances have come only from the AEC themselves," and that this was "like asking the tobacco industry to give an objective unbiased opinion on the hazards to health in smoking."

The next month Pruitt, Viereck, and Foote took the drastic action each had been contemplating for some time. They went public with the facts, opinions, and research findings that they hoped would destroy Project Chariot. The instrument they chose for this first salvo was the Alaska Conservation Society's *News Bulletin*. Working long hours in Ginny Wood's little log cabin, the group produced a thirty-page newsletter that essentially offered a summary of the Cape Thompson research to date. Pruitt wrote a review of the biological investigations at Cape Thompson; Viereck summarized certain botanical issues, and nearly the whole of his lengthy letter of resignation to President Wood was reprinted. Foote wrote a historical overview that questioned many of the AEC's basic assumptions. Ginny Wood and Celia Hunter worked on the editorial section.

Even though the connection between radiation and caribou was not within the scope of his contractual duties, in a brief section of the *News Bulletin* called "Implications of Project Chariot," Pruitt called attention to some interesting peculiarities of the Arctic ecosystem that, he thought, might bear on radiation issues. Pruitt did not go into great detail, but sent his material off to Barry Commoner, a plant physiologist at Washington University in Saint Louis who had organized the Committee for Nuclear Information (CNI). Commoner's group had covered the Chariot proposal briefly in its newsletter, and Pruitt had hope that CNI would expand its Chariot investigation, publicizing the potential ecological problems to a wider audience than the ACS reached.

Produced in advance of the AEC summary report—which the biologists expected would have a prodevelopment slant—the Alaska Conservation Society *News Bulletin* was essentially a "minority report" emphasizing environmental considerations. Overall, Viereck considered the treatment "extremely mild." The group strove to keep "the factual review section separate from the editorial section," he said, and to document their statements with citations to the literature. With considerable effort, 1,000 copies were mimeographed, collated, and mailed to conservationists and other interested parties all over the United States. Government officials in Washington were targeted in particular.

•   •   •

Though Les Viereck had been given his notice at the University of Alaska, he was "really being grand about the entire situation," according to Brina Kessel. He helped plan the botanical work for the 1961 field season at Ogotoruk Creek and wrote up a progress report on the 1960 season's work. By March, the Committee on Environmental Studies for Project Chariot had prepared a draft of its "First Summary Report" on the research at Ogotoruk Creek. The committee was disinclined to release to the public the progress reports turned in by the investigators; instead it preferred to offer its own characterization of the researchers' preliminary results. Also, for some unknown reason the committee's summary dealt only with the 1959 field season, even though the 1960 summer's work had been written up and submitted to the committee months before the summary's publication in April of 1961. Though its information was a year old and covered only half the research activity, the document claimed to set forth the research "to date." But at least the AEC was sending the draft around for the investigators' review.

When the University of Alaska biologists got a look at the document, they denounced it. In pages of written comments to Brina Kessel, they complained that the summary was "non-representative of the data in the original reports" and slanted to favor the detonation. Pruitt said that the section on land mammals was "so completely at variance with our results" that he had to rewrite the entire section. He pointed out many examples of what he took to be rhetorical tricks employed to put the data in the best possible light. For example, the report contained a species list that noted the occurrence of voles in the project area, but omitted the mention of moose, a species that might have greater emotional appeal to the public. Caribou hunting at Kivalina was said to benefit "twenty-five households," a number that some readers might take to be rather small. It was not stated that caribou hunting benefited every single household in the village, but that was the case: Kivalina consisted of twenty-five households, total.

Complaining about the summary report's continual references to studies conducted at the Nevada Test Site, Pruitt wrote: "We have emphasized that extrapolations from a temperate desert cannot be made to a coastal Arctic environment." The report also gave sweeping assurances as to Chariot's radiological safety: "It would appear that [radiation effects] would be negligible, undetectable, or possibly nonexistent in areas distant from the excavation." Pruitt protested that "there are absolutely no data to back up the statement that modifications of habitats will be non-existent beyond the limits of throwout." His own references to the possible concentration of radioisotopes in lichen and caribou had been ignored in the AEC report.

Les Viereck also wrote pages of comments and corrections, including the charge that "there is a definite attempt to belittle the snow data obtained by the botanists." The scientists had been reporting that Ogotoruk Creek valley was often largely bare of snow due to frequent high winds, meanwhile the AEC was claiming that the fallout would land on snow and be flushed out to sea with the spring melt.

There were other complaints, as well. Doris Saario, a graduate student who was working under the University of Alaska contract on the human ecology at Kivalina, found elements of the summary to be "misleading." Among her objections was what she regarded as an equivocal analysis of the interference to Eskimo hunting patterns that the detonation might cause. The summary had said that disruption of hunting activities "could" result in hardship to the Eskimo unless substitute food was provided. Saario responded:

> It seems fairly obvious that disruption of hunting activities *would* result in hardship to the Eskimo. . . . With regard to a "substitute," hunting is not only a means of subsistence—it is a "way of life." In these terms, I cannot conceive of a substitute.

Brina Kessel compiled the researchers' many corrections to the summary report and sent them on to John Wolfe, saying: "In general, all the investigators, including Miss Saario, felt that the report was biased in favor of

Project Chariot." But all the effort expended in rewriting and correcting the summary report was to no effect, as Wolfe wrote back in reply:

> Your commentaries on the report arrived in the same mail cart that carried the report to press. There was no hope of re-typing, but at the first up-dating we will consider every suggestion. . . . Your people should always keep in mind that their final reports will be published in full.

Don Foote had similar luck with John Wolfe and the committee. Upon receipt of the draft, Foote pounded out a 4,000-word reply. It was sent weeks before Wolfe's mail cart left with the report. But very few of Foote's dozens of detailed corrections made it into the document. For example, the report had concluded that Kivalina was maintaining itself on a "subsistence level," but that Point Hope was not. In Foote's view, however, Point Hope was actually a truer subsistence economy because its cash income derived from the sale of Native products, rather than welfare. And he was particularly irked by the committee's paraphrasing of his results having to do with interruptions to the Eskimo's hunting patterns. The committee had said:

> Disruption of hunting activities and attendant harvests without substitute in any month could result in hardship to the Eskimo . . . in the opinion of one of the investigators in human geography.

The editorial insertion of the phrase "in the opinion of one of the investigators in human geography" upset Foote for two reasons. It suggested that Foote was alone in his belief that an interruption to hunting would cause hardship, and that he endorsed the idea of providing substitute food as a way to mitigate the hardship. Neither idea was remotely true, and Foote made very clear his objection to the "substitute" notion:

> The statement . . . should be clearly defined as *not* included in my conclusions of June, 1960. I have never supported the idea of substituting anything for the traditional Point Hope Eskimo diet. I would be very interested in knowing how the

author of this section drew the conclusion that I did. [Emphasis in original.]

Despite the fact that *both* of the two investigators in human geography had declared in writing, in the strongest possible terms, that the summary statement did *not* reflect their views, Wolfe made no changes to it. Foote also pointed out a more basic flaw in the AEC's report. He noted that while the document claimed to "provide the major basis for biological cost estimates" should the detonation take place, it never mentioned any of the physical effects of the proposed detonation. It did not discuss seismic effects, air shock, or disposition of radioactivity either close in or as fallout. How, Foote questioned, could anyone estimate the biological cost of an activity that had not been defined? It was a scientific absurdity. The committee had produced a summary that described an environment, failed to describe a contemplated disruption to that environment, yet concluded that the effect of that disruption would be insignificant. With respect to broad ecological harm, the committee said that "the chance of significant biological costs at this level appears exceedingly remote." And without providing any supporting evidence, Wolfe's committee concluded that radiation risks would appear to be "negligible, undetectable, or possibly nonexistent in areas distant from the excavation."

If, when he started the Project Chariot research in 1959, Don Foote harbored any illusions about the inviolability of science, few remained by the spring of 1961. He wrote John Wolfe:

> One reason I took the research position of Human Geographical Studies was my conviction in the sincerity of your personality and your dedication to the principle that the wise use of nuclear explosives demands considerations of pre- and post-detonation effects on the entire biosphere.
>
> During the past year, however, my belief in you and the program has been shattered. To a great extent this was the effect of my politically immature mind awakened to the modern interplay of politics and science. It was caused by my realization that when science becomes directly involved with politics, the concept of scientific truth is modified, warped and totally abandoned.

Wolfe reminded Foote and the University of Alaska investigators that the present effort was only an interim summary, and that the AEC would publish their final reports in full. In fact, the publication of the full reports would not occur for five more years, and in the meantime the AEC would issue two summary reports, which simply ignored the bulk of the contributors' objections.

•   •   •

"Biologists in particular are a snarky bunch of bastards," says Bill Pruitt, "you get a half dozen biologists together and you're going to knock heads together pretty soon." Eventually, Pruitt and Kessel knocked heads—and not just because they were biologists. Their personalities clashed utterly. Kessel was born of academic parents into, as she says, "a very prestigious scientific family." Her parents were affiliated with Cornell University, and her grandfather, James McKeen Cattell, was a psychologist and prominent as a scientific publisher and editor. Among other publications, he had owned and edited *Science* magazine for nearly half a century, having purchased it from Alexander Graham Bell in 1895. Cattell was also a champion of academic freedom who'd been expelled from Columbia University for his pacifist opinions, and earlier from Johns Hopkins, where he had been regarded as a nuisance. He had also helped to found the American Association of University Professors. As one writer said: "In both his writings and his conduct, Cattell upheld the rights of professors against deans and university presidents." His granddaughter inherited his love of scholarship, but by some irony of genetics, it is Pruitt, more than Kessel, who seems the philosophical descendant of James McKeen Cattell.

As department head and acting dean at the University of Alaska, Brina Kessel tended to believe in authority and a defined chain of command. Not so Pruitt. He possessed the sort of strong personality the North often attracts: He was smart, self-reliant, and stubborn, not averse to challenging conventional wisdom or, for that matter, authority. The sort of administrator Pruitt admired was one who "didn't care whether school was kept."

When Pruitt went visiting from house to house in Kivalina or Point Hope, each time he ducked into someone's entryway he'd extract a little whisk broom from his parka and carefully brush every crumb of snow from his clothes before entering. Fastidious on the trail, he was less than dainty in the halls of academe. Returning from dog-team trips, he would disrupt scholastic serenity with, among other things, five-gallon cans full of rotting animal carcasses—his "specimens." In the fall of 1960, as Kessel wrote in a business letter, "Pruitt had 18 ground squirrels housed in Blazo [white gas] cans stinking up the physiology lab, but we got rid of them before school started." And to Kessel's intense annoyance, he liked to sit in the lab with the windows open and the icy air pouring in, contentedly working away in his parka and mukluks.

Despite his idiosyncrasies, Pruitt was a leading authority on northern mammals and a good teacher whom the students liked. Within weeks of President Wood's suggestion that the department might be better off without the dissenting biologists, though, Bill Pruitt's standing at the University of Alaska began to decline. He noticed that his relations with Kessel were souring. She wanted access to his correspondence and Pruitt regarded the issue as "only one of a series of rather petty (but damned annoying) incidents lately. I am making valiant efforts," he wrote Don Foote, "to look at all this without developing a persecution complex, but it is sometimes difficult."

By March of 1961, after he had helped get the Alaska Conservation Society *Bulletin* ready for duplication but before it was mailed out, Pruitt flew out to Kivalina and was once again traveling by dog team through the windswept landscape and over the frozen sea ice of northwest Alaska. During the springs of 1960 and 1961, he spent forty-six days conducting field research by dog team. He covered 700 miles in all with his Eskimo guide and field assistant, Lawrence Sage, from Kivalina. They took one team and alternated running and driving. They'd both push the sled up the hills, then each jump on a runner for the ride down. "We made trips by dogs from Kivalina to Chariot site—Ogotoruk—and then to Point Hope," Pruitt recalls. "Sometimes we had to turn back because the wind was strong enough that it was picking up gravel, and throwing it in our faces. And the dogs wouldn't face it." In February of 1960, Pruitt and Sage drove their team into winds exceeding fifty miles per hour and air temperatures to twenty below. The resulting "wind chill factor" approached ninety degrees below zero. Under these conditions, it takes between thirty and sixty seconds for exposed human flesh to freeze.

"Consequently," he wrote, "the greatest amount of one's time is spent in simply staying alive." One of Pruitt's reports noted his experience with a sort of ground blizzard:

> During this trip we encountered a snow-wind phenomenon frequent in the region. Nearing Cape Seppings, in clear air with only moderate winds, we saw ahead a solid gray wall which proved to be blowing snow. The edge of the *siqoq* . . . formed a sharply defined wall, so sharp that our lead dogs disappeared into it as if passing through a door.

On this trip, he was unable to visit all of the thirty-six one-acre plots that he had established miles back in the hills in order to trap and study lemmings and voles. But two months later, when conditions had improved to the point where, as he dryly observed in a scientific report, "traveling [was] possible but occasionally somewhat hazardous," Pruitt managed to visit all his plots. He whipped off his mittens and took a photograph of each site. He knelt and collected pellets and scats. He took the temperature of the air at the snow level and measured the thickness, density, and hardness of each of several layers of snow. Always he kept an eye out for caribou, wolf, and wolverine and recorded these and other mammal sightings.

It was while Bill Pruitt was doing his field research at Cape Thompson in March of 1961 that a stunned Brina Kessel received in the mail, as a dues-paying member of the Alaska Conservation Society, her copy of the ACS *News Bulletin*. Kessel felt Pruitt and Viereck's work on the publication represented a breach of protocol on several counts, as she wrote to Professor Al Johnson, her confidant in departmental matters, who was still away on sabbatical:

> I, of course, don't mind their opposition to Chariot, but I do mind the lack of ethics involved here—especially in Biological Sciences professional staff: 1) I knew *nothing* about this proposed report on Chariot until I received it in the mail, 2) They have published other scientists' data without permission (a violation of publication rights? I think so.), 3) the editorial is almost pure propaganda (written by Ginny, but Les is

President and Pruitt on the Board [of the ACS]), 4) The factual data in the summary may be accurate, but it is presented in a biased manner. The biased presentation is bad enough where their own data is involved, but inexcusable when they have used other scientists' data. They have done exactly what they have objected to the AEC doing.

Al Johnson, who was inclined to agree with his boss, thought "the factual material was presented fairly well, although, as you say biased." He said he "hadn't considered the ethical point that you raised, but on first thought am inclined to agree." The *Bulletin* articles had indeed made use of information contained in other Chariot researchers' progress reports to the AEC, which reports the AEC had distributed to all investigators. Perhaps, if the progress reports were not considered publications, an impropriety arose when the ACS authors made the first published use of other researchers' data. But they did acknowledge the original researchers in footnotes, and nothing in the Alaska Conservation Society archives suggests that the use of the data disturbed the scientists whose work was cited. The ethical point seemed of great importance to Kessel. She had also protested the AEC's "summary report," but not with as much vigor. And in the summary report, the AEC had characterized several researchers' work in ways to which the scientists had specifically and vehemently objected, in writing, in advance of publication.

Kessel also complained that she knew nothing of the article prior to its publication, but it is not clear why she would expect to be kept apprised of the political activities of her staff, especially since permission to publish Chariot research had been specifically granted by John Wolfe in a 1959 letter to all the investigators. Given all of this, the biologists suspected that their superiors were upset because of their criticism of Chariot, rather than the fact that they had used others' data; that if their advocacy had tended toward economic development rather than conservation—if they had been members of the chamber of commerce, for example, and used the data to bolster a "positive" review of Chariot—that such advocacy would have been tolerated. After all, John Wolfe had been making unrestrained use of the researchers' data for years—not just to issue summary reports, but also to shape public opinion by stimulating *favorable* articles in the press.

As far as Kessel was concerned, however, the Alaska Conservation Society *News Bulletin* was the last straw for one William Obadiah Pruitt, Jr. When the unsuspecting biologist returned to his lab in early April after two weeks behind a dog team, "all hell broke loose," he wrote. "I was guilty of being biased, untruthful, using data without permission. . . . Brina informed me that my contract was not being renewed with the University because of a long list of grievances. These ranged from 'not working hard enough,' 'uncooperative,' to 'wearing a lab coat of the wrong color' (honest, this was actually said)."

Pruitt "got fired up at all this," as he wrote in a letter to Don Foote, and "arranged an appointment with Wood." The meeting, which lasted an hour and a half, went just the way Viereck's interview with Wood had gone. Wood led the talk into many areas, drew Pruitt out on his ideas for future biological work at the university, and asked him to write up several of his proposals for Wood's review. "At the end," said Pruitt, "Wood said in effect, 'about those things Brina has accused you of, you're very close to the source, but as for me I don't bother with such things at all.'"

Like Viereck, Pruitt was hopeful that the new president, who seemed so interested in the future of the biological sciences, would help mend the rifts and set the department back on course. Both men spent a good deal of energy writing detailed proposals for Wood's consideration, which, Pruitt said, "we really put our hearts into." Viereck even headed off an incipient protest action from people upset at his discharge, telling his supporters that Wood had encouraged his hope of rejoining the university. But there was something that Pruitt and Viereck did not know as they waited for President Wood's reaction to their proposals. Namely, that even though Wood had said he didn't "bother with such things," their dismissal had been his idea.

In June, Pruitt was given a seven-and-a-half-month contract, enough time to write up his final report on the Chariot research. He was to be "phased out" on January 22, 1962.

●   ●   ●

Along with giving him a short-term contract, Kessel relieved Pruitt of responsibility for the caribou and ground squirrel sections of the Cape

For thousands of years, the residents of northern Alaska have practiced an economy based on animals. Here Inupiat Eskimos of Point Hope, Alaska, butcher a bowhead whale on an ice floe in 1960. *(Don Foote collection, UAF Archives)*

A traditional sod house was constructed of whale bone or driftwood, covered with sod, and heated with an oil lamp. Some of these houses, including this one were still in use at Point Hope in the 1950s when Eskimos abruptly entered the atomic age. *(Dan O'Neill)*

A Point Hope hunter glasses the hills for caribou as the dog teams rest. Hunters freighted into the village as much as 100,000 pounds of caribou meat per year. In 1959 and 1960, every single caribou taken by Point Hope hunters was shot within twenty-five miles of the proposed ground zero. *(Don Foote collection, UAF Archives)*

Edward Teller (*left*) and Gerald Johnson (*right*) with colleagues in the hole of the 1961 Gnome shot. The Plowshare detonation would accidentally vent a "high release of radioactivity." Alaskans said the atomic physicists seemed like boys playing with firecrackers. *(Courtesy of Lawrence Livermore National Laboratory)*

En route to the Project Chariot site on one of his several trips to Alaska, Edward Teller *(center)* changes planes at Kotzebue, Alaska, 1959. *(Courtesy of Lawrence Livermore National Laboratory)*

University of Alaska president Ernest Patty *(left)* and regent Elmer Rasmuson *(right)* confer an honorary doctorate on the "Father of the H-bomb," Edward Teller, in 1959. The citation noted Teller's "fearless endeavors . . . against the menace of tyranny." *(Courtesy of* Denali, *the University of Alaska yearbook for 1959)*

Shown in 1962, Daniel Lisbourne was president of the Point Hope Village Council and one of the Inupiat people's strongest leaders in modern times. A scientist working in the village told the AEC that Lisbourne was "well read and as concerned about fallout, etc., as any good New Englander." *(Don Foote collection, UAF Archives)*

Lawrence Radiation Laboratory's Edward Teller *(left)* and Gerald Johnson *(light suit)* promote the idea of "geographical engineering" to a receptive group of University of Alaska officials, including C. T. Elvey *(bow tie)* and Robert Wiegman *(right)* in 1959. *(Courtesy of University Relations Negative Collection, University of Alaska Fairbanks, Archives)*

In downtown Fairbanks in 1959, Teller mingled with civic leaders, members of the press, and the business community. Though initially skeptical of Teller's proposals, Alaskan business leaders led the campaign for nuclear excavation projects in the new state. *(Courtesy of University Relations Negative Department, University of Alaska Fairbanks, Archives)*

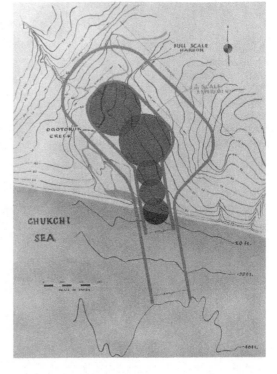

In the original Project Chariot design, a "full scale harbor," shown as the larger outline, would be produced with detonations totaling 2.4 megatons. Plans were later scaled down twice to call for 460 kilotons, shown here as the inner outline with five explosives and 280 kilotons. *(Lawrence Livermore National Laboratory)*

Scenes from a 1958 AEC film showing a stylized representation of the Ogotoruk Creek Valley near Point Hope, Alaska. The film was shown to Point Hope villagers in 1960. *(Photos enhanced by author for clarity; from Industrial Applications of Nuclear Explosives, courtesy of Lawrence Livermore Laboratory, 1958)*

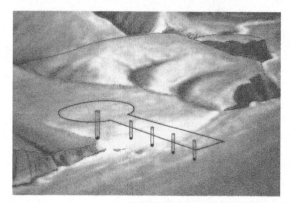

A:
An inline series of four 100-kiloton nuclear explosives are placed at a depth of 30 meters and a terminal shot of 1 megaton is buried at 50 meters.

B:
Simultaneous detonation. Cloud height would actually reach 30,000 feet (the cliffs are about 700 feet above sea level).

C:
The sea rushes in to fill the entrance channel and turning basin of the finished harbor.

Driftwood lines the beach at the proposed ground zero, the mouth of the Ogotoruk Creek, near Cape Thompson on the northwest coast of Alaska. In 1992, with the release of documents discovered in researching this book, a small radioactive waste dump was discovered not far from this spot. *(Dan O'Neill)*

William Ransom Wood, president of the University of Alaska from 1960–1973, defended Project Chariot, declaring that "if the United States government decides that the project is a safe one, there is no reason for concern." Two biology professors lost their jobs during Wood's administration, after they publicly criticized Chariot. *(Courtesy of University Relations Office, University of Alaska Fairbanks)*

Organized as an autonomous advisory body by the AEC, the Committee on Environmental Studies for Project Chariot was led by a chairman, vice-chairman and secretary who worked—directly or indirectly—for the AEC. Several Alaskan scientists were unhappy with the committee's characterization of their research data in early reports, which seemed to favor the AEC's previously stated objectives. *Left to right*: Maxwell E. Britton, Arthur H. Lachenbruch, Robert L. Rausch, Kermit H. Larson, Ernest D. Campbell (secretary), Allyn H. Seymour (vice-chairman), Norman J. Wilimovsky, and John N. Wolfe (chairman). *(Courtesy of U.S. Department of Energy)*

The AEC camp at Ogotoruk Creek had sleeping quarters for eighty-four people in Jamesway huts. The site has been used by Eskimos as a hunting and trapping camp for hundreds of years, at least. *(Don Foote collection, UAF Archives)*

Biologist William Pruitt packs caribou antlers back to camp for analysis as part of the AEC-funded environmental studies in the Cape Thompson region. In 1961, Pruitt was dismissed from the University of Alaska after his reports noted relatively high levels of fallout in lichen, caribou, and Eskimos in the Arctic. *(Courtesy of Les Viereck).*

Pruitt works on a caribou rack and skull at the field laboratory at Ogotoruk Creek. *(Courtesy of Les Vierek)*

Leslie Vierek presses plants at his field camp near Cape Thompson in the summer of 1959. Vierek resigned in protest from the AEC research program, charging that the agency persistently mischaracterized some of the researchers' preliminary findings. With that action, University of Alaska officials relieved him of his teaching position. *(Courtesy of Teri Vierek)*

Within weeks of his arrival at Point Hope, Don Foote got out into the country via dog team with his mentor, Point Hope's great hunter Antonio Weber. Here the team and drivers rest on tundra so windblown it is largely bare of snow. *(Courtesy of Berit Arnestad Foote)*

Foote attempts to winterize his drafty house on the Point Hope spit. One winter night he noticed that, with the furnace going full blast, a water bucket ten feet away froze solid. *(Courtesy of Berit Arnestad Foote)*

La Verne Madigan (*right*), head of the New York-based Association on American Indian Affairs, assisted Alaskan Eskimos in their struggle to stop Project Chariot and to assert aboriginal rights. Here she plans strategy with Dan Lisbourne of Point Hope (*center*) and Guy Okakok of Barrow (*left*). *(Courtesy of Tundra Times)*

The Point Hope Village Council in 1968, a few years after the Project Chariot controversy: *(seated, left to right)* Joe Frankson, David Stone, Rose Omnik, Daniel Lisbourne *(president)*, George Kingik, Clyde Howarth. *(Standing, left to right)* Andrew Tooyak, Sr., Ronald Oviok, Sr., Amos Lane, and Henry Attungana, Sr. *(Courtesy of Steve McCutcheon)*

Because the mainstream press failed to cover Native issues adequately, Howard Rock (*left*), an Inupiat artist from Point Hope, and Tom Snapp (*right*), a journalist from Fairbanks, founded the first statewide Native newspaper, *Tundra Times*. *(Courtesy of* Tundra Times*)*

Tundra Times was founded largely to give voice to the Eskimo people's opposition to Project Chariot. The paper became an important instrument of political cohesiveness for all Alaska Natives as they fought for aboriginal rights through the 1960s and early 1970s. *(Courtesy of Richard Veazey)*

The 100-kiloton Sedan shot at Nevada Test Site in July 1962 was designed out of "frustration" as an "alternative to Chariot" to answer many of the same questions. No preshot environmental studies were required at NTS. *(Courtesy of Lawrence Livermore National Laboratory)*

Sedan was the largest explosion to have occurred on North America up to that point. Its boiling cloud of radioactive dust rose "higher than had been expected," to more than two miles in the air, drifting across the United States and into Canada. *(Courtesy of Lawrence Livermore National Laboratory)*

The Sedan explosion ejected 24 billion pounds of irradiated earth. The resulting crater measured 1,280 feet across and 320 feet deep. Today, thirty-two years later, signs still warn NTS employees not to dig near the crater. *(Courtesy of Lawrence Livermore National Laboratory)*

William O. Pruitt (*left*) and Leslie Viereck (*right*) in 1993 on the occasion of receiving honorary doctorate degrees from the University of Alaska, Fairbanks. In part, the men were recognized for the very actions for which they lost their jobs three decades earlier. *(Courtesy of Cal White).*

Thompson study. That left him with only the section on the ecology of the other terrestrial mammals to write up. On December 22, 1961, a month before he was to leave, Pruitt submitted his final report to Kessel. No one suspected that this act would touch off a memo war that would lead to inquires and investigations by the American Civil Liberties Union, the American Association of University Professors, and the American Association for the Advancement of Science.

Kessel kept Pruitt's report until January 17, getting it back to him just five days before his last day of work. She had made numerous editorial revisions, as well as some changes "of a censorship nature," according to Pruitt, ". . . in that the deletions changed meanings and prevented the Atomic Energy Commission from receiving the full benefit of my experience of Arctic mammalian ecology." Pruitt said he accepted a number of Kessel's editorial changes even though most were foreign to his style of writing, but could not accept those modifications that he considered to be censorship. The next day, he passed on his revision directly to the steno pool, rather than returning it to Kessel for approval.

When Pruitt moved out of his office on the twenty-second, about half of the lengthy report had been transferred to mimeograph stencils, the final step before running off the 100 copies required by the AEC. These were not sent back to Pruitt for proofing, however, because Kessel took possession of them and made additional changes. On February 18, she again sent Pruitt an edited copy of his report. Sections of it had been cut apart, some portions had been moved or removed, and then taped back together. With this version she enclosed a reasonably pleasant "Dear Bill" letter outlining her objections. It was clear that Kessel had expected to have been given the manuscript for final approval before it had gone to the steno pool, and in any event had not had the opportunity to review Pruitt's "Recommendations" section.

Aside from some lingering issues with the prose, Kessel felt "that a major revision of the last half of your [Recommendations] section is necessary—not only because it is prolix and carelessly written, but primarily because the detail on the possible effects of radiation is definitely not pertinent to the University of Alaska investigations."

Pruitt had considerable writing talent and would in time publish widely beyond the scientific literature, including a popular book on northern animals and articles in *Harper's, Atlantic, Scientific American, Holiday, Audubon,* and other journals. Still, Pruitt's work could show the failings Kessel pointed out, especially because he had worked over and written up three years' work in four months' time. For her part, though she was to publish very little, Kessel was a gifted editor, according to colleagues. By this point, however, the working relationship between the two biologists had completely broken down. Pruitt fired off his reply two days later:

> I have received your letter of 15 February 1962, along with the mutilated and censored portions of the section on recommendations . . . I must remind you that I am the author of the report and that you do not have the authority to make such changes in it. You do not have my permission to do so.

Pruitt also asked to see his original report because he had no copy and could not be sure exactly how his words had been altered.

His letter remained unanswered by early March, and Pruitt decided to write President Wood. He did not go into detail about his differences with Kessel, but did point out the "ramifications" of her actions. "If the report does not agree with the version I submitted to the steno pool," he wrote, "the correct version will be published independently with detailed explanations. The resulting publicity would probably not reflect credit on those individuals responsible for the censorship or for their institutions." Though Pruitt threatened to publish his "correct" version, it would have been difficult to do because, as he had been complaining, his only copy of it was in the possession of Brina Kessel.

On March 29, Kessel sent Pruitt a clean copy of his final report as she had edited it. But she did not send him his original manuscript by which he might compare the two versions. Her tone was strictly formal:

> You are hereby given the opportunity to object to any part of this report. . . . If we do not receive your specific objections, suggested additions, deletions or changes, and stated reasons for each of these deletions, additions or changes, on or before 20 days from the receipt of this report, it will be conclusively presumed that you approve of this report in its entirety.

In conclusion, Kessel asserted that "the university itself is the one ultimately responsible for the content of all final reports" submitted to the AEC. "These are final contractual reports," she said, "and do not constitute publication in the scientific literature."

Again, Pruitt responded immediately:

> Your puzzling statement about my "opportunity" to object to any part of the report and your self-imposed time limit of 20 days exhibit a bizarre lack of understanding of the ethics of scientific writing. No amount of quasi-legalistic verbiage can obscure or justify censorship.
>
> I challenge your statement that the University is the one ultimately responsible for the content of all final reports. The ultimate responsibility for any scientific writing, whether intra-agency report, contractual report or open publication lies with the author. Your statement that the document under consideration is a final contractual report and not publication in the scientific literature imputes different degrees of accuracy or logical rigor.

And again Pruitt asked to see his original manuscript, which was the only way he could determine just what parts of the two-inch-thick document had been modified, and how.

What was in Pruitt's report to cause all the commotion? Obviously, the stylistic objections were not significant, though Pruitt might have profitably accepted more of Kessel's suggestions having to do with his prose style. In general, Pruitt had simply gone out of his way to emphasize issues of likely interest to conservationists worried about Chariot's effects. For example, he didn't merely note what he called the "endangered" status of the grizzly bear, he wrote four pages of commentary including each and every sighting in the region. For her part, Kessel did not stop with trimming the excessive coverage. She eliminated references to the bear's (and the wolverine's) "endangered" status altogether, and only permitted the listing of sightings that occurred in the previous year. She said other records for grizzly bear and wolverine had already appeared in earlier progress reports, while Pruitt quite reasonably thought a "final report" should include the full inventory of data collected.

Kessel also toned down Pruitt's emphasis on the economic importance to Eskimos of caribou and polar bear. For example, in his section called "Literature Cited," Pruitt listed approximately a dozen scientific papers dealing with the economic and ecological importance of the caribou. Kessel deleted them all, perhaps with the rationale that the university's caribou investigations were to be the subject of another report not written by Pruitt.

But the main rumpus centered, again, on Pruitt's "Recommendations" section. Pruitt had written that the land-mammal study "was specifically designed to provide data only on the kinds of mammals present in a restricted region and to delimit roughly their natural history and their ecological relationships." He stressed that his study "was not designed to produce data on the effects or relationships of the proposed nuclear blast and the mammals." But, he said, the role of land mammals in radiation transport had nonetheless become apparent during the course of his research:

> Although not officially concerned with radiation effects of Project Chariot, such aspects and their potential dangers have become evident during the study. . . . The important points to stress are our woeful lack of knowledge of the effects of such high concentrations of a long-lived radioisotope in an economically and ecologically "key" mammal [caribou].

Pruitt provided a table that drew on published research to compare levels of strontium 90 in organisms found at Ogotoruk Creek with those found at lower latitudes. He said: "As far as strontium 90 concentration is concerned we can only point out the potential danger of this situation and strongly recommend no further radioactive contamination until sufficient research has been conducted, published and evaluated." Kessel deleted the table and Pruitt's strong recommendation.

In a letter he sent to the Project Chariot Bioenvironmental Committee, Pruitt concluded:

> The deletions form a disturbing pattern. Almost all are sections on endangered species, potential ecosystem disruption or recommendations for further needed research. These are just the aspects of Arctic ecology and resource conservation

that are avoided by the powerful commercial and governmental exploiters and "developers."

Despite all his protests, Pruitt's original manuscript was not returned
to him during the twenty-day objection period. In fact, on April 11 and
13, when only about half of the period had expired, Kessel sent on to
John Wolfe and the environmental committee a hundred copies of Pruitt's
report. On the title page, the author was identified with the words:

William O. Pruitt, Jr.
as modified by
Brina Kessel, General Supervisor
University of Alaska Project Chariot Investigations

Kessel noted in her letter to the committee that the report did not have
Pruitt's approval and included a copy of his original "Recommendations"
section for the committee's information. She said that the section had
been "rather heavily edited and portions deleted" because the "possible
effects of radiation was not pertinent to the University of Alaska investigations." Kessel had substituted a paragraph for the last two and a half pages
of Pruitt's "Recommendations" section. But the paragraph, stripped of
Pruitt's indignant tone, seems a reasonable distillation of his call for the future studies he considered necessary. What was not allowed in Kessel's
version was Pruitt's comments on the "potential dangers" of the bioaccumulation of radioisotopes.

Pruitt said his excised comments did not relate to "possible effects of
radiation," and that he would not presume to make such comments because they were outside his area of competence. "The deleted statements," he wrote the Chariot environmental committee, "are concerned
with the ecological and environmental roles played by the species concerned, and on the need for clarification of the mechanisms of transport
of material and energy through the food web." In an interview years
later, Pruitt explained, "I was careful not to say anything about the effects of the radioactive contamination. My area of expertise is in following the food up through the food chain, showing that the contamination
is going to end up in the upper levels of the food chain, in the top con-
sumers [of caribou] which would be wolves and humans." While Pruitt

did not comment specifically on anticipated health effects, per se—he did not mention cancers or birth defects—he did cite data showing high levels of strontium 90 in lichen and caribou, and he used these data to "strongly recommend no further radioactive contamination [be introduced in the region] until sufficient research has been conducted." He did point out the "potential danger of this situation" and, therefore, did seem to be drawing conclusions as to the effects of radioactive contamination, as Kessel claimed.

*   *   *

Kessel showed the modified version of Pruitt's report to her two most senior faculty, Al Johnson and Gerry Swartz. Both men had essentially the same reaction. In a letter written a year later, Johnson recalled taking issue with Kessel's tactic: "I talked to her about it and objected to its unilateral nature as well as those aspects of academic freedom which I felt were involved." Kessel refused to discuss it, Johnson said, even though she had first brought the issue to his attention. "In summary," he wrote, "Dr. Kessel informed me that . . . she had the full support of Dr. Wood, Vice-President Elvey, as well as the University attorney, whose opinion had been solicited in the matter." Apparently the university attorney felt Kessel's actions were legal because the contract had been made between the AEC and the university, not between the AEC and Pruitt. Johnson says he persisted in his objections but received no satisfaction.

It was Johnson's view that the material on radiation deleted from Pruitt's report contained information relevant to the food-chain issues under study. He thought Pruitt was extracting from the published literature data concerning radiation levels and that he was pointing out the potential danger to Eskimos if the Chariot shot were carried out. And this danger, Johnson wrote in 1963, "has been confirmed most conclusively in the months that have followed early 1962." Though he had no proof, Al Johnson presumed that the information on radioactivity was deleted from Pruitt's report "so as to avoid antagonizing the AEC."

Gerry Swartz's recollection of the incident had not faded twenty-six years later. He felt considerable empathy with Kessel's review of certain parts of Pruitt's report, but he could not abide Kessel's preemptory modification of scientific work. "I had never seen a scientist treated that way before," said Swartz, recalling his reaction to the title page, "[I] felt it was kind of a monstrous thing to do and indicated my feelings to Brina."

Unaware that Kessel "had the full support of Dr. Wood," Pruitt again appealed to the university president. He reminded Wood that he had written four letters asking for the return of his manuscript "so that I could proof read the altered report. I have never received the manuscript. . . . The 20 days were not allowed. . . . In all my experience of publishing in scientific journals and commercial magazines I have never before had a manuscript impressed."

On April 25, having heard nothing from Wood, Pruitt wrote out a "Statement to Project Chariot Environmental Committee and All Concerned," in which he outlined the "censorship" issue as he saw it and informed the committee that he was "unable to vouch for the accuracy of anything in the mimeographed document sent to you as my final report." Wood finally responded by turning the matter over to his vice president, C. T. Elvey, a tough administrator who had helped secure big-dollar funding for the university's Geophysical Institute. "Instead of writing you an answer," Elvey wrote Pruitt, "I feel after a review of all correspondence that you and I should talk over the entire problem." Elvey's papers (he died in 1970) do not contain a contemporaneous written account of the two-hour meeting, but Pruitt's do:

> He was nasty—he knew just what he could do legally if I did not cooperate with the Univ. on the matter. I got my dander up and told him bluntly I thought the Univ. was way out on a limb as far as scientific ethics was concerned and if there was any sort of public hassle I felt sure the Univ. would come out second best.

In an interview decades later, Pruitt told the same story. "I really definitely remember the session . . . Elvey called me in and really raked me over the coals . . . Did I realize that I was endangering the financial situation of the University of Alaska, that they'd had all these threatening phone calls from the Atomic Energy Commission threatening to cut off the support?" Elvey said the university could not risk losing the funding for the sake of one individual, according to Pruitt, and therefore Pruitt had been, essentially, fired.

It is probably not possible to prove whether or not Kessel, Elvey, and Wood in their treatment of Pruitt and his writing were responding to direct AEC pressure, or even the anticipation of AEC displeasure with the university. But one document shows that the university administration's alteration of Pruitt's work followed an AEC complaint about it. Two months before Kessel's modification of Pruitt's report, John Wolfe wrote Kessel (with a copy to President Wood). In the course of answering various questions Kessel had raised, Wolfe said that the final payment from the AEC to the University of Alaska was "contingent upon the receipt of a satisfactory [final] report." In the same letter, he noted that Pruitt's recently submitted progress report was "not highly satisfactory" and that Kessel could "approach this subject with him with whatever tact or forthrightness" she deemed proper or ignore his remark altogether. Wolfe took particular exception with Pruitt's table showing levels of radiation in lichen and caribou. He said that Pruitt's work was "disconcerting," contained statements "that could be misleading," and that he would probably write Kessel again on the subject of Pruitt's report. President Wood sent his copy of this letter to Elvey; the paragraph dealing with Pruitt was marked in pencil, and Wood's penciled cover note suggested that Elvey look into the matter.

When the American Civil Liberties Union investigated Pruitt's firing, the chairman of its Academic Freedom Committee wrote to President Wood and raised the same issues as Pruitt had done. Concluding its questions of Wood, the ACLU asked: "Is it true that officials of the University declared that their dependence upon government grants and contracts necessitated the 'placating' of government agencies?"

President Wood apparently did not reply to the ACLU, though he did tell the campus newspaper that Pruitt had *not* been fired. It was merely a situation, he said, where "his contract came to an end and the University had no opening for a man with his qualifications and background." But on this very point Al Johnson wrote the ACLU that "money for a permanent position in Dr. Pruitt's specialty became a part of the biological sciences department budget in the [academic] year following Dr. Pruitt's 'dismissal,' and I assume that Dr. Pruitt would have filled this position had he been acceptable to the University administration." And when the ACLU asked Les Viereck to comment on the allegations that Pruitt was fired in order to preserve good relations with important funding agencies, he wrote: "From my own experience with the administration of the University of Alaska, I have little cause to doubt [Pruitt's charges]. I have had similar statements made to me." Brina Kessel, too, said she could believe that Elvey explained the facts of life to Pruitt, because Elvey was a savvy bureaucrat who knew how to get science funded at the national level. In a 1988 interview, even Wood thought it possible that Elvey had said the university could not afford to antagonize the AEC. And he would have been right about one thing, said Wood, "that we were very dependent on the contracts that were generated by the scientists."

But in the same interview, Wood denied that he had been involved in the modification of Pruitt's report and claimed to remember only hazily the controversy, and to remember not at all the lengthy meetings on the subject he had had in his office with Kessel and Pruitt:

> WOOD: I recall vaguely some controversy of that sort, but it never came into my office, that I know of.
>
> QUESTION: It never came into your office?
>
> WOOD: No. No. Not that I know of. The matter was handled with Brina. She may have come to the office to tell me about it, but I don't recall any detail of it.

* * *

On February 9, 1961, Don Foote wrote to his brother Joe:

> The Alaska Chariot scene is boiling and I expect trouble. Have reason to believe an FBI or CIA agent is on my track in Kotzebue-Point Hope. Going under name of [name deleted at publisher's request], 5'8", 175, balding, mustache, English accent, claims from Surrey, England, free-lance writer in New York.

The same day he wrote to Alan Cooke, his friend and former classmate from Dartmouth:

When in Kotzebue I met a [name deleted], about 5'8" tall, 170 lbs., balding, mustache and says he's an Englishman from Surrey. He claims to have just arrived as a temporary BIA teacher, having come from New York where he is a free-lance writer for such papers as the *Times,* writing on architectural history of famous New York buildings. . . . You have claimed [name deleted] is an FBI agent or perhaps CIA agent. If you have good or binding evidence of this, or if what I have repeated above is false, please wire me collect at the above address. I also need desperately some more information on him.

On February 13, Alan Cooke wired Foote: "He is agent. Letter follows."

In his follow-up letter, Cooke, who was assistant librarian in the Stefansson Collection at Dartmouth College, filled in the details:

I feel certain that your BIA teacher, [name deleted], is an agent for these reasons:

When he appeared in the Collection last winter, he told an interesting story of being "with a student organization" that "permitted me to travel widely." . . . His interest in the Stefansson Collection, he told me, was Ipiutak art. He had got interested in this, I gathered, in his visit to Point Hope the previous winter. . . . As I recall, I volunteered the information that Tom Stone [Foote's field assistant] was around and that I had correspondence from you. He expressed extreme interest in both facts—but I could find neither your correspondence nor Tom Stone. . . . I found Stone after he left and told him about this. He was interested in all I had to say about the funny little fellow and was looking forward to seeing him. But they never did connect on that visit—if I am to believe Stone. We often spoke about it for reasons I now give.

Tom Stone's roommate, whom he has known from grade school, Al somebody, worked in the Alumni Records Office. This is an extraordinary coincidence, and the only reason I know anything more of [name deleted]. Tom spoke of [name deleted] to Al who realized that this was the same person who came fairly regularly to the Records Office to check on the Hungarian students here. Al knew that on his last visit he had asked for not only the Hungarians' records but also for that of Stone. I scarcely think there can be any doubt of this—for Al knew his name and we know his name, it was the same time, etc. . . . Tom's roommate said he was an FBI agent and that this fact was well known in the records office.

Don Foote had noticed that his mail was being opened during this period. But his brother Joe told him: "As a government contractor you are always fair game for investigation. As a firebrand, more so." To "government contractor" and "firebrand," Joe Foote could have added the fact that the project involved nuclear bombs during the cold war—Don was a natural target for surveillance.

Replies to Freedom of Information Act (FOIA) requests filed with the FBI and the Defense Intelligence Agency indicate that those agencies claim not to maintain files on Don Foote today. The CIA, however, does. A year and a half after a FOIA was submitted, the CIA declined to release the (then) twenty-four-year-old document. An appeal was made within days, and two years and seven months after the original request (by law, the process should have taken twenty days), the CIA denied the appeal, citing the National Security Act of 1947. At the time of the denial, Don Foote had been dead for twenty-three years.

Foote's contract with the AEC expired May 31, 1961; it was not renewed. Viereck, Pruitt, and Foote, the three Chariot critics who had "gone public," were all relieved of their employment. But the pamphlet they had written with Ginny Wood and Celia Hunter, the Alaska Conservation Society *News Bulletin*, was creating a distinct ripple outside Alaska. And the ripple was widening. It was becoming, in fact, a wave.

14   A NATIONAL PROTEST

> More than likely, the moral and intel-
> lectual leadership of science will
> pass to biologists, and it is among
> them that we shall find the Ruther-
> fords, Bohrs and Francks of the next
> generation.
>
> —Sir Charles P. Snow, 1961

After World War II ended with the dramatic dropping of two atomic bombs on Japan, Barry Commoner, a plant physiologist then at Washington University in St. Louis, began to think deeply about what is generally called the social responsibility of the scientist. Obviously, the possibility of nuclear war was a preeminent concern, as was the danger inherent in testing nuclear weapons in the atmosphere. Just as obvious was the fact that reasoned public discourse on these political issues involved some understanding of scientific information. Unless citizens intended to abdicate their democratic prerogatives to government experts, some public education was essential. And Commoner believed that scientists had a responsibility to explain this technical information and thereby assist the public in making these science-related—but *political*—decisions.

None of the several associations of scientists that had sprung up after the war seemed to Commoner to fill the bill. Groups such as the Federation of American Scientists, the Pugwash movement, whose organ was the *Bulletin of the Atomic Scientists,* and the Society for Social Responsibility in Science certainly were all cognizant of the scientist's new role in public affairs. But they all seemed to project a view of the scientist as an expert either in the analysis of political matters or in the analysis of social ethics. And scientists had no more claim to be arbiters of the public good or mediators of moral philosophy than had any other human being.

The case of nuclear excavation illustrates how a technical proposal presents more than simply a scientific dilemma. Harbors and canals blasted by H-bombs could facilitate economic development to improve materially the lives of many people. But along with that potential, the projects inevitably carry some risk to human health, such as the risk of increased incidence of cancers and birth defects from fallout. Suppose, for example, that expert analysis concluded that Project Chariot would result in a quarter million dollars annually in increased economic activity in northwest Alaska. But suppose the analysis also projected that, with equal likelihood, a dozen Eskimos at Point Hope could be expected to have their lives shortened by a decade due to late-developing cancers. Experts—statisticians and epidemiologists, radiobiologists and ecologists, economists and politicians—might be able to assign a *probability* to these eventualities, but who could assign the *value* to be placed on economic advantage, or on human suffering? An Edward Teller might see the balance tipped one way, an Albert Schweitzer, the other.

Barry Commoner's notion was that in these matters it was not the scientist who was the authority, but—through the political process—the ordinary citizen. And the scientist's contribution, he said, ought to be more along the lines of a technical consultant's who could sift through the facts, outline scientific principles, discuss the limits of accuracy and alternate interpretations, then step aside and let the public determine its will.

Accordingly, in 1958 Commoner formed the Greater St. Louis Citizen's Committee for Nuclear Information (CNI), a pioneer group of scientists and laypeople whose mission was to provide information, not advocacy. "CNI does not stand for or against particular policies," the group's literature stated. "It presents the known facts for people to use in deciding where they stand on the moral and political questions of the nuclear age."

By 1960, Commoner chaired a Committee on Science in the Promotion of Human Welfare for the American Association for the Advancement of Science (AAAS). With his fellow committee members, one of whom was Margaret Mead, Commoner analyzed the role of science in the political affairs of the modern world:

> Having become a major instrument in political affairs, science is inseparably bound up with many troublesome questions of public policy. That science is valued more for these uses than for its fundamental purpose—the free inquiry into nature— leads to pressures which have begun to threaten the integrity of science itself.

Without mentioning the Atomic Energy Commission by name, the AAAS committee singled out ardent nuclear partisanship as a cause of the erosion of such scientific values as "objective, open communication of results; rigorous distinction between fact and hypothesis; candid recognition of assumptions and sources of error." And the result, said the committee, was that "the identity between science and an objective regard for the facts" had become clouded.

In connection both with his chairmanship of the AAAS committee and his leadership in the Committee for Nuclear Information, Commoner began to track Project Chariot. In the spring of 1960, his group wrote to John Wolfe asking to see any environmental data that might have been developed from the investigations at Cape Thompson. Wolfe felt that it would be improper for him to release any preliminary data, but that CNI could contact the investigators individually. [This CNI did, and as a result the group published in the summer of 1960 two issues of their newsletter, *Nuclear Information*, dealing with Project Chariot and Plowshare.]

The two issues "stimulated such an unprecedented volume of response from scientists and non-scientists here and abroad, and from government agencies," wrote the CNI editor to Alaskan investigators, "that we feel impelled to publish additional material." The CNI editor requested articles from the Alaskan scientists for an expanded issue on Chariot, noting many complaints that information about Chariot was not publicly available, and further noting that John Wolfe's letter appeared to say that the scientists were free to release their data.

Bill Pruitt considered that CNI's interest represented a tactical opportunity too good to pass up. "I have been very favorably impressed with their dispassionate approach to atomic energy," he wrote Don Foote, ". . . I suggest we use this outfit as a vehicle for introducing our attack, just the light artillery, especially the biological and scientific objections (the real heavy artillery will be the emotional, ethical and political push)." The vehicle for the "heavy artillery," Pruitt, Foote, and Viereck agreed, should be a mainstream national magazine.

Despite their interest, it was months before the busy Alaskan researchers—who were also working on their articles for the Alaska Conservation Society *News Bulletin*—were able to send their contributions to the proposed Chariot issue down to Commoner's group. Meanwhile, "the flow of inquiries and comments" on the first Chariot article continued to pour in, according to CNI. When the material from Alaska finally arrived in Saint Louis in February 1961, it sent Barry Commoner hurrying off to the library.

●   ●   ●

In the late 1950s and early 1960s, biologists were examining a puzzling phenomenon. Air currents in the stratosphere deposited fallout mostly in the Northern Hemiphere's Temperate Zone (where, incidentally, most of the world's population lived). Consequently, some measurements had shown that fallout levels in the North Temperate Zone were ten times higher than levels measured on the ground in Arctic regions. But for some unknown reason, the level of strontium 90 showing up in Alaska caribou was many times higher than that in domestic animals in the lower states. For example, the bones of grazing animals in the lower states averaged about twenty-five *micromicrocuries* of Sr90 per gram of calcium (or twenty-five strontium units), while the bone and antlers of caribou in the north contained 100–200 strontium units. Caribou meat was seven times more contaminated with strontium 90 than the meat of U.S. domestic animals. And the stomach contents of cattle raised on the fallout-rich Nevada Test Site was about 200 strontium units, while caribou stomachs were showing 1,264 strontium units. Furthermore, the caribou, and the Eskimos who ate the caribou, appeared to be higher in Sr90 content than any other group in the world.

Why were these two species, caribou and man, who lived in a region of low fallout, showing such high concentrations of radioactive contamination? Biologists believed that the differences in the efficiency of uptake must be attributable to differing biological processes in the two regions. And they focused on the unique biology of the caribou's major food, lichens.

Lichens are amazing testimonials to floristic cooperation. They are not like ordinary vascular plants, but are, rather, two distinct organisms aligned for mutual benefit. Algal cells intertwine with fungal filaments to produce a plant body that can take a scaly, leafy, or stalked and shrublike form. They can grow on rocks or soil or on tree trunks. Unlike ordinary plants, which draw water and nutrients out of the soil, sending this food up from their roots and transporting it to the plant's parts through vessels, lichens are rootless and derive their water and their mineral nutrition from the air. Rain, and the minerals dissolved in it—or even dust particles blown by the wind—land on the lichen body and are gradually dissolved and absorbed into it.

This arrangement was satisfactory for the lichen throughout all the millennia of its existence—until 1945. Then the rains began to bring down radioactive dust. Lichens are the ideal organisms to capture fallout. They cover wide areas and retain virtually 100 percent of the radioisotopes that land on them. Also, because they grow so slowly and are long-lived, they contain the accumulated burden of many years' fallout. If this contamination was bad for the lichens, it was worse for the caribou, and worse still for the Eskimos. When the caribou grazed on the lichens, their principal winter food, they gleaned the fallout from many acres and stored it in their tissues and bones. As Eskimos ate the caribou, they further concentrated into their bodies the radioactive strontium and cesium that were once dispersed over miles of tundra. So, because of the unusual biology of the lichens, because caribou have a predilection for lichens, and because Eskimo villages such as Point Hope might consume 100,000 pounds of caribou meat annually, the radioactive contamination was amplified at each successive level of the food chain.

As Barry Commoner sat in the library learning about lichen and caribou and Eskimos, an idea was driven home to him with great force and clarity. "It was my introduction to ecology," said the man who would become a leader of the environmental movement before the decade was out. "It was when I realized that the different ecosystem in Alaska deeply conditioned the outcome of this technological impact, that I realized that what we were doing in our work on radiation was really an aspect of what is now called environmentalism." This realization, and Commoner's ability to articulate the principle and to organize and lead concerned citizens, would help to broaden antinuclear activism into the environmental movement. But that would come later. For the moment, he had a magazine to get out.

•   •   •

Bill Pruitt had noticed the uniquely efficient lichen-caribou-man pathway for fallout more than a year earlier than Commoner. He'd seen references to a couple of scientific papers on lichens and, he says, whenever a caribou biologist hears about a paper on lichens his "eyes start to light up." Naturally, he sent away for reprints. One paper was by a Norwegian named Hvinden, and the other by a Canadian named Gorham. The Canadian research had shown that lichens sampled in that country were high in fallout radioactivity compared with other plants, while the Norwegian study showed that strontium 90 from fallout was utilized biologically by animals that grazed on lichens. Domestic reindeer in Norway showed twice the Sr90 concentrations in their bones as did sheep grazing in the same region.

Pruitt knew that the Cape Thompson region was not overrich in lichens, and even in places where lichens comprised more than 50 percent of the vegetative cover, the cover might have extended over only half the ground, the rest being bare soil. Still, when the winter-feeding caribou were not eating lichens, they were generally eating sedges, which, next to lichens, were showing the highest readings of radioactive contamination. And the Chariot fallout could certainly spread over wide areas of the North Slope of Alaska, where the lichens did grow abundantly. Reading these papers, Pruitt saw what he called the "ecologically obvious" and its implications for Project Chariot. He submitted a proposal to John Wolfe in February 1961 to study the "vegetation-caribou-radioactivity relationships," but funds "were specifically deleted," he claimed. In his progress report submitted in advance of the "First Summary Report," which was released in the summer of 1961, Pruitt noted that radiation might concentrate in lichens and caribou, but this observation did not survive John Wolfe's editing of the document. Finally, in a draft of his final report submitted in late 1961, Pruitt listed in a table the few published data noting the relatively high levels of radiation found in lichens and caribou. But Wolfe considered the data's inclusion "somewhat out of order." He said, "We are aware of these and other published analyses, and when they are useful in various kinds of comparisons, we fully intend to use them."

By then Wolfe was, as he said, aware of the lichen research. But it was Pruitt who seems to have been responsible for this enlightenment, and it apparently came as more of a surprise to Wolfe than his casual tone might suggest. Al Johnson, who was then in Norway on sabbatical, recalls the circumstances:

> Once Wolfe found out about that, he got very excited about it. And worried about it. In fact, I have a telegram in my file from him asking me for information about lichen growth rates and that sort of thing. . . . The Scandinavians first showed that impact on concentration in caribou, reindeer. And Pruitt was in touch with that. Wolfe and his group did not know about that at that time. So I really believe that neither John Wolfe nor anybody else would have been so foolish as to try to hide from that particular thing.

Strong evidence that this bioaccumulation hazard was not common knowledge at the AEC before Pruitt started bringing the international research to light can be found in a previously restricted AEC document that summarized the findings of the environmental studies:

> Special mention is perhaps warranted of the fallout/lichen /caribou/man food chain. This is an example of an environmental problem *identified in the course of the studies* which must be taken into account if the Chariot detonation were to be conducted. This food chain appears to be the mechanism

which has resulted in markedly high body burdens of fallout-borne Cs-137 that has been found in the Arctic peoples whose diet is heavily dependent on caribou or reindeer. [Emphasis added.]

Though Pruitt had not been able to sound the alarm about lichen-radiation interactions in the AEC's published report, he did cover the topic in the ACS *News Bulletin* and he alerted the Committee for Nuclear Information's Barry Commoner, who could address the issue as a plant physiologist. "There is almost no basis," Commoner wrote in the June 1961 issue of *Nuclear Information*, "for using experimental studies of Sr90 in pastures, cows and milk to predict what will happen in Alaska." The grassland ecology was entirely different. For one thing, blades of grass have waxy coatings on their upper side that resist the absorption of dust particles. For another, grass dies back to its roots after a short, seasonal life. Fallout, which sifts down to the ground and enters the soil in pasture-land, is "diluted" by the presence of minerals already there. Strontium 90, for example, is chemically similar to calcium. So in soil where calcium is abundant, Sr90 will have a necessarily lower statistical chance of ending up in a plant. Finally, there is considerable discrimination at the root wall against Sr90.

Commoner also questioned the AEC's willingness to extrapolate the Temperate Zone data to the Alaskan situation where man's diet was concerned. A great deal of scientific work had been done to show how fallout Sr90 moved through pasture soil, entered the fodder, then cattle, milk, and humans. Not only was the Alaskan pathway completely different, but the dietary habits and hence certain metabolic processes of the Eskimos were dissimilar as well. The Eskimos ate a very limited amount of starch but a great deal of fat and meat. Presumably, Commoner reasoned, the diet was low in calcium but very rich in vitamin D, which promotes calcium absorption. It was possible that a low level of calcium in the diet was offset by unusually efficient calcium absorption facilitated by high dietary levels of vitamin D. And, because strontium is chemically similar to calcium, this knowledge suggested to Commoner that "the relative efficiency of Sr90 absorption from the characteristic Eskimo diet may be very different from persons living on a diet typical of temperate zones."

Though the AEC frequently attempted to relate their "experience in Nevada" to the Arctic condition, Commoner bolstered the Alaskan biologists' argument that such extrapolations were scientifically improper.

•   •   •

Another article in CNI's Chariot issue questioned the reliability of the AEC's published estimates on the extent of Chariot fallout. Washington University physicist Michael Friedlander's analysis began with the observation that the AEC had conducted only four previous atomic detonations in Nevada where the depth of burial and the explosive yield resulted in a crater formed by ejected earth. And in none of these four was the depth of burial or explosive yield similar to those designed for Chariot.

Two of the shots, Jangle-U and Teapot-S, which both had only about a one-kiloton yield, were buried at seventeen and sixty-seven feet respectively. In both cases, the shallow burial resulted in the venting of nearly all of the radiation into the atmosphere (more than 80 percent in the case of Jangle-U, and 90 percent for Teapot-S). The third and fourth tests, Neptune and Blanca, were buried more deeply relative to their explosive yields (Neptune, about 1 kiloton buried at 99 feet; Blanca, 19 kilotons at 835 feet). Consequently, much less radiation was observed at the surface (estimated to be 1–2 percent for Neptune and less than 0.5 percent for Blanca). But the largest bomb in the proposed Chariot detonation (200 kilotons) was ten times larger than the largest previous cratering shot (19 kilotons). The other four Chariot bombs at 20 kilotons resembled the Blanca test, but they were not planned to be buried even half as deep as Blanca.

In other words, of the four previous shots from which the AEC might predict the venting of a Chariot detonation, two were shallowly buried and vented nearly all their radiation, and two were deeply buried and vented little radiation. Chariot was to be buried at an intermediate depth, so its vented radiation would fall somewhere between a lot and a little. But where? It all came down to what shape curve one thought might be drawn between the two data points at the upper left corner of the graph and the two data points at the lower right corner. The AEC drew a curve for Chariot that indicated that only 5 percent of the radiation would vent. But Friedlander pointed out that "there was no reliable way of knowing the steepness with which the curve descends from Teapot-S to Neptune—and this is the crux of the matter." He concluded that "the predicted value of 5% for the vented radioactivity of the proposed Chariot explosions must be regarded as so uncertain as to suggest, with equal probability any value ranging from 1% to about 25%."

●   ●   ●

Within days of the release of CNI's *Nuclear Information,* the Atomic Energy Commission released its "First Summary Report" on Project Chariot that, as detailed in Chapter Thirteen, avoided a discussion of the characteristics of the Chariot nuclear explosion; provided nothing to indicate the basis of its conclusion that the radiation effects would be "negligible, undetectable, or possibly nonexistent" beyond the throwout area; and failed even to mention the unique aspects of the Arctic food chain, in which lichen and caribou concentrated radiation. With both publications circulating in June of 1961, *Science* magazine offered its readers a review of the two conflicting reports. In an article entitled "Project Chariot: Two Groups of Scientists Issue 'Objective' But Conflicting Reports," *Science* writer Howard Margolis faulted the AEC document for the three omissions noted above. But having said that, he focused the bulk of his critical comments on the tone and "technical soundness" of the report from Barry Commoner's group. In particular, he questioned CNI's calculation of the quantity of Sr90 that might be vented from the Chariot explosion.

Some evidence suggests that Margolis relied heavily on the AEC's viewpoint in the preparation of his article. Margolis had been briefed on CNI's "errors" before he had even seen the report, according to Commoner. Several of Margolis's criticisms form the core of AEC comments prepared for use within that agency. John Wolfe conferred with Margolis about the extent to which Margolis's piece had quelled public concern roused by CNI, and reported (with obvious satisfaction) to the University of Alaska's President Wood that Margolis thought the controversy had "died down quickly" after the appearance of his review. (Wood thought the Margolis piece "very cleverly managed.")

But the issue did not die down, because the next month Commoner wrote a rebuttal letter to *Science* that dealt point by point with the Margolis/AEC criticisms and suggested that the author's technique did not seem to meet the professional standards of either science or journalism:

A few days after the CNI report had been made public, one of us received a long distance telephone call from Margolis. In this call he made several criticisms of the CNI report, and asked for comment on them. During this conversation Margolis acknowledged that he had not seen a copy of the CNI report. Accordingly, a copy of the report was sent to him immediately. After several days he called again. In this second conversation nearly all of the points which we have enumerated above (including an explanation of the so-called "technical error") were explained to Margolis at some length. We regret that they do not appear in his article. In particular, we believe that ordinary journalistic practice would recommend that the specific reply given to his query about the supposed technical error in the CNI report should appear in his article alongside his discussion of the AEC "complaint" about it.

●   ●   ●

Fallout from the publication of the Alaska Conservation Society's *News Bulletin*, CNI's *Nuclear Information*, and from the AEC's "First Summary Report" spread across the country. *Science News Letter* carried an article announcing the conclusions of the AEC's "First Summary" under the headline: AEC FINDS NO BIOLOGICAL REASON TO STOP 'CHARIOT.' The preferred reading of the article, which devoted one sentence to the CNI point of view, was that despite some rather sweeping claims of danger, the thirty-plus scientific surveys "have not revealed any biological reason" for stopping Project Chariot.

In conservationist circles, however, opposition to Project Chariot was building. The Sierra Club reprinted the entire issue of the Alaska Conservation Society *News Bulletin* (it had all been devoted to Chariot) in the May 1961 *Sierra Club Bulletin*. Far from their humble mimeographed origins, the ACS articles now ran alongside photographs of Alaska wildlife and comprised nearly the whole issue of the nationally distributed glossy. Introducing the article, the editor noted that the Sierra Club's board of directors "commends and supports the Governor of Alaska for his stand in opposition to Project Chariot . . . pending a more complete study of the total effects including damage to native people, wilderness and wildlife." The Sierra Club had not so much announced its opposition to Chariot as commended the governor for *his* opposition. And Gov. Bill Egan had not precisely come out against Chariot, either. He had merely stated that he intended that his government would be satisfied of the safety issues before any blast would take place. Writing to the Sierra Club, Egan asked, "I should appreciate your advising me on what that statement of my purported position is based." And when Alaska senator Bob Bartlett read a copy of the governor's tart reply, he commiserated with Egan, observing that the Sierra Club bulletin was a rehash of the earlier ACS newsletter and the CNI bulletin:

> I found myself gagging on making a third try and had to desist. The Alaska Conservation Society heroes and heroines may be 100 per cent right about Project Chariot for all I know (although I do not believe it), but I am pretty much inclined to the opinion that however dedicated they are (and they are plenty dedicated), they are essentially third-rate people with no more knowledge of all of this than you and I, and maybe not as much. Admittedly, my opinion is colored by the fact that these are the same people who so ardently support the Arctic Wildlife Range.

The Sierra Club's plucky junior colleague, the twenty-five-year-old Wilderness Society, had passed a blunt resolution the previous year urging the abandonment of Project Chariot, which it said "would unalterably destroy the wilderness area of the westerly end of the Brooks Range, our last great wilderness in the world not in a tropical region." The text was written by Lois Crisler, author of *Arctic Wild,* who distrusted Teller's judgment, particularly after reading that he thought Plowshare projects "would be a decisive victory in man's historic battle to shape the world to his needs." That point of view might have been acceptable a century ago, Crisler argued, but "the main fight now turns to saving enough nature to shape human beings."

The Wilderness Society's director, biologist, naturalist, and wildlife illustrator Dr. Olaus Murie had first tramped and run dog teams through the Alaska wilderness more than forty years earlier. For him, Chariot was "part of our national carelessness with what we do with our land." He objected to the popular view of wilderness places as suitable testing ground for potentially hazardous experiments:

> Why, then, are such areas chosen for this atomic plaything—the ocean, Nevada, and Alaska? Because they are considered waste places; places where it doesn't matter what damage is done to wildlife or the few people there.

Smaller journals with limited circulation such as the *Defenders of Wildlife News* and *National Wildland News* ran tough editorials opposing the Alaska project. The latter called for "thinking people everywhere to demand an end to nuclear detonations." The former published a critical account of Chariot, and the group's president wrote to President Kennedy, saying that the Defenders of Wildlife doubted that the AEC, "steeped in its own lore," could impartially evaluate the scientific findings:

> One of the purposes of the experiment is to define the effect of the blasts on the biota. At the same time we are assured that the Atomic Energy Commission will not proceed unless it has assurance that the experiment can be conducted without jeopardy to living things. This is double talk.

The National Parks Association, a private, nonprofit educational and scientific group founded to protect the national park system, said its 16,000 members thought that "the risks can in no way be justified." Writing to then AEC chairman Glenn Seaborg, the group's executive secretary said: "We can assure you that sentiment in this matter will in all probability be quite nearly unanimous and that Project Chariot will have the condemnation and opposition of most conservationists."

Despite these protests, for the critics of Project Chariot the support of conservation groups did not come as easily as they might have expected. But "Chariot was an atypical cause," as one environmental historian has written, because the soggy tundra of the Cape Thompson region was not the sort of magnificent landscape likely to draw crowds, and caribou and grizzly bears, while scarce elsewhere, were abundant in Alaska. The Massachusetts Audubon Society, for example, declined to pass a resolution condemning Chariot after board member Bradford Washburn, who had many years of experience in Alaska, shrugged off the threat of contamination. One director recalled, "there were so many thousands of square miles of tundra, [Washburn] said, it wouldn't make much difference to pollute a few." Chariot's threats hinged on ecological relationships that were not only unspectacular, but invisible.

Also, many conservationists in the early 1960s had hopes that nuclear power generation might forestall the damming of wild rivers and flooding of wilderness areas. Finally, conservationists were somewhat cowed before the technical complexities of nuclear science. One of its correspondents thought that "the Wilderness Society should be concerned to the extent of their knowledge of the subject." And the Sierra Club's secretary warned executive director, David Brower: "We must be careful not to get into genetic and other fields we are not expert in." But the Wilderness Society's Lois Crisler disputed this logic, calling it the "mystery-cloaking" by which scientists were sometimes able to avoid public scrutiny; "it is to be condemned," she declared.

Notwithstanding that the issue was an atypical one for conservationists, a broad assortment of large and small conservation groups declared their opposition to, or wariness of, Project Chariot. And sportsmen's groups—which sometimes allied themselves with conservationist causes—also expressed concern over the massive detonation in Alaska. In January of 1961, *Outdoor Life* printed a cautionary account of Chariot prompted by a letter from the Alaska Sportsmen's Council executive director, A. W. "Bud" Boddy, who was also the mayor of Juneau. Boddy had written that Alaska sportsmen were worried about Chariot's effect on the state's tremendously bountiful fish and game:

> Not only sportsmen, but every citizen of the U.S. should be seriously concerned with what effects this planned nuclear explosion might have on game and fish, two of Alaska's most valuable natural resources—to say nothing of their possible effects on our human resources.

Outdoor Life took a wait-and-see approach, but noted that the environmental studies had not been made available to the public as they should be, and that the stakes to sportsmen were high:

> A hundred miles or so to the east of the Chariot site, in the western end of the Brooks Range, there is some of the finest big game country left in North America—Dall sheep, moose, grizzly bears, and caribou are plentiful.

A national conference of peace workers held in Chicago discussed the Chariot scheme in the context of it being an AEC ploy to resume atomic testing during the moratorium. They considered Chariot as a possible focus of a national protest action by peace activists, as one of the organizers wrote to Les Viereck:

> It was agreed that should negotiations for a test ban break down in Geneva . . . we would mobilize all forces for a major action directed at preventing resumption of testing.

●   ●   ●

Meanwhile, Keith Lawton, the Episcopal priest at Point Hope, was doing all he could to hold the AEC's feet to the fire and protect his congregation from what seemed to him a dubious project. Needless to say, these efforts rankled the AEC. In a 1964 review of the Chariot "public information" effort, the AEC attributed much of its public relations trouble to Lawton. They accurately noted that although Point Hope and Kivalina were about the same distance from the site, Point Hope was "vigorous in opposition," while Kivalina "voiced no objection." The AEC said Point Hope's "unrelenting" opposition was "led by a local minister [Lawton] and an investigator for the Environmental Study Group [Don Foote]." The AEC concluded that the Eskimo people "had been furnished by outsiders with articles on the effects of the bombings at Hiroshima and Nagasaki, and from these some said they feared a similar fate."

Reviewing the disastrous March 14, 1960, meeting at Point Hope, the AEC public relations specialists criticized Lawton for hosting the program and for asking questions on behalf of his congregation. They leave no doubt as to whom they perceived as the Eskimos' chief ringleader:

> Questioning began within a critical atmosphere. The questioning was led by Keith Lawton, local Episcopal missionary, who has been reported agitating against the project. He tape-recorded the session and conducted the questioning as though it were a news conference. . . .
>
> Lawton led the discussion into danger of fallout; reports of sterilization of Japanese A-bomb victims; lingering and dangerous effects of radiation to humans, animals, seals, whales, and all forms of life. . . .
>
> The villagers also expressed fear of the blast effects on their homes and on the water and land in nearby areas. The emotional pitch exhibited indicates that opposition has been stirred up purposefully and that arrangements had been made for an almost staged performance of the opposition.

Of course, another interpretation of why opposition had coalesced at Point Hope might be that this village had made an effort to acquire information from sources other than the AEC. And if the Reverend Mr. Lawton had taken the lead in organizing the gathering and dissemination of information—which he had—he saw that involvement as part and parcel of his ministry. He responded hotly in a 1988 interview when read the above excerpts from the AEC report:

> If anybody at the Atomic Energy Commission wants to consider me—I was living in the village as priest in charge of a congregation that involved almost all the people of Point Hope—and if they want to consider me at that point an outsider, I violently object to that. I mean I *violently* object to that. I take exception to that. I was in the village as a resident in the village. I belonged to the village. I was where I was supposed to be and I was doing my job in the village. I was not an outsider.

Lawton said, "I'm glad they found us ready for them because we intended to be," but he says the idea that the opposition was staged was "ridiculous." He said that while he helped the people to understand technical issues related to nuclear explosions, the people had their own sources of information too:

> Well the Eskimo people weren't illiterate. They could read, and they could speak for themselves. . . .
>
> The Atomic Energy Commission was dealing with people who had been in the Pacific theaters of war in World War II. People who had traveled down to California, Hawaii. Many of the people who were in the National Guard were familiar with nuclear effects of the bombs dropped on Hiroshima and Nagasaki. They had been informed of this kind of stuff in their military experience. . . .
>
> It was spontaneous. I was very proud and pleased with some of the questions the Eskimo people asked. And it showed me they had plenty of sense about what was going on.

The priest's active interest—advocacy, even—on the subject of Project Chariot raised eyebrows within his own church, and among church administrators of various faiths. In the late 1950s and early 1960s, the question of Church involvement in social and political issues was a prominent one in the national consciousness. The Reverend Martin Luther King, Jr. led boycotts and mass nonviolent demonstrations against racial inequality, proclaiming: "Too often, the churches had a high blood pressure of creeds and an anemia of deeds."

As Lawton remembers it, his then superior, Bishop William Gordon, the famous "Flying Bishop" (who piloted his own plane to remote villages) and powerful leader of the Episcopal Church in Alaska, gave him absolute freedom to approach Project Chariot according to his conscience. But Gordon endorsed Chariot and employed the AEC's stock justifications for the blast in a letter to concerned church members:

> I must point out to you that similar explosions have been carried out in Nevada in proximity to people more cultured and civilized in the modern sense of the word than the people of the Arctic coast. This is not a new thing.

I also believe that we should explore the potentials for the good of atomic energy, that there are tremendous possibilities for mankind in the future from peacetime uses of atomic energy. . . . Certainly we must have some experiments in order to bring this about.

Gordon went on to say that he felt "very strongly" that "certain people" connected with the project "who knew little or nothing about atomic explosions" but "who have pretended to be experts" had harmed the project.

Gordon apparently squelched even the mention of concern over Chariot by fellow delegates at the Alaska Council of Churches' annual conference. Early in 1961, Rev. Richard Heacock, then a young Methodist minister in Juneau, drafted a resolution opposing Project Chariot, which he submitted to the Alaska Council of Churches' Committee on Christian Social Relations. But action on the measure was blocked, as Heacock wrote in a 1961 letter:

The truth is that my carefully written resolution didn't get out of the Christian Social Relations Committee. It seems that Bishop Gordon and Mr. Dutch Derr strongly vetoed it and no mention was made of the project in the committee report.

When I saw it was not included in the report I asked the chairman why and he told me. I went to Bishop Gordon to see what he knew. He would only say, "It's in good hands. We don't need to worry about it." When I asked what hands he said, "The AEC's hands."

Heacock felt the issue too important to allow it to be buried by so few men in committee. And though it was "pretty scary" because he was a young newcomer to Alaska, and because "everybody knew Bishop Gordon," he "screwed up enough courage" to address the full council meeting in open session. Heacock asked the privilege of the floor just before adjournment, stated his concern, and read the statement he had written. He moved that the statement be referred to the committee for study. "The motion carried," said Heacock, "but a subsequent motion by Bishop Gordon specified that the statement *not* be included in the minutes of the assembly." (Emphasis in original.)

Eventually, Bishop Gordon's confidence in the AEC diminished as he came to believe that the government scientists were "not much concerned about the people in the area." In an interview shortly before his death in 1994, Gordon credited Keith Lawton and Richard Heacock for broadening his view: "They helped me see [Chariot] in a larger light."

Stymied at the interdenominational level, Heacock nonetheless saw the passage of an anti-Chariot resolution within the Methodist Church in Alaska. As early as May 1960, his church's Woman's Division of Christian Service had drafted a resolution that said: "We continue to object to the AEC Project Chariot." and requested that the AEC "cease its plans." Statements of deep concern over American nuclear weapons development, testing, and Project Chariot in particular were also part of the Alaska Methodist Church's annual reports throughout the early 1960s.

Heacock and the Alaska Methodist women's society also spread the word about Chariot beyond Alaska by taking advantage of the Methodist Church's network of civic education programs, traditionally run by the women of the church. "It has been described as one of the largest and most significant adult education programs in the world," says Heacock, "because in almost every city and town in the United States and even in rural places you'll find United Methodist churches, and most of these have a United Methodist women's organization." By the spring of 1961, the national group, the Woman's Division of Christian Service, had taken specific action on the Alaska harbor "in response to expressed concerns of Women's Societies in Alaska related to the dangerous effects of Project Chariot." The group called on President Kennedy to delay the blast until its usefulness might be proved and the people's safety assured.

Heacock also had another idea. He asked a legislator from Kotzebue, Jacob Stalker, to have the mayor of Point Hope produce a tape recording, in English, of residents expressing their attitudes toward Chariot. When the tape arrived, Heacock thought it expressed the central problems with simple, powerful eloquence, as an excerpt illustrates:

DAVID FRANKSON (MAYOR): What do you think about this plan at Cape Thompson?

JOSEPH FRANKSON: Well, one thing, I don't want it to go off. I don't know much about the atomic bombs, but what I heard and what I know little bit about them, well, it's to me, just like taking a chance on our lives here at Point Hope. Well, on what I hear from people, and from people that know a little bit about atomic bombs, I know it can do a lot of serious damage to people's body. And I hate, I don't want to take a chance on my life. You know, anybody that's born anyplace always likes his home. I don't care where people are from. I know I like this place, Point Hope. So I like to keep living here.

DAVID FRANKSON: Without any disturbances like that?

JOSEPH FRANKSON: No. Well, we know they're taking a chance because I know they are doing a lot of studying. They must be taking a chance if they are doing a lot of studying here.

Heacock sent the tape off to the Woman's Division of Christian Service in New York and it was circulated around the country to women's groups within the Methodist Church. "And I'm sure it was related to opposition that grew to Project Chariot," says Heacock.

＊　　　＊　　　＊

Aside from group or institutional activism, individual opposition to Project Chariot was sprouting from the grass roots. In Norwich, Vermont, Alan Cooke, who had worked briefly with Don Foote on the Chariot studies, began "anti-AEC-ing." With the help of his wife and mother-in-law, who was a professional writer and editor, Cooke sent letters to some thirty members of Congress complaining that the AEC was preparing to go ahead with the explosion "in the face of reasoned and trustworthy warnings against it."

By far, the most energetic and dedicated grass-roots protest was the inspiration of a man named Jim Haddock, of Manchester, New Hampshire. "I imagine myself to be an average man," he wrote Les Viereck in early 1961, "Amherst graduate, employed by the local electric utility in scheduling production of power, load forecasts, etc. I have a better than average knowledge of mathematics and nuclear physics through special courses at

M.I.T. I have never been associated with any 'causes' before and would not be now if I had been able to find some one else to carry the ball."

Haddock had learned of Chariot through a purely chance encounter with Keith Lawton, when Lawton had exchanged pulpits with the rector of Grace Church in Manchester for the summer of 1960. Lawton spoke about Chariot before various groups and even appeared in interviews on radio and television. Dramatizing Lawton's message was the presence of Antonio Weber, a highly respected Inupiaq hunter from Point Hope, and his wife, Hilda, who traveled and hunted with him. The couple had come along with the Lawtons for the summer, and for Jim and Doris Haddock, they put a human face on the impending tragedy.* The Haddocks, together with another couple, Max and Elizabeth Foster of Ipswich, Massachusetts, and the Fosters' son Peter, comprised the ranks of what they facetiously called "the Campaign."

Like the Alaskan conservationists, the Campaign had a mimeograph machine and a typewriter. But they also had "connections," as they say in New England. Max Foster had been a partner in a Boston law firm and three of his classmates at Yale were in the U.S. Senate; Elizabeth Foster was a board member of the Massachusetts Audubon Society, while the Haddocks' son was engaged to the niece of a distinguished New England doyenne, Mrs. Joseph Lyndon Smith, "Aunt Corinna." One of the Campaign's more interesting attempts to thwart the onrush of Chariot involved presenting the case to Aunt Corinna who, as it was generally accepted, could "call up the President any time." Elizabeth Foster wrote an account of the Campaign, including this recollection of the tea given by Aunt Corinna for Doris Haddock, Jackie Lawton, and Hilda Weber of Point Hope:

> The party was awesome and not wholly satisfactory. Blistering weather, the unaccustomed grandeur of the surroundings, Aunt Corinna's gracious but daunting formality, were all too much for Hilda, who could not utter a word. Mrs. Law-

---

* In 1999, at the age of eighty-nine, Doris Haddock (better known as Granny D.) began a 3,200-mile walk across the country to demonstrate support for campaign finance reform.

> ton lost her poise when her small son upset his milk on the
> oriental rug, which left Doris [Haddock] to present the Char-
> iot problem without help from her chief witnesses [Hilda We-
> ber and Jackie Lawton].. . . The party returned home to
> Manchester without hope that anything had been accom-
> plished. Clearly Aunt Corinna did not intend to call up the
> President.

Not the president, perhaps, but Mrs. Smith also happened to serve on
the board of directors of the Association on American Indian Affairs, and
this connection would prove valuable both to the Campaign and to the
Inupiat people of Alaska. Meanwhile, Jim Haddock prepared and mimeographed a six-page paper outlining the case against Chariot. The project
could never be carried out, Haddock wrote, if it were proposed to be conducted on a cattle range in Texas, or on the fishing grounds of the Atlantic
Seaboard:

> It would be instantly rejected because of the immediate and
> convincing protest with which the people . . . would,
> through their representatives, through their newspapers, over
> radio and television networks, blanket the country. These
> 300 Eskimo men, women and children have no voice with
> which to make any protest.

Haddock also questioned the morality of exposing human beings to
risk when they "have been given no opportunity to express their acceptance or rejection of that risk."

To this memorandum, Haddock added Les Viereck's letter of resignation from the University of Alaska, the ACS statements opposing Chariot,
and the Wilderness Society resolution. He would make, he wrote Les
Viereck, "a serious and strenuous effort" to get the packet of information
"to every point of power possible." Routes of access were explored up
through the labyrinthine channels of eastern society. Secretary of the Interior Stewart Udall received the information "from a close personal friend
who is granddaughter of Groton's headmaster." Dean Acheson, Kennedy's
foreign policy adviser and former secretary of state, was a friend of Elizabeth Foster's brother and of Max's, and a full account was sent to him

"with a request to get it to Kennedy." Acheson replied that he had been advised "by one who knows" that he could say "categorically that nothing is imminent." Haddock interpreted this to mean that Acheson had spoken to Kennedy, that the president was not wedded to the experiment, and that it was likely that considerations dealing with test-ban treaty negotiations probably would control the future of Project Chariot. McGeorge Bundy, another of Kennedy's Harvard-trained advisers, had worked on nuclear disarmament policy and was also a distant relation of Max Foster's. Bundy, too, received the materials and passed them on to others on the White House staff whom he knew to share misgivings about Chariot. Ted Sherburne, a friend of Jim Haddock's and the new director of the American Association for the Advancement of Science, received a packet along with another one for him to give Glenn Seaborg, the new chairman of the Atomic Energy Commission. Max Foster also knew Henry Luce of *Time* and *Life*, who was encouraging but thought the time not yet ripe for a story. Jim Haddock wrote the Lawtons that he never dreamed Max Foster would use his connections, but that he was doing so "shamelessly." "He is unutterably tenacious once he is roused," Haddock wrote.

By the time the St. Louis Committee for Nuclear Information was getting to work on its Project Chariot issue, the Campaign had already made the AEC take notice. When Plowshare chief John Kelly met with CNI in the spring of 1961, "he told a tale concerning [the Campaign's] original mailing," according to CNI staffer Judith Miller. "It appears that 36 Congressmen, Governors and other official gentlemen who received that mailing all forwarded the copies to the AEC . . . where they were all piled on Dr. Kelly's desk for reply." By May, the AEC's then director of military applications, A. W. Betts, wrote to the commissioners that "the volume of incoming letters objecting to Project Chariot has increased greatly," and these, he said, came "directly or indirectly from a small group." Betts identified the small group: Don Foote and Keith Lawton led the list followed by the Haddocks and Fosters, Viereck and Pruitt, the Alaska Conservation Society and the Committee for Nuclear Responsibility. Betts said the Haddock-Foster pamphlet had been forwarded to the AEC by the White House, by about twenty congressmen, and by the U.S. ambassador to the United Nations. Betts advised AEC chairman Glenn Seaborg and the commissioners that the "attack on Chariot will continue and may even expand."

The Campaign was a small group, but a dedicated one. On July 3, 1961, its efforts began to look as if they might yield some success. That day, at eleven o'clock in the evening, the Haddocks got a telephone call from Aunt Corinna. Now, a year after she had hosted the "not wholly satisfactory" tea party, Aunt Corinna was calling to say that the Association on American Indian Affairs (AAIA) was going to send its executive director, La Verne Madigan, to Alaska. The group, at Mrs. Smith's urging, had appropriated money to "make available to Eskimos the legal and investigatory services long available to Indian tribes." She wanted to know if the Haddocks could introduce Miss Madigan to the Lawtons. A half hour later, Madigan called. She would be leaving at once for Alaska and would be accompanied by Dr. Henry S. Forbes, chairman of the association's Committee on Alaskan Policy. Their mission was to "assess the facts and protect [the Eskimo's] interests." When the Haddocks went to bed that night they had reason to hope that they had put out the light on Project Chariot.

15   DRUMBEATS ON THE TUNDRA

> These people are asking what right the Bureau of Land Management has to give the Atomic Energy Commission land they claim . . . that's a damned good question.
>
> —John Carver, Assistant Secretary
> of the Interior

> What happened in the villages in Alaska in the 1960s was a revolution.
>
> —Mary Clay Berry

In the spring of 1961, a new American president, John F. Kennedy, had his eye on several developing events. Alan Shepard would shortly become the first American in space, though Yuri Gagarin of the Soviet Union would nip him by twenty-three days for the honor of being the first human in space. And, in between those two historic space shots, a CIA-trained invasion force with some Americans participating would land on the south shore of Cuba at the Bay of Pigs and attempt to spark a general uprising of Cubans against the communist government of Fidel Castro. Unfortunately for the insurgents, Castro was more popular among Cubans than the previous dictator, American-backed Fulgencio Batista, and Castro's army secured the Bay of Pigs in three days. The political repercussions of the humiliating misadventure—it was not only a military defeat, but a violation of the charter of the Organization of American States—occupied Kennedy for a considerably longer period. He probably never saw a letter sent to him during this time by the women of the Point Hope Village Health Council.

If the Eskimo women seemed beyond the orbit of contemporary political affairs—living farther from Washington, D.C., than Washington was from the coast of Africa—their experience nonetheless strangely coincided with their young president's. They too were preoccupied with a struggle being waged in the shadow of nuclear weapons. And upon it, they also thought, depended the survival of a way of life. But in defense of their cause, the women did not raise the abstractions of geopolitics or cold war ideology. They marshaled only the simple realities of their daily lives:

> All the four seasons, each month, we get what we need for living. In December, January, February and even March, we get the polar bear, seals, tomcod, oogruk, walrus, fox, and caribou. In March we get crabs. In April, May and June, we hunt whales, ducks, seals, white beluga, and oogruk. In July we collect crow-bell [sic] eggs from Cape Thompson. . . . In the middle of September many of our village go up Kookpuk River to stay for the fishing and caribou hunting until the middle of November. In November we get seals again and we used the seal blubber for our fuel. . . .
>
> The ice we get for our drinking water during the winter is about twelve miles off from the village towards Cape Thompson. We melt snow also to drink and for washing. In spring, May and June we used ocean ice. . . .
>
> We are concerned about the health of our children and the mothers-to-be after the explosion.

The next declaration sent to Washington from Point Hope was drafted by Howard Rock, a fifty-year-old Inupiaq artist who had studied at the University of Washington, but who had more recently been sliding into alcoholic dissipation on Seattle's Skid Row. Rock had returned to his birthplace in the spring of 1961 to "make my peace with everyone and then go out on the ice to die," according to his biographer. But returning to Tikigaq, to the elemental and difficult life of his people, and to a land steeped in the memories of his childhood, had a salutary effect on the painter and he took up his brushes again. By the time La Verne Madigan of the Association on American Indian Affairs arrived in the village in July, she saw in Rock his potential as an articulate spokesman for not only the Point Hope people, but for all Alaska Natives.

AAIA, the New York–based charitable organization, was headed by Pulitzer Prize–winning author Oliver La Farge, whose novel, *Laughing Boy*, portrayed American Indian life in the Southwest. Dr. Henry S. Forbes, a wealthy Massachusetts physician and head of AAIA's Alaska committee, accompanied Madigan, the executive director, to Alaska. The group was at great pains to say that it had become involved in Chariot *after* it had received word of the project from Les Viereck, and that they were simply "responding to the bewildered requests" for help from the Eskimos, in particular, a "specific request for help" from David Frankson, the mayor of Point Hope. But Viereck had not contacted AAIA. Rather, Viereck's letter detailing his run-in with the AEC was passed on to the AAIA by Jim Haddock. And Frankson's letter seems less a "request" for assistance than an acceptance of assistance already offered by AAIA: "We deeply thank you for your wish to stand for us over this Project Chariot." No doubt Madigan was anticipating charges from white Alaskans (which would indeed come) that "outside" agitators were stirring up trouble among the Natives.

●   ●   ●

When the AAIA party arrived in Point Hope, David Frankson called a special meeting of the village council. Dan Lisbourne came, as did Antonio Weber, Patrick and Elijah Attungana, Laurie Kingik, Bernard Nash, Guy Oomituk, and Joseph Frankson, until the table was ringed with jovial, round-faced men in plaid work shirts. Howard Rock sat in, too. They all shook hands with the blond woman from New York, fished out their cigarettes, and sat back to listen as La Verne Madigan outlined a legal tactic that she thought might thwart even the Atomic Energy Commission. The Eskimos had legal rights to the land they used, she said.

The Organic Act of 1884, Alaska's first land law, provided that Alaska Natives "shall not be disturbed in the possession of any lands actually in their use or occupation or now claimed by them, but the terms under which persons may acquire title to such lands is reserved for future legislation by Congress." While the law did not permit Alaska's 33,000 Natives to acquire title—as it did the territory's 430 white residents—at least it protected the Natives' continued occupancy. And the Statehood Act of 1958 reaffirmed the promise to protect the Natives' lands by disclaiming any interest in lands where Eskimos, Indians, or Aleuts already held a right or title. But again, a definition of the Natives' "right" was left for future congressional action. And by 1961, nearly 100 years after Alaska's purchase, Congress had taken no action to secure title for Alaska Natives based on their possessory rights.

Nevertheless, Madigan said, the Eskimos had never signed a treaty relinquishing their claim to the land. They had never sold it or lost it in war. They still retained their rights, and the government was obliged to see to it that they "shall not be disturbed in the possession" of the land until those rights were more clearly defined. If the council would write a letter in protest to the secretary of the interior, Madigan said, AAIA would hire lawyers on behalf of the village. Then, if the legal argument prevailed, the Eskimos could throw the AEC off their land. The council agreed, and Howard Rock was enlisted to draft the letter because he was "the only one who could write letters pretty well," as he said later. Rock also mentioned that he thought it might be wise to involve other Eskimo villages in the struggle, and to produce and distribute some sort of newsletter or tape recordings to cement the coalition.

Believing that the fate of Tikigaq depended on his force of logic and the strength of his prose, Rock labored on the letter for a week. He wrote and rewrote, then supervised the typing and retyping, saw it signed by the members of the village council and sent to Secretary of the Interior Stewart Udall. "As any reasonable and loyal citizens of the United States," it read, "the people of Point Hope are not against experimentations of the peaceful uses of the nuclear explosions. . . ." But to protect their way of life, Rock continued, "we are forced to make declarations in the face of this impending event." As the Village Health Council had done, Rock named the long list of land and sea animals upon which the people's life depended. Then he raised the land issue:

> At this point, Mr. Secretary, we would like to pose a question that has been foremost in our minds. The clarification of this inquiry would be most enlightening to the people of Point Hope. Why was the Bureau of Land Management, which we understand is under the Department of the Interior, given any right to allow the Cape Thompson area of land to be utilized by the Atomic Energy Commission for the explosion of Project Chariot? . . . We, the people at Point Hope, have a claim to this land and we consider your office, the Department of Interior, to be the protector of our rights.

If necessary, Rock said, the Eskimos "will, and must, resort to legal channels to sue for our incalculable heritage and our age old aboriginal rights to hunt on our land. It is saddening to us that such declarations have to be made, but we feel deeply that our way of life in the village of Point Hope is being dangerously threatened."

Howard Rock's letter would have more effect than he possibly could have imagined.

●　　　●　　　●

Four days after Rock finished his letter, another meeting took place at Point Hope. The AEC had taken seriously David Frankson's request that one of the five AEC commissioners be sent to Point Hope to explain the project to the people. Late in the afternoon of July 29, 1961, Commissioner Leland Haworth, accompanied by other AEC brass (John Kelly, chief of the AEC's Peaceful Nuclear Explosives Branch, and James Reeves, assistant manager for testing at Albuquerque), and Rodney Southwick, the public relations man, met with the Point Hope Village Council and about 150 assembled villagers. For four hours the AEC men sought to assure the villagers that a decision to proceed with the blast had not yet been made and that, should the shot take place, interruption to the people's hunting would be minimal. But Keith Lawton, David Frankson, and Howard Rock persisted with what John Kelly called "quite a number of penetrating and well-thought-out questions." Kelly suspected that the people's fears had been allayed by the presentation. But Southwick recognized that, while the Point Hopers' attitude was improved since the disastrous March 14, 1960, meeting, "there was still an atmosphere of considerable doubt and some opposition expressed."

Haworth and his group also visited Kivalina, where the people "did not appear to have any strong feeling one way or the other about Chariot"; Fairbanks, where Dr. Wood professed his regret that some of those opposed to Chariot were associated with the university; Juneau, where they met with Governor Egan; and Anchorage, where a state senator asked for an explanation of a number of dead radioactive walrus that were found off the coast of Alaska in the early 1950s. (On this last matter, Kelly later obtained confirmation that "there had been such walrus found but that the Air Force Intelligence people had assumed jurisdiction of these walrus and had treated the matter as classified information. . . .")

Altogether, the visit by Commissioner Haworth achieved mixed results. As Southwick correctly observed, from the standpoint of public relations, "Point Hope is still a difficult problem."

●   ●   ●

There was trouble brewing, too, at Barrow, Alaska, the northernmost city in the Western Hemisphere, located some 500 miles north of Fairbanks. Historically, the Natives of Barrow had hunted ducks when they migrated north in the spring. After a long winter subsisting on whale meat, the Eskimos craved a change of diet, and the arriving ducks and geese—and the fresh meat they represented—lifted everyone's spirits. If the village had not taken a whale the year before, then the migratory birds represented more than dietary variety—they represented survival. What the Eskimos didn't know, however, was that a treaty signed with Mexico and Canada in 1916 restricted hunting to a season beginning about September first. The law made sense for the thousands of hunters at lower latitudes: The birds had the chance to rear young, unmolested, in the spring and summer, and the hunters could shoot a limited number during the fall migration. But at Barrow, Alaska, it would be a brave duck still swimming the tundra ponds into September, as the ice began to form.

The Migratory Bird Treaty law had never been enforced in Arctic Alaska until May 20, 1961. On that date, a federal game warden arrested a Barrow resident at Meade River, sixty miles south of the village, for shooting three geese out of season. Nine days later, the warden arrested another Inupiaq hunter for possession of an eider duck. The warden, who was staying at the Top of the World Hotel, awoke the next day to find outside his door a line of hunters, each with a duck in his hand. All 138 hunters demanded to be arrested. They signed statements that they had taken the ducks illegally—though some of the ducks were really last year's, which the people had dug out of freezers and distributed—and the warden stuffed 600 pounds of feathered evidence into sacks. Later, 300 Barrow residents signed a petition and sent it to President Kennedy asserting their right to subsistence hunting. A Presbyterian missionary, John Chambers, summarized the situation:

> As it is now the federal government won't let the natives take ducks and geese while the ducks and geese are in the Arctic, but after the birds go south, where they are killed by the thousands by white sportsmen, the government says to the hungry native, "Now you can hunt."

The whole episode left the patriotic Eskimos deeply troubled by the souring relations with the state and federal government. But bureaucratic agencies didn't seem to understand what the people needed for their survival, nor did the Eskimos have any voice in the political affairs that affected them. Seeking an advocate, Barrow's Guy Okakok wrote to La Verne Madigan at the Association on American Indian Affairs:

> We did not know that we can't shoot ducks when they come through our shore. Anyway, please come so I can let you know everything. It's a long story. . . . Our country here don't compare to their countries southside. I mean the climate. Oh, I wish I'd tell you everything.

After meeting with the people of Point Hope about their opposition to Project Chariot, La Verne Madigan flew on to Barrow to hear about the people's struggle over hunting rights. As a result of these two visits, AAIA agreed to provide the Eskimos with legal and investigatory services and underwriting for a conference on Native rights to be held that coming fall at Barrow. Madigan and Guy Okakok spent the latter part of the summer visiting several Eskimo villages to hold preconference planning sessions.

When the Native rights conference finally opened in Barrow on November 15, more than 200 Inupiat men and women crammed into the meeting hall: the men leaning forward, wide-eyed and alert, eyebrows arched as if in mild surprise; the women sober-faced, their fancy outfits dominated by fur-ruffed parkas and accessorized by bandanas and the harlequin glasses popular in the 1960s. Down in front, in their best parkas and mukluks, the delegates sat cross-legged on benches and smoked.

Madigan worried that they might fail to establish an orderly agenda. But her anxiety passed as chairman Guy Okakok simply followed Eskimo custom and gave each delegate a chance to speak, and common concerns rose steadily to the surface. "The Eskimo leaders" she wrote later, ". . . conceived the Conference, planned it, conducted it in their own way and largely in their own language." And they wrote a statement of policy, the opening lines of which are hauntingly reminiscent of the cries for aboriginal rights voiced by the American Indian a century earlier:

> We the Inupiat have come together for the first time ever in all the years of our history. We had to come together in meeting from our far villages from Lower Kuskokwim to Point Barrow. We had to come from so far together for this reason. We always thought our Inupiat Paitot was safe to be passed down to our future generations as our fathers passed down to us. Our Inupiat Paitot is our land around the whole Arctic world where we Inupiat live, our right to hunt our food any place and time of year as it has always been, our right to be great hunters and brave independent people, like our grandfathers, our right to the minerals that belong to us in the land we claim. Today our Inupiat Paitot is called by white men aboriginal rights.

With rising anger the Eskimos charged that they had been misled about their rights:

> We were told that if the government reserved our aboriginal land for us, we could not be citizens of the United States—could not vote—would be tied to the reservation like a dog—could not have businesses on our land or sell products of our

land. That was a lie told to us Inupiat to take away our aboriginal land and mineral rights.

Brainwashed was the term La Verne Madigan used. She believed that
"the misinformation was systematically spread, for the same lies have
been current in all Eskimo villages." But none of the Eskimos realized
what was happening, they said, "because we never before had a chance
to talk to each other."

The delegates sorted out their problems into two broad classes. One
concerned aboriginal land and hunting rights, and the other had to do
with economic and social development. In regard to the latter category,
the Eskimos defined their condition in a state-of-the-union sort of way,
and recommended steps the state and federal governments could take to
consider adequately the circumstances of the Inupiat people and their
needs. To attend high school, for example, northern Eskimo kids had to
travel to Sitka, more than 1,000 miles away. If a centrally located school
could be built at Kotzebue, the delegates said, "the students' morale
would improve knowing that they are not too far away from home." They
offered similar suggested reforms in the areas of housing, employment,
health, transportation, and so on.

When the delegates turned to the issues that had occasioned the conference, aboriginal rights, they made clear that they had "two problems of
our rights which are special. One is the Migratory Bird Treaty with
Canada and Mexico. One is Project Chariot, the proposed nuclear explosion at Point Hope." On Chariot, they put the Inupiat position into the
kind of bluff language that would become a hallmark of civil agitation
over the next decade: "We deny the right of the Bureau of Land Management to dispose of land claimed by a native village." They called on the
BLM to revoke the AEC's permit before Chariot went any further.

●   ●   ●

The only journalist covering the historic Barrow conference was the *Fairbanks Daily News-Miner*'s Tom Snapp. Though he was relatively new to
Alaska, Snapp had tried to give serious, hard-news coverage to Alaska Native issues. Previously, the *News-Miner*'s interest in Native affairs was
generally limited to running "personal items," anecdotes, and gossip sent

in from the villages. The paper printed these reports without the benefit of editing—the imperfect syntax and grammar effected a kind of pidgin English. Between the less than dignified presentation and the soft-news content, the newspaper did practically nothing to illuminate the substantive issues faced by roughly 20 percent of the state's population.

The *News-Miner* almost didn't allow Snapp to fly north to cover the conference. And it was only the likelihood that Undersecretary of the Interior John Carver might say something newsworthy that persuaded Snapp's editors to relent. It was a tough assignment, though, because the proceedings were conducted in both Inupiaq and Yupik, the two languages of the Alaskan Eskimo people, and Snapp spoke neither. Howard Rock interpreted what he could understand of the Inupiaq, though Snapp said Rock was pretty "rusty," and a man from Unalakleet helped him when delegates spoke in Yupik. But toward the end of the conference, Snapp kept hearing two words popping up in the testimony that didn't require any translation—his own name:

> I could hardly wait 'till there was a break 'till I could find out why they were mentioning my name. . . . And what it was is that somehow they didn't take that I was a reporter, and were appointing Howard and I to a committee to start a newspaper because their issues were not being covered.
>
> Well [laugh], . . . there's just very little advertising out there in the bush. And I said, "Oh, you mean a newsletter don't you?" Because I could see, you know, a newsletter. You could put that out with not too much trouble. But a newspaper? Oh, no, [they wanted] a newspaper! First they wanted a daily newspaper, which, you know, that was just out of the question. And then finally they said that they would be satisfied with a weekly newspaper. But they wanted us to come up with the funding.

The conference's final statement called for the establishment of a "bulletin or newsletter," but Snapp and Rock understood that the conferees wanted a substantial publication, and back in Fairbanks they set about looking for funding to start a regular biweekly newspaper. Printing costs were high in Alaska and, as Snapp noted, advertising revenues would likely be low. Then there was the sheer size of Alaska, a fifth the size of the contiguous states, which made travel almost unaffordable. The two would-be publishers drove out to the University of Alaska seeking advice in applying for a grant, but they were referred to thick volumes in the library where the application procedures required by large foundations seemed far too cumbersome and time-consuming. "We finally decided to cache it," says Snapp, "and just try to do it our own method. So our method was we called Miss Madigan from that Association on American Indian Affairs and asked if they could give us the names of their five richest members." Madigan didn't think the tactic proper, but promised to give the idea some thought and talk more with them in Fairbanks shortly. When she came to town in June of 1962, she took Rock and Snapp to eat at the Model Cafe and there handed them the names of two of the association's wealthy patrons.

One was Dr. Henry Stone Forbes, who chaired AAIA's Committee on Alaska and whom Howard had met in Point Hope the year before. Rock and Snapp knew that Forbes had attended Harvard Medical School with Alaska's U.S. senator, Ernest Gruening, and that he had a relative living on a homestead on the outskirts of Fairbanks. Forbes looked like their man, and the burden of selling the proposal to him fell to Tom Snapp:

> I sat down and typed all one day and all night and the next day—this huge letter that was just real thick. What it was was telling the reasons why a newspaper like that was needed. And I was an authority. It was mostly that their issues were not being covered. And who would know better because I was the one supposed to be covering it for the *News-Miner?*

It was a while before Forbes got around to reading the forty-seven-page dissertation, but as soon as he had finished the last page, he called back. He said he would guarantee $35,000 the first year and some additional funding in succeeding years. There was just one string attached, Snapp remembers. Howard Rock was to be the editor, and Snapp, the assistant. Snapp would teach Rock the newspaper business—one on one. Snapp had already made arrangements to return to graduate school and he was reluctant to postpone his plans. He believed in the newspaper, but he wanted to avoid the sort of relationship he'd seen elsewhere when white folks assisted Native people—a relationship of dependency. Also, he wanted to finish his graduate education and return to mainstream journalism. In the end, he agreed to stay for a year and a day.

Forbes's call came in mid-September and Snapp knew that they had to get an issue out soon to take advantage of the current election season and the paid advertising it could bring. If they hustled, they could publish three biweekly issues before the November 6 elections. That meant they had only two weeks to get the first issue out. The first step was to sell some advertising and, as they knocked on doors in the Fairbanks business community, Rock and Snapp found out that not everyone was enthusiastic about Native empowerment. Lael Morgan, Rock's biographer, writes that after "a couple burly whites threatened to give Rock a beating and Snapp was bitterly assailed by Natives who believed *News-Miner* editorials that the AAIA was trying to move them to reservations, the two men decided to solicit advertising as a team."

Despite the obstacles, 5,000 copies of the *Tundra Times* rolled off the presses on October 1, 1962, and onto the streets of towns and villages all over Alaska. The lead sentence of the top story of the first issue zeroed in on the struggle for settlement of the Natives' land rights. Covering the secretary of the interior's trip to Alaska in September, the *Tundra Times* reported that Stewart Udall had "cited the settlement of historic rights and claims as the most important problem facing Alaska natives today." Udall was quoted as saying bluntly that the problem of settling Native rights and claims "has been delayed too long and it is time to look it squarely in the face." But it wasn't time, apparently, for the *Fairbanks Daily News-Miner* to look the issue in the face. Though five reporters from the *News-Miner* attended the secretary's news conference, no mention whatever of Udall's strong opinions on Native rights was made in the paper's coverage. Instead, the *News-Miner* characterized Udall's remarks as "emphasizing the need for vast resource development in Alaska." The value of an alternative to the mainstream press could not have been more dramatically illustrated, as Snapp and Rock were pleased to point out to their East Coast backer.

The first issue of the *Tundra Times* also contained two articles related to Project Chariot, the controversy that had given rise to the newspaper. The first item was reported with cautious relief, the second with a great deal of sadness.

16   SPIKING THE WHEELS
         OF CHARIOT

> It is, therefore, important that all concerned with this program be careful that their actions or words are not such that they can be interpreted as indicating that the Commission has canceled Chariot.
>
> —John S. Kelly, Director,
> Division of Peaceful Nuclear
> Explosives, 1962

Sitting at his plywood worktable in his drafty cabin on the Point Hope spit in January of 1961, Don Foote sifted through Chariot newspaper clippings from the previous year. His brother Joe had sent clips from the East Coast, and Les Viereck had forwarded others from Alaska papers. "There is a distinct change in the tenor of the press during the past few months," he wrote Viereck, referring to Lower Forty-eight coverage, which was beginning to examine Chariot skeptically. "This is significant. . . . *What is wrong with the Alaska press?*"

Joe Foote, a Harvard Law School graduate who was himself a newspaperman, thought that "privately, most people in the press and radio field in Alaska seem to have serious doubts about Chariot, but they realize that it is still a sin to oppose a large federal project in the state." Even so, by 1961, some elements of the Alaska press finally began at least to cover the opposition to Project Chariot. In February, the *Fairbanks Daily News-Miner* quoted extensively from an Alaska Conservation Society news release calling for AEC assurances of safety. The next month a writer at the *Anchorage Daily Times* reported that Project Chariot was "running into heavy weather in Alaska," and quoted the March ACS *News Bulletin's* conclusion that "sufficient grounds exist to question seriously the advisability of the proposed nuclear explosion." While the *Times* was willing to report opposition sentiments, the headline writer labeled these views "roadblocks" to the project.

And the *Times* editorial position was made clear (if it was ever in doubt) later in the summer in a piece entitled ALASKA TEST IS NEEDED FOR PROGRESS OF MAN. In it, publisher/editor Robert Atwood continued to insist, despite the AEC's public recanting, that the excavation would produce a "harbor" that "would create an economic value" from resource development. Chariot would also stimulate better living conditions "in an area now on the fringe of civilization." Ignoring the fact that Chariot fallout would be transported great distances by wind, the *Times* assured its readers that "the only danger foreseen is that of radiation, and scientists expect it to be confined within a radius of about one mile." It was Atwood who in an earlier editorial likened AEC scientists to the "family doctor" who knew better than we what was best for us.

But the reportage that really excited Don Foote and the others campaigning to stop Chariot came from the paper of record, *The New York Times*. On June 4, 1961, just after the release of Barry Commoner's *Nuclear Information* issue on Chariot, the paper carried an article headlined CARIBOU'S FONDNESS FOR LICHENS MAY BAR ALASKA ATOMIC BLASTS. In what Jim Haddock recognized as "the first dash of cold water the project has ever received in the columns of *The New York Times*," the article gave widespread exposure to the idea that "in the process of blasting out the harbor, the commission might also contaminate the food chain in the Arctic region so that radioactive strontium 90 would pass from plants into animals and thence into the bones of Eskimos." By late summer, a major article in the *Christian Science Monitor* predicted "a strong public debate over the merits and dangers of the Chariot program."

Even the *Fairbanks Daily News-Miner* published a series of articles on Chariot, which, as Joe Foote noted, "showed a desire to probe the facts of the project which was wholly lacking in previous press treatment." The four-part series, which ran in August of 1961, was the work of Tom Snapp, shortly before he resigned from the *News-Miner* to help found the *Tundra Times.* Snapp reported what had been ignored during the years of adulatory coverage. He exposed the Eskimo opposition to the blast, and in a complete departure from standard practice, he described that opposition from the viewpoint of the Eskimo, rather than from the perspective of the development interests. Snapp also covered the concerns of the Associa-tion on American Indian Affairs and the scientists such as Les Viereck who objected to the AEC's public statements.

By the end of 1961, the *Christian Science Monitor* framed anti-Chariot activism as a David-and-Goliath story headlined ALASKAN ESKIMOS BUCK AEC. And the same month the widely read and respected *Bulletin of the Atomic Scientists* ran a piece outlining the case against Project Chariot, drawing on the publications of Don Foote, Viereck, and the AAIA.

"Here's the big chance, Don," Bill Pruitt wrote Don Foote in the fall of 1961, explaining that conservation writer Paul Brooks had been commis-sioned by *Harper's* magazine to write a piece on Project Chariot. "Paul will do a good job if he has the pertinent data." Pruitt said, ". . . I suggest you clue Paul in on all phases of Chariot that you think should be aired." Brooks was editor-in-chief at Houghton Mifflin Company and well enough connected with the Sierra Club that its executive director, David Brower, authorized funding of the writer's reconnaissance trip to Alaska. Don Foote immediately wrote his brother Joe to pass on the news of Brooks's commission, and within a month Joe had joined forces with Brooks on the *Harper's* article. Joe Foote sifted through all the documentation on Chariot that his brother Don had gathered, and prepared lengthy and de-tailed background materials for Brooks. In April, 1962, "The Disturbing Story of Project Chariot" appeared in *Harper's* as "the answer to prayer," as one Chariot opponent put it.

In their article, Brooks and Foote documented the policy shifts and inconsistencies that had dotted Chariot's history: "Is the plan to create a harbor, as originally announced, or isn't it? Is it an economically self-supporting venture for the people of Alaska, as originally announced, or is it simply an experiment in nuclear excavation?" And can the experiment, as the AEC claimed, "be conducted without jeopardy to the local inhabi-tants" while at the same time testing the effect of a nuclear explosion on living things? Wading through a "morass of scientific technicalities," the authors determined that a definitive assessment of risk was not possible. The scheme had always been experimental in nature, they said, and "the essence of experiment is uncertainty." In conclusion, Brooks and Foote drew in the wider implications of Project Chariot—the effect on the American democracy of an agency unaccountable to the people:

> Few of us are in a position to judge the ultimate scientific value of an experiment like the Chariot explosion. But it *is* up to us to know what is going on in that far corner of the United States. And to realize that another scale of values is also involved: not the precise relations between depth of burst and crater characteristics, but the precise relations between unlimited power and the awesome responsibility that goes with its use.

From Jim and Doris Haddock and Max and Elizabeth Foster running "the Campaign" to stop Chariot on the East Coast, to Barry Commoner and his Committee for Nuclear Responsibility in the Midwest, to Celia Hunter and the Alaska Conservation Society in Fairbanks, to the Eskimo people at Point Hope, anti-Chariot activists began to hope that the incisive article would mean a happy ending after all, that public outrage would reduce Edward Teller's bizarre vision of "geographical engineering" to a footnote in American nuclear history.

The *Harper's* article certainly created an immediate stir within AEC offices. Ernie Campbell of the San Francisco Operations Office wrote a thirteen-page rebuttal and sent it off to AEC headquarters in Washington. Campbell picked at minor details in the article, but finessed the more material issues. He objected to the article's implication that AEC San Francisco worked for the Lawrence Radiation Laboratory: "We still prefer to think that LRL is a SAN contractor, rather than the converse as implied in the article." "Prefer to think" was indeed the appropriate expression. Everyone within the atomic energy establishment knew that Livermore operated with virtual autonomy from its nominal contractor across the bay. As Gerald Johnson, Livermore's associate director, has made clear, "San Francisco Operations was a contract body. They managed our contract. They didn't tell us what to do. They had no authority over technical program at all, and if they tried to exercise it, we'd tell them to get lost. Which they did. They got lost."

Campbell complained that the authors had unfairly implied that Cape Thompson had been selected as the venue for Chariot at a time when Teller was soliciting siting suggestions from Alaskans. "The decision to locate Project Chariot at Cape Thompson was reached only after on-site reconnaissance surveys in the summer of 1958," wrote Campbell. In fact, LRL had asked the AEC to initiate a classified application to the Department of the Interior to withdraw 1,600 square miles of land around Cape Thompson two months before the "reconnaissance surveys" Campbell cites.

And when Brooks and Foote pointed out the AEC's on-again, off-again position that the harbor must be economically self-supporting, Campbell wrote that he could find no reference to such statements in AEC files. Had he done a bit of newspaper research, however, Campbell would have found that nearly every Alaska newspaper article concerning Chariot printed in the summer of 1958 quoted Livermore representatives as insisting that the project must meet the test of economic viability.

●   ●   ●

As AEC officials in Washington evaluated the impact of the *Harper's* article, they were dealt yet another blow from within the ranks of the federal government. Secretary of the Interior Stewart Udall and his inner circle of advisers "had little sympathy for Chariot," as the AEC was to discover. Udall had been lobbied from several angles—from the "granddaughter of Groton's head master," to Oliver La Farge. The Point Hope people had written La Farge, president of the Association on American Indian Affairs, and a personal friend of Udall's, and La Farge had in turn written the secretary. He described the "earnest and detailed protests" he had received from the Eskimos and urged Udall to looked into the matter to "make sure that everything is done that ought to be done for the protection of these natives, their hunting grounds and fishing waters, and their food supply."

Shortly after these contacts, Udall's confidential assistant, Sharon Francis, wrote Les Viereck, whose letter of resignation from the University of Alaska she had read in the Campaign's packet. Francis was moved by

Viereck's "inspiring and thought-provoking" statement, and disturbed by what seemed to her a gratuitous disruption to the Eskimos' lives. She became an anti-Chariot partisan within the department's highest ranks and a conduit of information from the Alaskan dissenters (Viereck, Foote, and Pruitt) to the secretary. "I have the complete moral support within this Department," she wrote Viereck, "of the science advisor to the Secretary, Roger Revelle . . . ; Assistant Secretary for Public Lands, John Carver; and Solicitor Frank Berry." Currently, she said, her work "involved exploration into every bureau within this Department in order to find any possible flaw or question through which we could deal with" Chariot.

If Udall needed an additional nudge, it came in the form of Howard Rock's letter for the Point Hope Village Council citing the Eskimos' proprietary rights to the land around Cape Thompson and appealing for the protection of their "incalculable heritage." The "heartwarming" letter "touched the Secretary profoundly," said Francis. And, she wrote Viereck, the land-claim issue could well be a deciding point:

> Right now the Solicitor is looking into the validity of these claims. If they hold water, the Secretary will be unable to grant the AEC a land withdrawal for the Chariot explosion. . . .
>
> One wonders why the Bureau of Indian Affairs has never looked into this before; one wonders why BLM [Bureau of Land Management] checked mineral rights but not Indian rights on these lands; one wonders how a bureaucracy can be so uncoordinated within its own ranks. . . .
>
> The Secretary will be writing the Point Hope Council shortly, and we hope to be able to assure them that we can defend their rights. . . . If at this point we are not successful, we will have to continue trying elsewhere.

One other point Sharon Francis mentioned was that in 1958, after the AEC requested withdrawal of lands around Ogotoruk Creek, the Bureau of Land Management published notice of the proposed land withdrawal in the Federal Register. The notice referred to the tract as "containing approximately 40 square miles." In fact, the area was about 1,600 square miles, nearly the size of Delaware. Eventually, a correction was published, but the Department of the Interior decided to limit the AEC to a land use permit in order to conduct the studies. If and when the blast was approved, the matter of a "withdrawal" would be considered by Interior.

On January 30, 1962, Secretary Udall wrote to Glenn Seaborg, chairman of the AEC. Udall said that his department, which had expertise and responsibilities "in the fields of Indians and wildlife," wished to participate in the review of the Cape Thompson environmental studies once they were completed. Furthermore, he said, Interior wanted all future dealings between the AEC and the Native people of the Cape Thompson region to include Department of the Interior participation. "In our analysis of the environmental report," Sharon Francis wrote Don Foote, "you can be sure that human and cultural aspects will be scrutinized with particular care."

The upshot of Udall's letter was a meeting on the first of May between two AEC officials and about a dozen of the most senior officers of the Department of the Interior. The AEC team was led by John Kelly, director of the improbably named Division of Peaceful Nuclear Explosions, a newly established division created to separate administratively the Plowshare program from the not very peaceful-sounding Division of Military Applications. "It was apparent," said Kelly in reference to the meeting, "that the Interior representatives had little sympathy for Chariot." Roger Revelle, Udall's science adviser, pointed out that "the AEC could neither successfully nor honestly fulfill the multiple roles of 'investigator, judge and jury' in safety matters," according to Kelly. And Assistant Secretary John Carver and BLM director Karl Landstrom voiced "serious doubts as to whether the land could be taken from the Eskimos and made available for this experiment." It would take an executive directive from the president or congressional approval to withdraw the land, Interior felt.

Of course, these were not insurmountable obstacles for the powerful AEC, provided it could cultivate in Kennedy—as Edward Teller and others had cultivated in Eisenhower—the political will to keep nuclear testing a paramount national priority. The situation had changed, though, because the influence of "hawk" scientists under the leadership of Teller gave way when Eisenhower established the President's Science Advisory Committee following the launch of Sputnik. As an official AEC history notes, the new formal apparatus finally "offered direct presidential access to scientists fundamentally antithetical to [the AEC's] Teller, Lawrence and Strauss." Consequently, the shape of internal debate on nuclear issues changed, and Kennedy's election further ensured the ascendancy of scientists who, like the new president, worried about atmospheric testing and were open to disarmament initiatives.

Less than two weeks after the meeting between the AEC and the Department of the Interior, a story surfaced in *The New York Times* that took Chariot critics completely by surprise. Under the headline A-BLAST TO DIG HARBOR MAY BE DEFERRED, *Times* writer Lawrence Davies declared that "Project Chariot may well be dead, killed by adverse publicity about its effects on Alaskan Eskimos and their hunting grounds." Davies, whom John Wolfe had used to report his "no biological objections" conclusion nearly two years earlier, wrote that it was the "educated guess" of unnamed scientists who were reviewing the environmental studies that the AEC might "hold the project in abeyance." And even though "scientists conversant with the situation" thought radiation was "no hazard," the AEC still "worried over the publicity about Eskimo opposition," Davies said. "The biggest problem" from an environmental standpoint, he said, again quoting anonymous sources, was the fate of birds, which, of course, was "of great concern to birdwatchers." The reporter also published—without supplying evidence or corroboration—the accusations of unidentified sources who charged conservationists "with keeping the Eskimos stirred up and 'giving direction' to the opposition to Project Chariot." Had Davies bothered to interview any Inupiat people, he might have concluded that they did not so neatly fit the racial stereotype of easily manipulated innocents, and that in any event they had strong leaders of their own in Dan Lisbourne and David Frankson.

But biased news reportage aside, Davies's piece was astonishing because it had obviously come from AEC sources and it was predicting Chariot's demise. As Jim Haddock said when he saw the article, "no self-respecting public relations man [would have] let this story out if life remained in Chariot—he must be preparing public opinion and explaining away three million bucks." A recent look at formerly classified documents shows that Haddock was exactly right.

•   •   •

"The pioneers have rocky roads," the AEC's Ernie Campbell liked to say. And by the summer of 1961, the road Chariot traveled was becoming increasingly bebouldered. As the publicity battle began to shift in favor of the Eskimo people's position, another factor was beginning to create problems for the AEC and Livermore: the Project Chariot environmental studies. Besides the revelation of the lichen-caribou-man pathway for radiation uptake (which had boosted the Eskimos' widely publicized opposition), the AEC was finally learning something about the winds at Ogotoruk Creek. And though the commission wasn't talking to the press about it, the data were not encouraging.

"I have become deeply concerned," wrote James Reeves, assistant manager for test operations at the AEC's Albuquerque Operations Office, "that during the months of March and April it is highly probable that there would not be an acceptable shooting day with the fallout directed toward the land." A landward disposition of fallout was important because, as the AEC had publicly stated, a primary reason for conducting Chariot was to determine how much radiation would be released by a shot at cratering depth, and to trace how that radiation would move through the environment. To do that, the scientists had to be able to collect and analyze samples of the fallout. Obviously, directing the fallout over land meant that the logistics of the experiment would be simpler. They could drive overland in track vehicles, sampling soil, vegetation, streams, and ponds, and monitor the movement of the radiation up through biotic systems over time. But if the fallout was blown to sea, the collection could only be done for a limited time, and then on a shifting ice pack.

March and April winds were the focus of Reeves's concern for two reasons. First, the AEC required a minimum of six hours of daylight for detonation and various postshot operations. Second, they needed the presence of sea ice so that, if fallout did drift over the sea, it wouldn't be diluted in the water before samples could be collected. The only time of the year that offered both enough daylight and sea ice was March and April.

Publicly the AEC had repeatedly claimed that fallout over sea *or* land was acceptable, and John Wolfe had admitted that sending the radiation toward the sea was the safest course from a biological standpoint. Privately, however, the AEC had no intention of abandoning the experimental possibilities afforded by sending the fallout over the tundra. In fact, though the AEC had consistently sought to reassure the public that the level of vented radiation would be kept as low as possible, internal memoranda from Livermore reveal that, up to a point, Chariot planners favored more, rather than less, radiation released. In a 1961 memo startling for its candor, five of Gerald Johnson's advisers investigated the possibility of reducing the yield of the Chariot shot from 280 kilotons to 200 kilotons. Discussing the "advantages and disadvantages" of such a change, the group listed the smaller shot's reduction in radioactive fallout among the "disadvantages" because that would mean "less radioactivity out so post-shot work will be based on smaller doses." Considering political ramifications, though, the writers felt that a lower dose delivered to northwest Alaska would confer political "advantages" due to "much less activity release (although you can't say this publicly)."

As one LRL scientist put it, Livermore wanted "a meaningful radioactivity experiment." And measuring the movement and effects of released radiation was viewed as of "signal importance . . . not only to Chariot, but to the future of nuclear excavation." At a meeting held in 1958, scientists from the AEC's Division of Biology and Medicine had pressed for studies to establish biological baselines in connection with Chariot because "the opportunities to study the ecology in relation to whatever degree of contamination results should not be lost." Of course, humans were part of the ecology of the Cape Thompson region, though studies of the effects of radiation from Chariot on people were not specifically mentioned in the minutes of the meeting. However, at the same meeting, AEC scientists reviewing "the human environmental problem" of nuclear explosions, generally, did advocate experimenting on human beings, specifically studies in which "very large numbers of individuals" would be exposed to low levels of radiation in order to "ascertain the effects." The scientists said they could not predict whether the experiment would yield "positive or negative findings," and they were particularly concerned that such a study be conducted with "the desired degree of sensitivity."

So it was no wonder that Chariot planners had always assumed a landward disposition of fallout. "All our efforts with regard to analysis of wind conditions have been on this basis," wrote Reeves. But as the meteorological data came in, they showed that the winds at the Chariot site almost always blew to sea. In his letter to AEC San Francisco, Reeves wanted to know if "there is any possibility that serious consideration is being given to directing the fallout pattern toward the sea." He was worried about "the damage that would incur to the Plowshare Program if we were to announce in advance, which we would have to do, that we were going to detonate something during March or April and subsequently find no suitable weather." If no favorable winds materialized by the time the ice went out, the shot would be delayed a year while five nuclear bombs sat in the ground at the mouth of Ogotoruk Creek. It was beginning to look as if the AEC either had to abandon one of the principal stated objectives for the shot—the radiation experiment—or risk some serious PR tarnishing.

Another issue that emerged in the course of the environmental study involved the geological suitability of the chosen site. The frozen siltstone and mudstone tended to turn into a "saturated muck" with very low shear strength when allowed to melt. The saturated areas and the adjacent permafrost soil would certainly melt as the relatively warmer seawater transferred heat to the crater walls. "As the melting progressed," wrote LRL's John Foster in early 1962, "the resulting change in the shear strength of the debris would undoubtedly result in serious deterioration of the crater slopes." This would mean that the harbor might tend to fill itself in as the shoreline slumped, so construction of dock facilities on unstable ground would be problematic, to say the least. AEC public statements sometimes insisted that experience gained in permafrost soils would be transferable to more temperate climates—such as Panama—where the AEC hoped to do nuclear excavation. The internal view, however, was that "it would be difficult if not impossible to relate this experience to any other medium or location." The immediate effect of water action on the crater walls could be observed at Chariot, but any long-term observations as to slope stability would be so complicated by permafrost melting that such information would be "of little value."

·   ·   ·

By showing that the Ogotoruk winds nearly always blew to sea, the environmental study was doing its job all too well. Instead of smoothing the way for Chariot, the data were revealing obstacles. Another factor serving to rein in Chariot was the development of technical information from other nuclear tests. In August 1961, the Soviet Union broke the moratorium on nuclear testing that had been in effect since 1958, and the United States resumed detonations the following month. As the controversy over Chariot mushroomed and the project languished "in a standby status," these nuclear tests, and others that were already designed and scheduled, began to undermine the need for the Chariot experiment. For example, the Danny Boy test at the Nevada Test Site in March 1962 apparently confirmed LRL's models on radiation release as a function of depth. And by that date, an "orderly series of experiments to obtain cratering data" was being developed jointly by the AEC and the U.S. Army Corps of Engineers. The first shot of that program, the 100-kiloton Sedan shot at the Nevada Test Site set for July 1962, was designed out of "frustration" as an "alternate to Chariot" to answer many of the same questions the Alaskan shot had been designed to address, says Livermore's Gerald Johnson. In particular, Sedan would refine scaling laws in the 100- to 200-kiloton range. Subsequent shots would aim "to specifically investigate the many problems associated with excavation, such as row charges, megaton cratering laws, effects of media," and so on, according to AEC documents.

In effect, while Chariot was conceived as an integral experiment designed to answer several technical questions at once, Livermore now realized the efficacy of designing several discrete experiments, each to test separate phenomena. As an added incentive, this "orderly series of experiments" could take place in Nevada, away from the critical eyes of dissenting scientists. Finally, even though vented radioactive debris might contaminate downwind communities off the test site, at the Nevada Test Site the detonations could be conducted without the bothersome necessity of an environmental assessment.

In Washington, the AEC commissioners were also considering the international picture. The minutes of their meeting of February 14, 1962, mention "intervening international political events which could effect [sic] national policy in this area." The commissioners were probably referring to the implications of so-called peaceful nuclear explosions (PNEs) in the protracted negotiations over a nuclear test-ban treaty. The United States and the U.S.S.R. had been wrangling over terms of an agreement at talks in Geneva since 1958, and one significant sticking point was the American PNE program, Plowshare. Edward Teller and the AEC insisted that Plowshare projects should be exempt from treaty provisions that otherwise banned such detonations. But the Soviets knew that weapons-related advantages would be gained from such testing and insisted that there be no PNE exemption. The AEC commissioners who expressed doubt over Chariot's fate in February 1962 knew that the previous fall, following the breaking of the moratorium, President Kennedy had offered the Russians a treaty banning all tests in the atmosphere, including PNEs. There was no doubt that the Kennedy administration viewed progress on arms control issues as too important to be derailed by an unproven nuclear excavation technology.

Though the AEC never said so publicly, internal documents from Livermore indicate that presidential authorization to execute Project Chariot had been sought in 1960–61. By February 1962, approval had not been forthcoming, and AEC commissioners questioned the wisdom of appropriating additional funds for Chariot "in view of the uncertainty of its future." AEC commissioner John S. Graham emphasized "the need to establish a firm date that beyond which, in the absence of a Presidential authorization, the project will be terminated." The commission agreed, and a letter was drafted for President Kennedy reflecting the AEC's position that, unless the AEC received "a Presidential authorization for Project Chariot prior to September, 1962, the experiment will be closed out." In the meantime, the AEC recommended that it "continue to hold Chariot in a stand-by status for possible execution" at a later time.

When the AEC solicited Livermore's position with regard to Chariot's future the next month, LRL director John Foster (Teller had resigned as director in mid-1960 so that he could publicly promote the merits of developing antimissile weapons) replied: "We recommend this project be canceled." Foster outlined the technical problems with Chariot: the wind data, which showed directing fallout over the land to be unlikely; the unsuitability of a permafrost site to learn about the effect of water on crater-wall stability; the fact that other nuclear tests would provide some of the same information expected from Chariot. Foster thought Chariot should be canceled, but he wanted to avert bad publicity. The laboratory was sensitive about any policy shift that "looks like another retreat to mollify local demands." So Foster urged a carefully managed public announcement:

> Such an action could have repercussions which would adversely affect the whole Plowshare program. In an effort to minimize repercussions, we suggest that the decision to abandon Chariot be included in your announcement of the future Plowshare program. Hopefully it would be possible to include

an important practical excavation such as a large harbor in
Australia or construction of a sea level canal.

Writing back, the AEC's John Kelly agreed that the public relations issue was crucial, and suggested that the decision to abandon Chariot *not*
be announced, even as he directed that the San Francisco office "prepare
plans for terminating all Chariot activities at the end of this field season."
Kelly said it was "important that all concerned with this program be careful that their actions or words are not such that they can be interpreted as
indicating that the commission has canceled Chariot."

By July 1962, John Kelly had prepared a staff paper for the AEC commissioners summarizing the "factors both for and against including Chariot" in the AEC's excavation program. Kelly listed three reasons why
Chariot should be conducted. First, the agency had spent almost $4 mil-
lion on surveys of the blast area and wanted a return on its money. Second, although the experiment was four years old and "obsolescent," it
would still supply data on nuclear row charges and the effects of flooding—both of which the AEC wanted. In short, Kelly said, "an experiment
comparable to Chariot is essential to the excavation program and the selection and safety evaluation of an alternate site will be very difficult."
And third, there was the need to manage the public's perception and
"since Chariot has been vigorously criticized from this standpoint of
safety . . . its cancellation will contribute to the skepticism on the safety of
nuclear excavation."

But Kelly's arguments *against* continuing Project Chariot were also
powerful. Other tests more cheaply conducted in Nevada could provide
most of the needed data. Also, the Alaskan conditions required that the
shot be timed with weather and biological cycles, requiring an eighteen-
month lead time after a decision to proceed. Then there was the fact that
Stewart Udall's Department of the Interior was disinclined to give to the
AEC land the Native people claimed. "This could lead," Kelly wrote, "to
lengthy inter-agency and legal discussions which would be adverse to the
Plowshare program." Finally, though he did his bureaucratic best not to
say so plainly, Kelly admitted that the environmental studies did raise
safety concerns. After first predicting that the chance of harm to local inhabitants would be exceedingly remote, he conceded that "there are some
uncertainties within the parameters involved in some of these predictions

however that can only be resolved by proceeding with an experiment." In other words, as a reason *not* to continue with Chariot, Kelly mentioned safety "uncertainties."

Kelly concluded that "Project Chariot should be canceled," but before wrapping up his brief, he dealt with the public-opinion consequences of canceling the project. He wrote an "Information Plan," the objective of which was "to minimize, insofar as possible, adverse criticism of the commission's decision" to cancel Chariot by providing a "publicly convincing explanation" of the commission's reasons. Kelly said the sort of explanation he had in mind would be "carefully timed to coincide with a convincing event," which would corroborate the technical reasons given for canceling Chariot. "Project Sedan will be such an event," he said. First the AEC would announce that Sedan had been a success and had provided useful cratering information. Then, in the second postshot press release, it would announce that, consequently, the Chariot experiment was no longer needed. Even before Sedan had been fired, Kelly provided press release language that concluded that Sedan's results had "largely obviated the need for Project Chariot."

•   •   •

The Sedan bomb exploded at 10:00 A.M. July 6, 1962, at the Nevada Test Site. Within three seconds, a blister of earth measuring 800 feet across and 300 feet high bulged above ground zero. At that point, incandescent gases burst through, shooting the desert alluvium 2,000 feet out in a great hemispheric fountain of rock and dirt before the heavier material showered back to earth, partially filling the crater. A base surge rolled out radially, expanding to a distance of about two and a half miles. An hour later, though it was too soon for any thorough analysis to have been performed, the AEC felt comfortable estimating that "95 percent of the radioactivity produced by the detonation was trapped in the ground, or in the earth that fell back promptly." And of the remaining 5 percent, the AEC said "most" landed "close to the crater and within the test site."

The truth was that this, the largest explosion to have occurred on North America up to that point, was substantially dirtier than the press statements would lead anyone to believe. Twenty-one years later, an AEC report acknowledged that the boiling cloud of radioactive dust, which rose "higher than had been expected" to more than two miles into the air, "deposited nearly five times as much fallout on and near the test site than had been predicted." This fivefold error was in rather precise agreement with a prediction offered by Barry Commoner's Committee on Nuclear Information a year earlier. Referring to the Chariot shot, which Sedan replaced, Commoner's group argued, "it seems not at all unreasonable that the [vented] yield could be five times" the value predicted by the AEC. Once aloft, the winds caught Sedan's radiation and carried it north at a rate of twelve miles an hour. And, because Sedan's ground zero was located about ten miles from the northern boundary of the Nevada Test Site, the radioactive cloud was probably already off NTS by the time the AEC delivered its first reassuring press release.

At 2:45 P.M., the AEC announced to the press that the cloud had crossed State Highway 25. The Nevada State Police closed the road, and a road maintenance station was evacuated, as were people from nearby ranches. The AEC advised others to remain indoors while the cloud passed. By 3:40 P.M., the dust cloud had crossed U.S. Highway 6, lowering visibility. The dust was so thick at Ely, Nevada, 200 miles away, that the streetlights had to be turned on at four in the afternoon. Five days later, after the cloud had passed over Utah, Colorado, Wyoming, Nebraska, and the Dakotas, tanker trucks and fire engines moved in to "wash" about seven miles of the still-blockaded Highway 25. With pressurized water, "radiological safety personnel" hosed most of the radioactive fallout off the road top and into the adjacent ditches. Two years later, AEC chairman Glenn Seaborg admitted that Sedan fallout had probably crossed into Canada. More than four decades later, a radiation warning sign at the Sedan crater lip warns that no digging or excavation is permitted in the area without special authorization.

A shot the size of Sedan, fired at Ogotoruk Creek, could have dropped radioactive fallout over the entire length of the North Slope of Alaska, or penetrated 1,000 miles into Siberia, depending on the winds and differences in the geological character of the blasted rock. The Chariot shot, at its smallest configuration (280 kilotons), would have been nearly three times as powerful as Sedan. At its largest configuration, Chariot would have been twenty-four times larger.

The Sedan bomb had been buried at 635 feet. It lifted 10 million tons of earth into the air, about half of which fell back into the hole. But 6.5 million cubic yards of light gray-brown dirt had been thrown to the sides of the resulting crater and covered an area five miles in diameter. From the air, AEC personnel saw the Sedan hole gaping a quarter mile across, dwarfing the other craters that pockmarked the Nevada Test Site. The conical canyon measured 1,280 feet across and 320 feet deep. To the AEC, this was "a significant contribution to earth-moving technology." It was proof that Plowshare had tremendous potential; it was just the sort of event that John Kelly thought ought to precede the "publicly convincing explanation of the reasons for canceling" Project Chariot.

The next month, on August 24, 1962, the AEC did issue a press statement on the fate of Project Chariot. It said the project was being "held in abeyance," and that "some of the data originally planned to be obtained from Chariot are now available or may be developed from other experiments." The statement left open the question of just what "role the Chariot experiment could play in the overall program," but said the AEC had "decided to defer, for the present, any recommendations to The President on whether to conduct the experiment."

Interestingly, when *The New York Times* writer Lawrence Davies had suggested the idea of Chariot's impending cancellation, he had used the very same terms as eventually appeared in the AEC's official press release three months later: *deferred* and *held in abeyance.* Perhaps the identical word choice was coincidental. Or perhaps it was the AEC writers who had borrowed Davies's language. But one might also speculate, as Jim Haddock had done, that the journalist had floated a trial balloon for the AEC to help the agency gauge public reaction.

Like most AEC public statements from the Chariot era, the details of the August 24 press release suffered when compared with the facts. If Sedan and future tests in Nevada were being developed to answer questions originally intended to be addressed by the Chariot experiment, why not simply conduct Chariot? And if, as the AEC now claimed, economic considerations favored Nevada, why had an Alaska venue been proposed to begin with? Because one of Chariot's main stated purposes had been to resolve uncertainties relating to excavations in a marine environment, explosions in the desert were hardly equivalent. Addressing this issue, Livermore's Gerald Johnson asserted that technical objectives not satisfied with shots at NTS would be met by a promising harbor project in Australia. But as Australian historian Trevor Findlay writes: "In 1962, however, the Australian project was only a vague possibility and cannot have been a factor in Chariot's demise." In short, the proposition that Chariot had simply been overtaken by events ignored the chronology of those technical developments and belied the internal analyses that recognized the political and public relations influence of certain "small but very vocal groups."

The AEC's "held in abeyance" strategy denied Chariot opponents a clear-cut victory and allowed the agency to continue insisting that the project would have been harmless to the local people. Sharon Francis, the anti-Chariot groups' insider at the Department of the Interior, saw the tactic as "the self-righteousness of desperately small men. . . . How much preferable forthright honesty would have been, and even now there is the lurking threat that the project has only been deferred, not canceled." Writing to Don Foote, Francis provided a glimpse—from Interior's perspective—of the interagency skirmishing:

> Those involved with the project at the AEC and at Livermore knew that Interior was prepared to exercise severe judgment in the matter of Project Chariot. Rather than have us issue a counter-report to the President which we would have publicized, and which would have been most embarrassing to the AEC, they withdrew early in the fray, and in the most face-saving manner.

Though not officially canceled, Francis felt reasonably sure that with the pace of technical advances, "Chariot will soon be obsolete."

•   •   •

PROJECT CHARIOT CALLED OFF, trumpeted a banner headline in the *Fairbanks Daily News-Miner.* Both the *News-Miner* and the *Anchorage Daily Times* ran the same Associated Press story out of Washington, D.C. The AP writer Raymond Crowley framed the government's capitulation to protesters as a defeat for science. In the process, Crowley trivialized the Inupiat people's opposition to the project and lampooned Native American ethnicity in general:

> Alaskan Eskimos won a victory over atomic science today.
> Their great white father isn't going to order any time soon, if
> ever, a big nuclear boom on their happy hunting grounds.
>
> The Atomic Energy Commission has shelved long-laid
> plans to blast out a new harbor above the Arctic Circle, near
> Cape Thompson in northwest Alaska. These plans—known
> as Project Chariot—had disturbed the Eskimos no end.

Crowley passed along the AEC's claim that much of the technical information "has been gathered in other experiments," and noted the expectation of protests from the Soviet Union if the project were carried out. "But," he wrote, "it was understood the Eskimos were more influential in bringing about today's decision."

Closing his article with another bit of fun-poking, Crowley managed to offend one more segment of society. He mentioned that "human geographers" were among the scientists employed on the project, and explained that "human geography is not what your eye naturally scans when a pretty girl trips down the street. It has to do with the interrelation between the earth's features and human life."

Another, somewhat more professional story out of Washington, run by the *Anchorage Daily Times* a few days later, agreed that the AEC's surprise decision "answered protests of the Eskimos" and other Alaskans concerned about potential environmental damage. This writer also included in his analysis the efforts of the St. Louis Citizens' Committee for Nuclear Information, but suggested that the Natives' opposition had been even more influential: "Whether the argument of the Eskimos or the citizens . . . helped the AEC to decide to give up four years of preparation . . . is not clear at this moment. Washington sources, however, indicate the Eskimos won the day." Musing over what must have seemed a puzzling new political reality, the reporter raised the question of what would happen now to such visionary proposals as the new Panama Canal, given "the fact that the protest of a few hundred Eskimos in Alaska has called a halt to an experiment to explore the further peaceful uses of nuclear energy."

When the first issue of the *Tundra Times* came out more than a month later, it contained two articles dealing with Chariot. The first briefly summarized the AEC's statement that Chariot would be deferred and promised a fuller treatment in later issues. The second item reported that La

Verne Madigan of the Association on American Indian Affairs, who had traveled to Alaska to help organize the Eskimo people, had met with an accident while horseback riding in Vermont. "The greatest champion for the Natives and fearless fighter for their rights," the paper said, was dead.

●   ●   ●

Back in New England, the Campaign was celebrating. After two years of evenings writing letters, of mass mailings and long-distance phone calls— all on their own time and at their own expense—Jim and Doris Haddock and Elizabeth and Max Foster relaxed. The U.S. Atomic Energy Commission had just blinked; an onrushing AEC project with millions of dollars of momentum had been derailed by a ragtag collection of citizen activists. And *that* called for a drink. Max filled the champagne glasses at the small dinner party and stood. He lifted his glass to Jim Haddock, the Campaign's unofficial leader, who had described himself as "an average man" who'd "never been associated with any 'causes' before":

> *I propose a toast to Jim.*
> *The reason why I want to propose this toast to him,*
> *Is that Jim is a man who,*
> *When he sees an unfair blow about to be struck*
> *Will step in to parry it;*
> *When he is wedded to a good cause,*
> *Boy does he really marry it;*
> *When he detects a double-crosser,*
> *He's not afraid to call him Judas Iscariot;*
> *Jim is the man who, when passed the ball,*
> *Can be relied on to carry it;*
> *Here's to Jim,*
> *Who spiked the wheels of Chariot!*

Of course, Jim Haddock wasn't *the* one who stopped Chariot. But he and a relative handful of committed people like him and a small Eskimo village had done so. Whale hunters and bureaucrats, bush pilots and church ladies, log-cabin-conservationists and the granddaughter of Groton's headmaster—against seemingly overwhelming odds, these citizens had brought about the first successful opposition to the American nuclear establishment. And something larger than Project Chariot had been knocked off course. Bogged down also was Edward Teller's headlong rush to establish Plowshare as a highly visible affirmation of the value of fission. Indeed, the civilian application of nuclear energy—other than for electric power generation—never regained its momentum.

17　BLACKLISTING

> If you can't speak your mind at the
> university, where can you speak it?
>
> —Ernest Patty

After losing his job at the University of Alaska, Bill Pruitt reported being blacklisted from working in his field elsewhere. In February 1963, he spoke on Project Chariot and scientific integrity at a national conference. Several members of the faculty of Montana State University (MSU) in the audience were "greatly impressed" and suggested he apply for a year's position there, filling in for a colleague who was to go on leave. The faculty of the Zoology Department, who were fully informed of Pruitt's run-in with the AEC and with the University of Alaska administration, nonetheless voted unanimously to recommend Pruitt for the job. The dean of the college concurred in the recommendation. But just at this point the president of MSU, Harry Newburn, ran into the University of Alaska's President Wood at a meeting in New York. The two men had been acquainted for more than thirty years. Wood advised Newburn that before hiring Pruitt he ought to talk to the Atomic Energy Commission. Newburn's second in command, Frank Abbott, made the call to John Wolfe of the AEC and, predictably, Abbott received an unfavorable review of Pruitt. Wolfe said, according to Abbott, that Pruitt was an excellent mammalogist but, when he got outside his subject, he would cause trouble. Radiation in

particular was an area about which Pruitt knew nothing and that had become a "crusade" with him.

Wolfe also suggested that Abbott contact a Bently Glass of Johns Hopkins who had investigated a complaint filed by Pruitt against the University of Alaska for the American Association of University Professors
(AAUP). On the basis of his conversations with Wolfe and Glass, Abbott
explained to the zoology faculty, he was rejecting Pruitt's application.

Pruitt had indeed filed a grievance with AAUP, the organization Brina
Kessel's grandfather had helped to establish in order to uphold "the rights
of professors against deans and university presidents." In fact, several
groups with an interest in academic freedom besides the AAUP had ex-
pressed concern over the University of Alaska's handling of the Project
Chariot affair. They had included the American Civil Liberties Union, the
American Association for the Advancement of Science's Committee on
Science in the Promotion of Human Welfare, and the Society for Social
Responsibility in Science. Even *Time* magazine covered the story of
Pruitt's firing as an instance of apparent political repression.

The AAUP's committee had found that the Pruitt case hinged on a legal issue, namely that the AEC had contracted with the University of
Alaska—rather than with the individual scientists—for the research.
Therefore the university's "general supervisor of this project was account-
able for the accuracy of the facts and findings and for the opinions and
conclusions derived from these findings." The statement seemed to side-
step the traditional confidence placed in the professional integrity of the
scientist, and Pruitt confessed he was "floored." But the AAUP's view was
that the university might properly edit and modify the report, and that academic freedom was not infringed by this action.

Though the AAUP had judged the University of Alaska within its rights
to modify Pruitt's work, that was not the same thing as endorsing the university's actions. What Abbott had neglected to tell the MSU zoology faculty was that Glass had stated that, were the decision his, he would not
have hesitated to hire Pruitt. Several members of the faculty fought the
MSU administration's decision publicly, bluntly charging that the incident
amounted to "blacklisting Dr. Pruitt because he disagrees with the AEC."
The professors' efforts did not prevail, however, and Pruitt's appointment
was blocked.

From his cabin in Fairbanks, Pruitt continued to seek work, especially at distant universities. "I frankly think," he wrote Don Foote, "that there is no chance of you, Les or I ever earning a living in our respective fields of specialization in Alaska. I also believe there is also no chance of any of us ever getting research money from the U.S. government. I at present am prepared to emigrate anywhere in the North I can find research support." While his résumés circulated, Pruitt subsisted by writing articles for popular magazines. When he finally landed a position, he packed up his family and belongings, asked a neighbor to rent out his cabin, and said good-bye to his friends. But he didn't say where he was going. The *Tundra Times* reported the "courageous scientist" as saying: "Sorry, if I told anyone where I am going I am sure the university would see that I didn't get that job too. And I want to teach and work. I have a family to support." Even after he crossed the border into Canada, Pruitt kept his destination a secret. As one friend remembered, "he visited me in Edmonton en route to a temporary job and he refused to tell me where he was going for fear that the AEC might learn of it and have him blackballed before he arrived."

Pruitt's destination was the University of Oklahoma, where he had been invited on a visiting professorship. He had not been at Oklahoma long when, he says, word reached him that the university president had been visited by representatives from the AEC and told that, really, Pruitt was not the sort to have on one's faculty. The Oklahoma president, according to Pruitt, threw the AEC men out of his office. Pruitt thinks he got the story from the president's secretary, but neither the eighty-six-year-old former president nor his also elderly former secretary (both nearly deaf) were able to recall such an incident when contacted in 1993. Pruitt also says that Alaska Methodist University "apparently had pressure put on them" when he applied there, and that his aged mother, back on the family farm in Virginia, was visited and questioned by the FBI.

It is worth noting that AEC officials did interfere during this period in the internal affairs of American universities in furtherance of a political agenda. A particularly disturbing example has been reported by journalist Peter Metzger. In 1969, AEC commissioner Francesco Costagliola wrote to the presidents of Stanford University and the Massachusetts Institute of Technology warning them that he would yank $40 million in AEC grants if the two universities were to "cave in to campus dissidents."

Costagliola also wrote to Cornell University, The Johns Hopkins University, and the University of Minnesota, even though student activism at those schools was not particularly conspicuous. He pointed out the level of financial support each institution was receiving from the AEC, then enclosed a "reprint which reflects some of my views regarding the academic community."

There is also an account of AEC pressure at the University of Alaska. In a paper presented at an AAUP conference years after he left Alaska, Al Johnson recalled an incident that took place during the 1959 spring semester, just after the outspoken university biologists had pressed the AEC to fund environmental studies: "President Patty called me in to discuss the AEC project, and concluded by saying that he had been asked to 'shut me up,' but, he said, 'I told them no—if you can't speak your mind at the university, where can you speak it?'"

Lending strength to the notion that Pruitt was actively blacklisted is the fact that the AEC kept up a cozy correspondence with President Wood concerning the activities of the Chariot critics, both while they were still on the staff and after they left. "We thought you might be interested," wrote the AEC's John Philip to Wood in 1961, "in the attached [CNI] bulletin. . . . This issue is devoted to Project Chariot and contains contributions by University of Alaska staff members." And months after Les Viereck had been let go from the University of Alaska, the AEC's John Wolfe was keeping President Wood advised as to where Viereck was seeking research support and noting—with barely concealed self-satisfaction— Viereck's lack of success: "A former employee of yours has asked the Ecological Society of America to support him in his efforts re Chariot. It will not and has so advised Viereck, using its status as a tax exempt organization as a basis for not participating in such things." When Pruitt applied for a job with the U.S. Fish and Wildlife Service and the AEC's John Wolfe was asked by the agency to provide information on Pruitt, he wrote back: "It is suggested that you send the form to Dr. William R. Wood, President, University of Alaska. I am sure that he or someone on his staff is more qualified than I to provide the information desired." That Wolfe enjoyed his joke of directing Pruitt's prospective employers to the one man least likely to recommend the biologist is clear from a penned-in postscript to the copy sent to Wood: "Bill, I suspect you know where my tongue is but don't you think I'm gaining stature as an executive?"

All in all, Pruitt felt that to be able to work in his field, he had to leave his country. He was prepared to immigrate to Norway. It turned out he didn't have to go quite that far—though nearly. Memorial University of Newfoundland at St. Johns, Newfoundland, Canada, took him in.

●   ●   ●

Les Viereck was not rehired to teach at the University of Alaska, though that was his longtime ambition. He formally applied for a position that opened up in the mid-1960s. Not only was he passed over, but the university never acknowledged his application or replied in any way. His wife, Teri, who was a part-time instructor in physiology at the University of Alaska during 1959 and 1960, found that her contract was not renewed the year following her husband's banishment. Les found work at the Alaska Department of Fish and Game in 1961. He reports that Alaska state politicians lobbied the department to fire him, but that the agency management resisted the pressure. A high-ranking Fish and Game official did order Viereck, as a matter of department policy, to resign from the presidency of the Alaska Conservation Society and, later, to resign from the ACS board of directors as well. About the same time, Fairbanks legislator Warren A. Taylor wrote to President Wood to ascertain if the University of Alaska had used any public money to support the Alaska Conservation Society. Taylor wanted it understood that a state-sponsored institution should not be involved in misleading propaganda in an attempt to sway the legislature.

After two years, Viereck landed a position at a U.S. Forest Service laboratory in Fairbanks.

●   ●   ●

Don Foote left Point Hope in 1962, returning to McGill, where he finished his Ph.D. and joined the teaching faculty. Almost immediately, he began inquiring about positions in Alaska, in particular at the University of Alaska's Institute of Social Economic and Government Research (ISEGR). When the director of that institute mentioned to President Wood that he wanted to hire Foote, he remembers Wood as saying that Foote would work at the university over his (Wood's) dead body. But the institute had substantial autonomy from the university administration. In 1968, Foote was offered a temporary position at ISEGR, and he took a sabbatical leave from McGill to come north. That led to a permanent job offer, which Foote accepted.

After working late on the night of Tuesday, February 26, 1969, Don Foote, his graduate assistant Don Prozesky, and friends Jim and Bonnie Babb drove in to the Nevada bar in Fairbanks. The Nevada was "sort of like the faculty club," says Jim Babb. They didn't drink a lot, according to the account they gave the police, maybe four beers apiece. Heading back the four miles to College at about 2:30 in the morning, February 27, Foote's vehicle apparently slid out of control, crossed the center line, and collided, nearly head-on, with an Alaska state trooper's patrol car. The temperature had risen to twenty degrees above zero, and, consequently, road conditions were "icy and extremely slippery," according to reports. The trooper vehicle hit on the driver's side door; Foote was driving.

No evidence suggests foul play. The trooper, as well as Prozesky and Jim Babb, reported that Foote's vehicle swerved across the center line several times. A blood sample taken at the hospital within an hour showed Foote's blood alcohol content to be .138 percent, high enough for him to be considered driving while intoxicated under current Alaska law. But several of Foote's friends find the circumstances suspicious. The trooper sustained only minor injuries; he was treated and released. Foote was hospitalized in "fair" condition with a crushed chest and a broken arm. Doctors apparently advised university officials that Foote's injuries were relatively minor, and those officials in turn telephoned Foote's family that he would need to postpone his scheduled trip to the East Coast for a few weeks. It was Bonnie Babb whom doctors worried about; she lay in a coma with a broken pelvis and a broken arm. When Jim Babb returned to the hospital the next day (he had only broken an arm and some ribs) to check on Bonnie and Don, he remembers that while doctors were equivocal as to Bonnie's prognosis, they thought Don looked pretty well. He himself thought Don seemed to be recovering satisfactorily. Les and Teri Viereck tried to see Don, but hospital staff told them he could not receive visitors. On the following day, Babb visited his comatose wife again, then he went up to Don's room. But he found the patient had been moved. In the stairwell he encountered a doctor, who "looked really shook." The

doctor said Don Foote had died. "And I said why? How? He looked OK."
The doctor, according to Babb, said he just didn't know. The cause of
death was listed as "acute pulmonary edema—cardiac failure" brought on
by the chest injury.

Automobile accidents happen all the time when the roads are icy in
Fairbanks. And doctors are likely to appear "shook" after losing a patient.
One detail obtained through the Freedom of Information Act, however,
seems likely to exacerbate the suspicions of Foote's friends. Just a few
months prior to his death, the U.S. Central Intelligence Agency was inter-
cepting and copying Foote's mail. The agency acknowledges that it still
has in its files a photocopy of Foote's correspondence from November 2,
1968. Though he had been dead twenty-three years at the time of the
FOIA request, the CIA refused to disclose the document.

Another coincidence is that Foote's death came just days before he was
to lead a symbolic "research halt" at the University of Alaska. As a
scholar/activist at the University of Alaska just as large-scale development
plans were being drawn up in response to the discovery of oil at Prudhoe
Bay, Foote was in a good position to agitate for environmental safeguards
for the people and the ecosystems of the North Slope. The research halt
was meant to focus attention on how the "misuse of scientific technological knowledge presents a major threat to the existence of mankind," and
was to be concurrent with a similar protest at the Massachusetts Institute
of Technology. A day after the *Fairbanks Daily News-Miner* carried the
notice of Foote's death, however, a short article said that without Foote,
plans for a local protest had dissolved. Though many of Foote's friends
suspect or believe that his death was not accidental, the available evidence does not sustain the conjecture.

For a memorial service held on campus, Les Viereck composed a brief
eulogy. Today he can't remember if he stood and read it or if, because he
is shy when it comes to public speaking, the paper stayed in his pocket. In
any event, it was a fine tribute. He spoke of how Don and Berit entered
into the life of the Point Hope Eskimo people, and not just for the purposes of an academic study, he said, but because they valued the people's
skills and wisdom. "But Don was always the scientist," he said. "There
was always the stop watch, the notebook, the scales, and the tape
recorder. He amassed great amounts of valuable data of a kind seldom, if

ever, obtained by others." Concluding, Viereck sought to define Foote's character as it was revealed during "the Project Chariot times," which he said were both "inspiring" and "troubled":

> Don was uncompromising in scientific integrity; he saw information being misused, suppressed or ignored. He saw his own scientific integrity questioned—and he fought back. . . . And he was successful, partly because of his well documented statements and reports, and partly because of his undying belief in the freedom—no, necessity—to speak out on his own convictions. I say undying belief because I think that Don has done much to set an example for others.

EPILOGUE

The objector and the rebel who
raises his voice against what he believes to be the injustice of the present and the wrongs of the past is the
one who hunches the world along.

—Clarence Darrow, 1920

William Pruitt and Leslie Viereck

In 1969, Pruitt left Memorial University in Newfoundland for a job at another Canadian school, the University of Manitoba in Winnipeg, where he became recognized
among students and colleagues as "the father of North American boreal ecology." For
twenty-eight years, Pruitt exiled himself from the United States, eventually becoming
a Canadian citizen. He took sabbaticals and attended conferences at centers of northern study throughout the circumpolar world, but not in Alaska or elsewhere in the
United States. Nor would he publish in U.S. scientific journals. In 1989, Dr. Pruitt received the Canadian government's Northern Science Award and Centenary Medal
for significant contribution to understanding of the North. He retired in 1996 at the
age of seventy-four, whereupon the title Senior Scholar was conferred upon him.

Les Viereck continued to work at the Institute of Northern Forestry in Fairbanks
from 1964 until the U.S. Forest Service closed the institute in 1996, and he retired.
Following the closure, a co-operative unit emerged at the University of Alaska Fairbanks, and Viereck remains affiliated there as a Forest Service Emeritus Scientist. He
is also Affiliate Professor of Forest Ecology at the University of Alaska Fairbanks and

periodically works on projects for several of the university's research institutes. His classification of vegetation in Alaska is considered the "bible" for researchers and scientists, and his books on Alaska's woody vegetation (particularly *Alaska Trees and Shrubs*) are regarded as indispensable guides. Viereck is internationally known as Alaska's leading forest ecologist.

Over the years, both Edward Teller and former President Wood had been awarded honorary doctorate degrees at the University of Alaska, as have several other erstwhile Chariot enthusiasts. In 1992, thirty years after Project Chariot's cancellation, dozens of Pruitt and Viereck's friends and colleagues in Fairbanks campaigned to persuade the university administration to recognize the two biologists similarly. Distinguished citizens and scholars from several countries sent in strong letters of support for Viereck and Pruitt, nearly all of them mentioning the scientists' courageous defense of scientific integrity as an accomplishment on a par with their outstanding scientific attainments. The faculty nominating committee voted its unanimous endorsement, and the university administration, initially reluctant to do anything that might embarrass the still-living former president, approved the popular initiative. When the citation was written, however, university officials sensitive to public relations edited out any mention that the two men, at great professional risk to themselves, had stood up for the truth of their findings and for principles of academic freedom. As a practical matter, it was more politic simply to recognize the men's scientific contributions and finesse the touchier academic freedom issue. But the faculty committee reconvened and, led by two senior professors who had been on the faculty during Chariot days (Rudy Krejci and Jack Distad) insisted that language similar to the original, committee-passed version be restored. And it was.

On May 6, 1993, a short, round, gray-bearded man stooping under a big pack stepped off a Canadian airplane on to the tarmac at Fairbanks and walked through Customs. A shout went up from a small mob of equally gray old friends—pals who had been young in Alaska together, many of them now in their seventies. Les and Teri Viereck, Celia Hunter, Ginny Wood, and others rushed to embrace Bill Pruitt and his wife Erna. Red-faced and absolutely speechless, his eyes filling with tears, Pruitt recognized some of the people by voice alone. Three days later, the University of Alaska awarded honorary doctorate degrees to William Pruitt and Leslie Viereck. In part, the university had conferred its highest honor in recognition of the very actions for which the men had lost their jobs thirty years earlier.

Former President Wood did not attend the commencement exercises for the first time in thirty-two years.

As an additional honor, the Alaska state legislature passed a resolution honoring Viereck and Pruitt that plainly acknowledged how the scientists came to lose their positions at the state's university. Concluding, the citation reads: "Gentlemen, you were right, and we, the people of Alaska, owe you a debt of gratitude for holding strong to your principles. The members of the Eighteenth Alaska State Legislature humbly thank you."

Don Foote

Though he had only a short professional life, superlatives with respect to Foote's scholarship come from every quarter. Sometimes called "the new Stefansson," he remains a major figure in Arctic human geography. As one colleague said: "Many ideas, theories, and techniques in these fields that were unknown or were at odds with conventional wisdom forty years ago, and which Don did much to pioneer, have become conventional and routine today."

William R. Wood

Wood presided over the University of Alaska (now called University of Alaska Fairbanks) from 1960 until 1973, a period of unprecedented growth. As one backer enthused, ever since his first day on the job, "Dr. Wood has had back hoes and bulldozers running over that campus." One of these projects was to construct a new power plant and, for a time, he was hopeful of replacing the aging boilers with a nuclear reactor. During his thirteen-year presidency, Wood aided in the establishment of several research institutes, tripled student enrollment, saw the value of the campus plant increase nearly tenfold, the budget rise by a factor of seven, and the university system expand statewide. But during Wood's reign, the campus was rocked by "four separate incidents of arbitrary firings," according to one former professor who has written on the subject. Some of these, including the Pruitt and Viereck case, were multiple firings. As one philosophy professor at the University of Alaska said: "Philosophical differences cannot be handled here because we do not have . . . what I call culture—culture in the exchange of ideas. . . . The atmosphere here is not conducive to academic freedom as it is generally understood."

For more than thirty years, Wood supported every large development project proposed for Alaska. As Project Chariot ground to a halt, Wood climbed aboard the bandwagon to dam the Yukon River near the village of Rampart. He served on the

Rampart Economic Advisory Board and later as vice president of Yukon Power for America, a booster group formed to promote the dam. In the late 1960s, when controversy erupted over whether to build the trans-Alaska oil pipeline and conservation groups rallied in opposition, Wood charged that the criticism was "anti-God, anti-man and anti-mind." It was anti-God because the Bible directed man to "fill the earth and subdue it." It was anti-man because the critics' worldview seemed to doubt the doctrine of mankind's supremacy over the earth and other living things. And it was anti-mind in that pipeline opponents seemed to lack sufficient commitment to material progress through human ingenuity.

After retiring, Wood stayed on in the community (unlike his successors, who all hastened south). Into his nineties he remained a revered civic leader and prominent Alaskan. He served a term as mayor of Fairbanks, was named "Distinguished Citizen" by the local Boy Scouts and "Alaskan of the Year" by the statewide chamber of commerce. The *Fairbanks Daily News-Miner* gave Wood a regular column, in which for more than a decade his views were featured as the apparent voice of reason in a world overrun by environmental extremism. He died in 2001.

Celia Hunter and Ginny Wood

After 1962, Hunter continued to serve as executive secretary and principal spokesperson of the Alaska Conservation Society. The remarkable success of this tiny, grass-roots organization—and the feisty intelligence of Celia Hunter—drew the attention of national conservation groups. After serving on its governing council and then as council president, Hunter left Fairbanks in 1976 for Washington, D.C. to become the national executive director of the 50,000-member Wilderness Society. For many years, she and Ginny Wood shared a log cabin outside Fairbanks. The hand-crank mimeograph machine was sent to the shed, and the manual typewriter gave way to a computer, but both women wrote regular columns and remained among Alaska's most articulate commentators on environmental issues. Celia died suddenly in 2001.

Conservation

The defeat of Project Chariot represented the first successful opposition to the American nuclear establishment and one of the first battles of the new era of "environmentalism." Here the rationale for caution was not the old logic of conserving a magnificent landscape or endangered species. Rather it was based on a more holistic concept of environmental protection, one premised on the realization that insidious degradation was possible because of the invisible connectedness of things. "Looking back on my career in environmentalism," said Barry Commoner in a 1988 interview, "it is absolutely certain that it began when I went to the library to look up lichen in connection with the Chariot program. That's a very vivid picture in my mind." Chariot led Commoner into environmentalism, and Commoner led others into what became known as the environmental movement. By 1970, Commoner was on the cover of *Time* magazine as the apparent leader of the new movement—the "Paul Revere of Ecology," as the *Time* headline put it, noted for his "political savvy, scientific soundness and the ability to excite people with his ideas." Elsewhere he has been called "the dean of the environmental movement," and "the father of grass-roots environmentalism." "I think," said Commoner in 1988, "in so far as I had an effect on the development of the whole movement—which I did, I have to admit—Project Chariot can be regarded as the ancestral birthplace of at least a large segment of the environmental movement."

●   ●   ●

The environmental study that the Alaskan biologists had demanded, and which John Wolfe had shepherded through the AEC, was published by the AEC as *Environment of the Cape Thompson Region, Alaska* in 1966. It was said to be the most comprehensive bioenvironmental program ever done at that time. "Nowhere in the free world at least," wrote John Wolfe in 1961, "and probably nowhere else, has such an assemblage of bioenvironmental researches been carried on simultaneously in the same location encompassing both land and sea environments." Because the Chariot environmental study was, at least ostensibly, intended as a planning tool, it can be regarded as the first de facto environmental impact statement (EIS). Begun ten years before the EIS formally existed, the Chariot study served as a model for some of the early EISs produced under the National Environmental Policy Act of 1969.

Alaska Natives and the *Tundra Times*

The Native land claims issue would become a central focus of the *Tundra Times* for the rest of the 1960s, as the paper served to politically sensitize and unify all Alaska Natives. In the summer of 1962, the Athabascan Indians of Alaska's Interior met for the first time since 1913 and followed the Barrow conference model to organize in defense of aboriginal rights and land issues. It was the beginning of a new era of political accomplishment for Alaska Natives, which would reach its apex in 1971 with the passage of the historic Alaska Native Claims Settlement Act. The 40 million acres of land that would be conveyed by that act would exceed all the land held in trust for all other American Indians. And the nearly $1 billion in compensation for lands already given up would nearly quadruple the total amount won by all other Indian tribes from the Indian Claims Commission over its twenty-five-year history.

Project Plowshare

After the Sedan "success," the Atomic Energy Commission's Plowshare program continued to expand in anticipation of excavating a new Panama Canal with nuclear explosives. By 1968, the program drew $18 million in annual federal funding. But by then, President Johnson was being pressed to reduce spending in light of rising inflation and an escalating Vietnam War. When the AEC requested a boost in Plowshare funding to $24 million for 1969, Johnson administration budget cutters approved only $14 million, a cut of $4 million. Richard Nixon, during his first run for the presidency in 1960, had promoted Plowshare in campaign speeches as "new and imaginative," as "having great potential," and as offering "tantalizing glimpses of great new vistas of future achievement." But according to Plowshare historian Trevor Findlay, when Nixon ascended to the presidency in 1969, he seemed to have no particular affinity for Plowshare. By then the program was largely identified with the Johnson administration, which had funded it amply for six years. To the AEC's request for nearly $30 million in annual funding, Nixon responded with less than $15 million. Congressman Craig Hosmer, a Plowshare supporter on the Joint Committee on Atomic Energy fumed: "In addition to the assorted professors, scientists, lawyers and literati who whine over Plowshare for philosophical reasons, a hard core of Plowshare opponents seem to have developed within the Executive Branch of the government itself."

The truth that boosters like Hosmer ignored was that besides being expensive, there just seemed to be little need to trade the enormous risks involved in multiple, high-yield nuclear detonations for the opportunity to correct "a slightly flawed planet." When the Atlantic Pacific Inter-oceanic Canal Commission delivered its recommendations to Nixon on the idea of a new sea-level Panama Canal, it rejected the use of nuclear excavation. The commission said the plan's "technical feasibility"—a term that surely included radiation safety and other environmental concerns—had not been proven. Besides the scientific issues, there was a political consideration as well. The canal project would have required signatory states to agree to an amendment of the Partial Test Ban Treaty (PTBT). As Findlay writes: "The most incredible aspect of the entire nuclear excavation saga had been the air of unreality about the chances of the PTBT being amended. . . . To start with, no one seemed to have any idea whether the Soviet Union would agree or not."

After the passage of the National Environmental Policy Act of 1969 (NEPA), environmental impact statements were required to be prepared for all Plowshare projects. NEPA could not require the disclosure of the bomb's characteristics, but it did force the release of more information as to the risks associated with the detonations. After NEPA, says political scientist Richard Sylves, "Plowshare could no longer proceed as elite or concealed policy-making." As the public's knowledge of civil works detonations grew, so did people's uneasiness with the idea. Plowshare's nuclear excavation program ended in 1970, while other Plowshare applications of nuclear explosions continued to be funded for a few more years. "The story of Project Plowshare," writes Sylves, "is one of sustained and futile scientific and engineering optimism in the face of a world public becoming increasingly intolerant of such nuclear ventures. . . . The termination of Plowshare was the reluctant admission that a nuclear utopia was not imminent."

In the Soviet Union, however, a similar program continued, and nuclear blasts were used for oil stimulation, creating underground storage cavities, extinguishing gas fires, conducting deep seismic sounding, and excavating. On February 10, 1992, Alaska's U.S. Senator Frank Murkowski, a member of the Senate Intelligence Committee, cited published and U.S. intelligence reports that indicated the possibility of "widespread nuclear contamination" in the former Soviet Union as a result of peaceful nuclear explosions. According to his information, Murkowski said, the Soviets had used nonmilitary devices for civil works projects during the 1960s, 1970s, and 1980s. Murkowski called for an intelligence investigation. Four days later, he received a report classified "Secret" from the Defense Intelligence Agency. According to his press release, the report gave "credence to his initial fears." Murkowski referred to accounts of Russian scientists who estimated that the Soviet Union may have released into the atmosphere a billion curies of radioactivity from 126 so-called "peaceful nuclear explosions."

In August, speaking at a special meeting of the Senate Select Committee on Intelligence convened in Fairbanks, then Director of Central Intelligence Robert Gates echoed Murkowski's distress over the Soviet PNE program, saying, "These [Soviet] crater-producing explosions produced widespread contamination." None of the more than two dozen experts testifying at the day-long hearing mentioned that radioactive contamination of the Arctic would have been a U.S. legacy as well if the AEC had been allowed to conduct Project Chariot. No one mentioned that just five years earlier in Anchorage, Edward Teller had spoken to a prodevelopment association and lamented that Project Chariot "should have been done. With no good reason, the environmentalists stopped us. . . . The Soviets are doing it; we are not."

Nothing if not persistent, Teller continued to advocate the peaceful use of nuclear explosions into the 1990s, revelations of the USSR's disastrous experience with PNEs notwithstanding. In late 1992, in response to the Senate hearings, Teller insisted that the bomb would yet be redeemed as a peaceful tool: "In the future it certainly will be used, and it can be used safely. The Soviets have been much too careless. We must find safe ways. When we find them, programs of that kind, including the kinds of things we tried to do in Alaska, will be practical and safe."

As the Soviet economy deteriorated, a business called International Chetek Corporation was founded in Moscow to offer nuclear explosions to anyone in the world with enough cash. "We are willing to entertain all ideas," said a Chetek marketing representative in Montreal, Quebec, bringing to mind Edward Teller's epigram of thirty years earlier: "If your mountain is not in the right place, drop us a card." "It doesn't matter who, where or when," said the Chetek salesman, "we have all the technologies and they're going to be used." The first offering at this "yard sale at the end of history," as one strategic analyst calls it, was Chetek's proposal to incinerate chemical and radioactive wastes, chemical weapons, retired warheads, and the like in an underground nuclear explosion. Chetek appears to have gone out of business within a year or so of its formation, so perhaps it will not be necessary to deal with the strategic and environmental repercussions of cash-starved Soviet scientists launching have-bomb-will-travel enterprises.

Edward Teller

Teller continued to be at or near the center of American defense and nuclear policy for decades, with an apparently life-time appointment as de facto science adviser to the president during Republican administrations. He was unwavering in his support for weapons development and testing, and equally steadfast in opposition to any attempts at controlling the arms race. Repeatedly, when his cold war agenda was threatened, Teller undertook intensive lobbying campaigns, promoting some not-yet-existing technology with an optimism that was later shown to be scientifically insupportable. At the end of World War II, when scientists began to head back to their universities and the Los Alamos laboratory began to shrink, Teller demanded that a crash program be launched to build a model of the H-bomb in which he proclaimed great confidence. In support of his model, he presented calculations that later proved to be wrong. The H-bomb only emerged when an unrelated model was suggested by a colleague. Nevertheless, Teller succeeded in keeping a vigorous nuclear weapons program alive. And he did so even though it meant "leading the laboratory, and indeed the whole country, into an adventurous program on the basis of calculations which he himself must have known to be very incomplete," as Hans Bethe, a Nobel Prize winner and Teller's boss on the Manhattan Project, has said.

In the 1950s and 1960s, Teller helped to kill the chances for a comprehensive test ban by insisting that the "clean bomb" was "well on its way to success." The clean bomb—a key to Plowshare miracles—was *not* at hand at that time, nor is it today after more than thirty years of experimenting. "During the moratorium," says Teller, speaking candidly about his greatest challenge as Director of LRL, "the problem was to keep the [Livermore] lab together and to keep people working on nuclear designs even though we could not test." The promise of a clean bomb assisted with that objective; the laboratory stayed in business, and treaty negotiations with the Soviets—which Teller feared would cause Americans "to turn away from nuclear weapons"—fell short of a comprehensive test ban. Instead, a "limited" or "partial" test ban treaty was signed, which, while banning nuclear weapons tests in the atmosphere, oceans, and space, permitted them underground.

In the 1980s, a new and powerful threat to the arms race emerged in the form of a "nuclear freeze movement." The initiative garnered widespread support as people questioned the need to continue stockpiling weapons capable of annihilating adversaries 1,000 times over. But again Edward Teller outflanked the arms-control advocates and repackaged the business of weapons production. He saw that nuclear weapons R & D—not to mention a broad wish list of other military projects—could be sold to the American people (and funded by Congress) under the rubric "defensive shield." Teller had convinced politicians of the feasibility of a fantastic technology, namely that not-yet-developed X-ray lasers, based in space, could knock out incoming intercontinental ballistic missiles (ICBMs). In March of 1983, President Ronald Reagan unveiled the Strategic Defense Initiative (SDI), later dubbed "Star Wars," which, he said, could make nuclear weapons "obsolete." With Reagan embracing Teller's vision, Star Wars became, by far, the most expensive military project in the history of the world. But the laser hardware never materialized. Thirty billion dollars were poured into the benign-sounding research that many scientists described as "a fraud" and "impossible to accomplish." The nuclear freeze initiative essentially disappeared. The weapons labs stayed in business, flush with money. Again, Teller had led the country into a fantastically ambitious and expensive military project on the basis of a

technical optimism that was neither scientifically supportable at the time nor validated later.

In each case—with his first model of the H-bomb, with the clean bomb, with
Plowshare, and with Star Wars lasers—Teller passionately insisted that he had
the key to brilliant new technologies if only nuclear weapons R & D were allowed to continue. In each case, the promised breakthrough failed to come about
as advertised, while progress on arms control was thwarted and the agenda of
the weaponeer advanced. "Time will reveal," wrote I. F. Stone in his 1963 book
The Haunted Fifties, "that those who campaigned for a nuclear test ban were
acting in the best interests of United States security while those who opposed it,
as Edward Teller and the AEC-Pentagon crowd did, were acting against our
country's best interests."

●   ●   ●

In June 1987, Edward Teller returned to Alaska. Once again he came to promote a
defense-related, high-technology project to be based in the Alaskan Arctic. He proposed installing SDI laser weapons on the North Slope. Speaking before Commonwealth North, a prodevelopment business group, Teller evoked the rhetoric of the
Chariot debate of thirty years earlier. He offered Alaskans, whom he again called the
most "reasonable" of U.S. citizens, an opportunity to "make a great contribution to
the nation's defense." He applauded the fact that the barren Arctic wasteland had
been validated now with oil development. Before Prudhoe Bay, he said to the laughter of his audience, "the North Slope was not yet there." Once again negating several
thousand years of Inupiat Eskimo culture, he continued, "Or to be more accurate, nobody was there *except* the North Slope."

The Alaskan businessmen attending the breakfast meeting pressed Teller to use
his political clout to bring the big federal project to the state: "May we presume," one
questioner asked, "that you will advise the President and those instrumental in SDI
development that Alaska and the North Slope presents the ideal locale for placement
of laser and particle beam defense staging?" Himself beaming, Teller replied: "You
may certainly presume that I will . . . advise the President's advisors. I have the experience that our president [Reagan] is a very good listener. . . . I am very hopeful that
you will see action."

Star Wars lasers were never installed on the North Slope or in space. But, with
Teller's involvement, the idea has been reconfigured and repackaged into a succes-

sion of anti-ballistic missile schemes, the latest of which has been built in Alaska. What has not changed in forty years is that billions of dollars continues to be transferred from public coffers to the arms manufacturers, that no meaningful security from intercontinental ballistic missiles has resulted, and that the United States continues to move closer to the weaponization of space. In 2001, when President George W. Bush abrogated the 1972 Anti-Ballistic Missile treaty with Russia as part of a continuing effort to deploy anti-missile systems, Teller offered a two-word reaction: "High time!" Bush awarded Teller the Presidential Medal of Freedom in 2003, praising his "long life of brilliant achievement and patriotic service." He died the same year at the age of ninety-five, one of the most politically influential scientists of the twentieth century.

Missile Defense

The history of missile defense has followed a roller coaster trajectory, led by gung ho champions during Republican administrations and at least acquiescent apostates during Democratic ones. In the 1960s, the military was able to send a missile launched from Kwajalein atoll in the central Pacific to within a mile and a quarter or so of a missile fired from Vandenberg Air Force base in California. In theory, that proximity was close enough to destroy an incoming missile, if the interceptor detonated a nuclear warhead. But such a crude antiballistic missile system (ABM) was known to have major flaws. Each incoming missile could contain multiple warheads and multiple decoys, and U.S. detection equipment might not distinguish between them. If only one warhead got through, an entire city could be destroyed. With any missile defense system, therefore, there is no margin for error. A ninety-nine percent success rate in defending San Francisco, say, could amount to total failure.

The task of discriminating between decoys and actual warheads might be made easier by withholding the intercept until after reentry, by which time any decoys (lacking heavy shielding) would have burned up. But this "terminal" phase lasts only a very short time—perhaps one or two minutes—and the available technology could not process the data fast enough. Besides, such a relatively low altitude intercept would mean detonating one nuclear bomb—and blowing apart another one—in the atmosphere above American cities, plastering our own population with radioactive fallout. These problems notwithstanding, Lyndon Johnson and later Richard Nixon pressed ahead with ABM. In 1975, at a cost of $6 billion, an ABM installation called Safeguard became operational in North Dakota, defending a single battery of American nuclear-tipped ICBMs. The facility was shut down just months later, Congress having decided its maintenance was a waste of money.

Funding of research for ABM systems continued at the Pentagon into the late 1970s, however, to the tune of about a billion dollars a year. When Ronald Reagan ran for the presidency in 1980, a plank in the Republican Party platform called for "vigorous research and development of an effective anti-ballistic missile system." Following Reagan's election, and with Teller promoting as a virtual certainty a new X-ray laser weapon, "Star Wars" emerged. Tens of billions of dollars later, no such weapons existed. And as the USSR imploded, even the Soviet threat went away.

With the election of Bill Clinton in 1992, the public was more concerned with a sputtering economy than with fantastical weapons systems. The enormously expensive Star Wars program appeared to be dead. But the arms manufacturers (who would make billions whether a missile defense system worked or not) were solidly allied with political conservatives (who seemed less concerned about an efficacious missile defense than about beginning the effort to put weapons into space). A new rational emerged whereby a missile defense system need not be expected to protect the country from the sort of full-on missile attack the former Soviet Union might have launched. Rather, it would defend against the odd missile that might be lobbed at the United States by a technically advanced (and apparently suicidal) "rogue state." Missile defense became part of the "Contract with America," a campaign stratagem that helped the Republican Party gain control of the Congress in 1994, and missile defense was once again back on track.

If President Clinton did not favor the scheme, now called National Missile Defense (NMD), he still signed off on continued funding. In 1999, with a presidential election looming and political considerations uppermost, Congress passed and Clinton signed the National Missile Defense Act. It simply stated: "It is the policy of the United States to deploy as soon as technologically possible an effective national missile defense." After three years of testing, and just before the 2000 election, Clinton said he did not have confidence in the NMD system. Rather than cancel it, however, he said he would let the next president decide whether or not to deploy NMD. In this matter, George W. Bush was pleased to be "the decider." The new secretary of defense, Donald Rumsfeld, had been an early and ardent advocate not only of missile defense, but of putting American weapons in space.

What military analysts had been calling a "rush to failure," now rushed even faster. Testing resumed, but when independent scientists such as the Union of Concerned Scientists looked at the experimental designs, they found that the Pentagon had created an array of artificialities that simplified the inceptor's task. For one thing, many of the tests were repeats of earlier tests, which placed the target array in a tight cluster, in the same place in the sky, at the same time of day (hence always illuminated by the sun in the same way). For another, the interceptor boosters used in the tests were slower, two-stage rockets that did not replicate the greater heat, vibration, and shock forces of the three-stage rockets that were intended for the system but that were not yet built. Slower rockets meant artificially slower closing speeds, so that the kill vehicle had more time to adjust to its target.

Test planners also made it artificially easy for the kill vehicle to discriminate between the mock warhead and any decoy that might be flying alongside. The decoy always had a very different electronic appearance to the kill vehicle's sensors. The geometry of the trajectories was so arranged that the target cluster (warhead, decoy, final stage of the rocket) was kept within a narrow field of view with respect to the kill vehicle. That meant that the kill vehicle's sensors "saw" all the objects at once. In a more realistic situation, the kill vehicle would have to rotate itself so that its sensors could gather data on several more widely separated objects. Artificially eliminating that problem gave the kill vehicle more time to rule out the decoy and steer toward the warhead.

Much of the critical detection apparatus, as well as the central computer system that would coordinate detection data with guidance instructions to the interceptor, were not yet built and so were simply not part of the tests. Instead, the defense was given in advance the incoming missile's flight plan, and was able to point the interceptor at the precise place in the sky where the rendezvous was planned. To make things even easier, a beacon was installed on the mock warhead for the interceptor to home in on.

Even with all these advantages, most tests were acknowledged failures. When the Pentagon claimed successes, independent scientists, engineers and other analysts often demonstrated the claims to be overstated or inaccurate. The Bush administration responded by impounding data from future tests. Henceforth, test data would be classified and peer review impossible.

To President Bush and Secretary Rumsfeld, the fact that the technology failed to prove out in testing was not a reason to delay its deployment. In 2002, Bush announced a crash program to field a rudimentary missile defense system by 2004. Fort Greely, Alaska, was selected to host a battery of interceptors, with two additional silos to be constructed at Vandenberg Air Force Base in California. Initially, the

Pentagon said forty interceptors would be installed at Fort Greely, but later it said it would keep secret the number of missiles deployed. The first missile was loaded into a silo at Fort Greely in the summer of 2004, and the site was declared operational on June 20, 2006. The very next month, however, heavy rains flooded seven silos, causing $38 million worth of damage (silos were filled to up with water sixty-three feet deep). When Secretary Rumsfeld visited Fort Greely in August of 2006, he acknowledged that the system upon which he was spending approximately $10 billion per year, had a limited capability, that it had never been tested end-to-end, and that he would need to see a full test done successfully before he had confidence that a defensive capability actually existed. Between Rumsfeld's admission and Ronald Reagan's unveiling of his Star Wars vision in 1983, American taxpayers had paid approximately $100 billion for a missile defense system. Still, the administration did not gainsay their decision to build a flawed system first, then test and refine it later. Baffled observers call this strategy "Fire, aim, ready." The catchphrase preferred by the Pentagon is "Something is better than nothing," as if it would make more sense to buy, rather than not to buy, an umbrella made of fish net.

The People of Point Hope

In the fall of 1992, the people of Point Hope painfully revisited the Project Chariot controversy. In August of that year, I passed on to an official of the Point Hope village corporation, and to the Alaska Military Toxics Network, documents uncovered while researching this book. The letters and memoranda showed that before abandoning the Chariot camp, government scientists had buried nuclear waste near the site. Shortly, banner headlines in the *Anchorage Daily News,* Alaska's largest newspaper, proclaimed, NUCLEAR WASTE DUMP DISCOVERED: ARCHIVES REVEAL '60s CHUKCHI TEST SITE; OFFICIALS HUSTLE TO DETERMINE HAZARD. The thirty-year-old documents described how the Atomic Energy Commission had contracted with the U.S. Geological Survey (USGS) to conduct experiments with radioactive tracers at the Chariot site. And they show that when the experiment was finished, the scientists illegally buried quantities of certain radioisotopes 1,000 times in excess of federal regulations. According to the documents, the AEC had asked the USGS to submit a funding proposal for the experiment, which was not specifically related to Project Chariot.

Just before the Sedan detonation at the Nevada Test Site in July 1962, scientists from the USGS placed collection pans on the ground a mile from ground zero. Sedan threw 24 billion pounds of irradiated soil out of its crater. Seventeen and a half of those pounds were collected in the pans and the next day transferred to a cloth sack.

This fallout contained the complete array of fission products, dozens of radionuclides, including the radiological poisons cesium 137, strontium 90, and traces of plutonium 239. Six weeks later, the USGS scientists flew with the sacks to Ogotoruk Creek. They also brought north from their Denver laboratory three other bags of sand mixed with three pure radioactive isotopes: cesium 137, strontium 85, and iodine 131. These three sand mixtures totaled twenty-six pounds, and together with the Sedan fallout, the total weight of all the contaminated soil flown to Alaska was 43.5 pounds. The total amount of radioactivity, as reported in the documents, was twenty-six milli-curies (twenty-six thousandths of one curie). However, as much as five curies of radiation might have been transported to Ogotoruk Creek, since the USGS had asked the AEC for permission to transport that amount, and permission had been granted. Five curies would represent a third of the radiation that was said to have vented in the worst nuclear accident in U.S. history, the Three Mile Island mishap.

On August 21, 1962, the scientists loaded their radioactive dirt and other tools and equipment into a "weasel" track vehicle and headed up the Ogotoruk Creek valley to a tiny tributary on the west bank called Snowbank Creek. An armed guard accompanied the team to ensure that Russian agents did not steal the fallout and, through radiochemical analysis, determine various characteristics of the Sedan bomb. Three-quarters of a mile from the Chariot camp, the men staked out twelve small sites on the treeless tundra: Eleven were terrestrial plots, one was a reach of Snowbank Creek. They wrapped one-by-six boards with polyethylene sheeting and imbedded them around the perimeter of the eleven plots, which averaged about three feet on a side. A notch was cut into the lowest board so that when the area was saturated, runoff would flow from the notch so that samples might be collected in plastic bottles. Into the turf at the head of each plot, the men hammered wooden plaques identifying the isotope in use. Pictures taken at the time show the plaques and their bordered plots looking uncannily like diminutive tombstones marking tiny graves.

Next, the scientists sprinkled the radioactive soil, the "tracer," onto the ground and the vegetation within the plots. With a gasoline-powered pump set up at the creek, and a garden hose, they simulated rain by spraying the plants and soil within the plots until they were drenched and water ran out the notch and into the plastic bottle. These samples were labeled and taken back to the laboratory in Denver for analysis.

The purpose of the experiment was to determine the extent to which water passing through irradiated soil would dissolve the fallout radionuclides and transport them to aquifers, streams, and ponds. That would indicate whether a detonation like Chariot might tend to concentrate radioactive contaminants in water supplies used by people and animals. The mixed fission products from Sedan were especially suitable stand-ins for the expected Chariot fallout because the Sedan bomb was a similar device to those planned for Chariot. It is noteworthy, however, that the experiment that could give the most definitive answers about Chariot's potential to deliver radiation to man via water was not part of the Project Chariot program of environmental studies organized by John Wolfe. It was apparently not funded through Project Chariot or Plowshare accounts, and was not commenced until late 1962, at a time when the AEC had already decided to cancel Project Chariot. Apparently the study was of interest to the AEC's Albuquerque Operations Office, which was concerned with the safety considerations of nuclear testing operations, generally.

At the twelfth site, the men put five and a half pounds of contaminated soil, 3.2 millicuries of radioactivity, directly into the creek. Then they collected samples of the water at twenty, forty, and sixty feet downstream to show the dispersal of the suspended particles and to measure the "resulting wave of radioactivity that passed downstream."

To decontaminate the plots after the experiment, the men excavated the soggy soil and vegetation down to a depth where only background levels of radiation were detectable. This amount of contaminated earth, which now totaled about 15,000 pounds, they loaded into fifty-five-gallon drums and hauled to a spot midway between two of the plots. They dumped the dirt out of the drums and threw the contaminated boards into the pile. This heap measured about 4 feet high and covered about 400 square feet. The material could not be buried in a pit because, even by August, the soil had only thawed to a depth of about two feet. Instead, one of the equipment operators brought up a bulldozer from the camp and pushed about four feet of clean dirt over the top of the waste pile. Then the scientists left.

Of course, the 3.2 millicuries put into the stream was not recovered. Presumably the "wave of radioactivity" flowed down Snowbank Creek, into Ogotoruk Creek, and on down to the sea. Other scientists working at the Chariot camp were unaware of this fact, even though the camp's water supply—for drinking, cooking, and washing— came not from a well, but from Ogotoruk Creek.

The radioactive material lay buried and forgotten for thirty years, almost exactly to the day. The site was not fenced or labeled or marked as off limits. No one monitored it over the years even though the porous nature of the uncompacted mound "could have allowed the radionuclides to leach out with rainfall," according to a 1993 scientific review. No one had bothered to consult with the people of Point Hope about dumping nuclear waste on land they claimed. And no one told them of the dump's existence after the fact.

The disposal contravened the Code of Federal Regulations, which limited the quantities of specific isotopes that may be buried in soil. Specifically, the cesium 137 and strontium 85 "exceeded one thousand times the amounts specified" in the law, according to federal regulators. Also contrary to the federal code, said regulators, was the fact that "no records were maintained of the byproduct materials disposed of by burial." Finally, the disposal was in direct violation of the Department of the Interior permit that allowed the AEC to occupy the Cape Thompson region. That permit stated unequivocally that "nothing in this permit shall be construed to authorize the contamination of any portion of the lands."

*   *   *

The shock waves that rippled through the Eskimo communities of northwest Alaska in 1992 upon the disclosure of the radioactive dump site reverberated around the world. NORTH SLOPE DECRIES NUCLEAR LIES, read a headline in the *Anchorage Daily News* soon after the story broke. Reviving after briefly folding for financial reasons, the *Tundra Times* revisited the story that had launched the paper exactly thirty years earlier. Atop its page one, the headline read: ATOMIC ARROGANCE: FEAR, ANGER RUN DEEP IN POINT HOPE.

Picked up by the wire services (Associated Press, Reuters, *Los Angeles Times*, and McClatchy), articles appeared in papers all over the United States, as well as in Canada, Britain, and Japan, at least: ESKIMOS FURIOUS ABOUT NUCLEAR DUMP; ESKIMOS UN-COVER 30-YEAR SECRET: NUCLEAR WASTE DUMPED NEARBY; ESKIMOS FEEL BETRAYED BY NUCLEAR EXPERIMENTS. The *Washington Post* gave the story prominent play, the *Los Angeles Times* ran it on page one, and *The New York Times* ran a major story in its Sunday edition with several photos.

Representatives of the Alaska state Department of Environmental Conservation (DEC) and the Army Corps of Engineers flew to the site on Thursday September 10, 1992, four days after the initial news story appeared. The first snowfall of the season preceded them. A mile from the site they saw a herd of musk oxen grazing. From the air, following maps I had found in the archival documents, the officials located the waste mound not very far inland from a spot on the beach where a whale carcass was being devoured by six grizzly bears. The mound proved to be easy to find from the air because swaths of disturbed tundra radiated from it like a starburst. A bulldozer had created the mound by scraping up dirt from every direction toward the center. Though thirty years had passed, the disturbance was still plainly evident.

The first operation, a test to see if there might be a high level of radiation at the site, involved getting the pilot to fly as low as he dared, while one of the men held a Geiger counter out the window. Nothing registered. Later, the men landed and hiked to the mound. They brought along a gas-powered ice auger borrowed from a Point Hoper, but the ground was frozen below a shallow depth. They resorted to using sledge hammers to pound an eight-foot steel rod into the frozen mound. After reaching a depth of about two and a half feet, they pulled the rod out and held the Geiger counter up to its tip. When the meter registered a slight increase in activity, the Corps of Engineers man declared that he'd seen enough. The investigators filled in the hole, marked the mound as off limits, and left.

"I'm scared now more than ever," said the mayor of Point Hope. "I'm just glad three of my people were out there and saw that meter. I trust them." As a news story explained, Point Hopers were "desperately seeking an explanation for the high number of cancer deaths in their village." Even though the Geiger counters had confirmed what they considered an inexcusable betrayal, Point Hopers felt a sense of vindication and relief. As they saw it, they were finally able to point to the cause of their medical troubles, and the discovery of the radioactive waste mound at Ogotoruk Creek seemed to confirm long-standing Eskimo tales of suspected buried poisons. But three paper studies conducted by the Centers for Disease Control over the previous decade all concluded that cancer rates on the North Slope were not out of line with the rest of the country, and that dietary factors and smoking were more likely causes of the Eskimos' cancers than environmental pollutants. Even though the cancer rate at Point Hope was once reported to be 38 percent higher than the national average, epidemiologists could not conclude the difference was statistically significant in a sample size as small as the population of Point Hope (600). But, as news reporters discovered, the Point Hopers "have never believed that." North Slope Borough mayor Jeslie Kaleak told an audience in Point Hope, "I have a personal interest in this. I have lost [members of my family] to cancer that must have been caused by Project Chariot." Even a veteran public-health worker in Arctic Alaska ascribed cancer deaths to the waste mound at Chariot camp:

> My government has made me an unknowing partner in its duplicity. . . . They used me to promulgate a lie that was killing people. They assured me over and over again that the people of Point Hope were imagining their problems. They had dreamed up their tales of waste being buried near their village.

Actually, the Native Alaskans' suspicion that they were sometimes used as unwitting participants in a radiation experiment was not without some basis in fact. Project Chariot was intended, as Livermore officials said, to be a "meaningful radioactivity experiment." A billion and a half curies of radiation would have been vented, much of it dumped on the Eskimo people's hunting ground, and scientists would return afterward to monitor its effect on the biota, including people. Whether Chariot was planned in the vicinity of an Eskimo community because the people there were few in number, nonwhite, not uniformly proficient in the English language nor highly educated in Western subjects—in short, because they lacked any sort of political clout—is not known. But it is known that Alaska Natives, including people from Point Hope, had been subjects in radiation experiments carried out by the U.S. military in the same time period, the mid–1950s. More than 100 Eskimos and Indians from six villages in northern Alaska were given radioactive iodine as part of an experiment conducted by the Arctic Aeromedical Laboratory at Ladd Air Force Base in Fairbanks. Subjects were given single doses of up to sixty-five microcuries (sixty-five millionths of one curie) of iodine 131 in an attempt to evaluate the role of the thyroid in human acclimatization to cold. Many subjects were dosed more than once. These doses were many times more than is considered appropriate under present medical standards, where only about six to ten microcuries are administered for diagnosis of thyroid anomalies. And that is only done when an illness is present or suspected, Healthy people, of course, are not dosed at all.

The obvious question is to what degree these Native people could have given informed consent to their participation in the study. At that time, many of the subjects in remote villages would not have spoken any English at all. Their own language would not have included a word for radiation. For many, the concept would not have been in their experience or education. The medical doctor who conducted the study remembers employing a bit of technological magic to secure one village chief's cooperation:

> We had with us a portable x-ray machine which generated enough x-rays so that we could take an x-ray picture of the chief's head with his brain and everything showing. And he was so impressed, of course, that . . . he joined immediately. And we were then permitted to live with them, share everything with them and make a very thorough study of their metabolism, their nutrition, behavior, hunting and all that sort of thing.

Surviving participants say they had not been informed that they were involved in an experiment that had the potential to affect their health.

*   *   *

As soon as the existence of the Ogotoruk Creek dump site was confirmed, Alaska's Senator Murkowski flew back to Alaska from Washington. It was September in an election year, and Murkowski had been spending a great deal of his campaign energy detailing the legacy of Soviet nuclear contamination in the Arctic. Here, all of a sudden, was an instance of *American* nuclear pollution, and though the amount was insignificant by comparison to Soviet dumping, it had turned up in the senator's own backyard. Joining Governor Walter Hickel, Murkowski flew out to inspect the Chariot site and to attempt to reassure the people of Point Hope. *The New York Times* reported that an elderly woman spotted the senator in the village and threw herself at him shouting: "You have poisoned our land!"

Both Hickel and Murkowski demanded that the federal government clean up the site. Murkowski met with Secretary of Defense Cheney and Energy Secretary Watkins, and Governor Hickel asked President George H. W. Bush to give cleanup of the site the highest priority. Next, Alaska's Senator Ted Stevens secured $1 million for site cleanup through the Defense Appropriations Committee. By February 1993, the Department of Energy put the cleanup figure at $3 million, and the agency said it would spend the money even though they were sure the mound posed no risk.

In March 1993, Stevens convinced the Senate Governmental Affairs Committee to launch a Government Accounting Office (GAO) investigation to identify other radioactive waste sites in Alaska. The GAO study initially was restricted to "abandoned" military sites, but it was pointed out (by the author) that such a limitation would likely miss the most contaminated sites in the state, excepting Amchitka, namely those at *active* military bases such as Fort Greely. Also, an investigation limited to federal land would miss federally licensed disposals on state land, such as the burial of low-level radioactive wastes by the University of Alaska Fairbanks on its campus and at the "Musk Ox Farm," a popular tourist attraction in Fairbanks. In apparent response, the study was broadened to include non-federal sites, active military bases, and some other sites about which it received information.

Issued in July 1994, the slim report is far from an aggressive investigation. Essentially, the GAO asked the principal polluters—the Department of Defense and the Department of Energy—to identify and characterize their dump sites, then it incorporated into its report whatever information the agencies had cared to present. There were no terrestrial surprises, though at the time of the report's publication, the army had yet to check 138 defense facilities for radioactive waste. But offshore, the report acknowledged that between 1958 and 1969, at a point in the ocean 350 miles southwest of Ketchikan, Alaska, the Boeing Company of Seattle (with the AEC's permission) dumped 294 55-gallon drums of radioactive waste (including strontium 90 and radium 226).

As the Chariot site nuclear-waste dump controversy escalated, lawyers, environmental consultants, bureaucrats, and reporters flew, phoned, or faxed north like prospectors in a high-tech gold rush. The North Slope Borough expected to spend $150,000 on their consultants, and they wanted federal reimbursement. The Department of Energy preferred to leave the material buried to decay to harmless levels over the next century, but said it would spend the $3 million in cleanup costs in order to calm the Eskimo people. By May 1993, the estimate had jumped to $5 million, by June, $6 million, but the agency was still inclined to spend the money. Plans were drawn up for the establishment of a fifty-one-man camp at Ogotoruk Creek to accommodate cleanup workers. If the government spent more than $1 million removing the contaminated soil from Ogotoruk Creek, said one DOE contractor, "it could easily go down in history as the most expensive and least cost-effective radiation protection measure ever undertaken in the history of mankind."

While the disposal of radioactive waste near Point Hope was carried out in an insensitive, secretive, and illegal way—and while it illuminates the AEC's motivations at that time, and its management of nuclear wastes since then—it is hard to reconcile the frantic public reaction when one considers Nevada. Not quite half of the radioactivity buried at Ogotoruk Creek consists of 17.5 pounds of irradiated dirt collected from the Sedan test at the Nevada Test Site. This material (perhaps a shovel's worth) lay buried under four feet of fill, thirty miles from the nearest dwelling (at Point Hope). That leaves twenty-four billion pounds of the same Sedan ejecta sitting on top of the desert floor in Nevada, blown about by the winds for forty-five years. If it was reasonable to demand that the government spend $6 million hauling out the comparatively microscopic Chariot contamination, what ought the people of Nevada to demand?

Or, nearer to home than Nevada, Alaskans might take a minute to consider how the Ogotoruk Creek disposal ought to rank alongside the serious radioactive contamination at Fort Greely, just outside the town of Delta Junction, Alaska (population 840). At the reactor site there, the military left behind—by the army's own reckoning—70,000 curies of radioactivity. That figure is roughly 2.5 million times greater than what was buried near Point Hope, and it does not include additional low-level radioactive waste that the army deliberately pumped into a well on the base, or that which they dumped into Jarvis Creek, both on the outskirts of Delta Junction.

"Remediation" programs such as the one performed at Ogotoruk Creek may become the pork-barrel projects of future decades. If so, it seems reasonable to ask where the megadollars will end up. Probably not with the people whose lands were violated, but rather with the weapons laboratories and their contractors, newly reconstituted as decontamination specialists—back to the geniuses who put the stuff there to begin with.

Perhaps even more worrisome is the fact that a near-hysterical public reaction to such disclosures will tend to validate government policies of deception and secrecy. If the public appears incapable of a rational evaluation of risk, then the agencies can be expected to extend a program of propaganda and disinformation that, while it may dampen excess anxiety, evades public accountability and corrodes democracy.

Alaska

It was not surprising that the Atomic Energy Commission chose Alaska as a venue for Project Chariot. Like the moon (which Edward Teller had also suggested as a target of experimental nuclear explosions), Alaska was often regarded as a barren wasteland, suitable for extracting mineral resources or as a laboratory for testing potentially hazardous technologies. At various times, the AEC considered other Alaska projects, such as blasting an instant harbor at Point Barrow with a five-megaton shot, or dredging Bering Strait with nuclear explosions.

In 1962, the first "portable" nuclear power plant to be built in the field attained criticality at Fort Greely, Alaska. Under emergency conditions, the army said, it would be possible to fly the components of such a plant to a remote site. The reactor operated for ten years as low-level radioactive waste was pumped into nearby Jarvis Creek and into a well that was drilled for that purpose.

As the Chariot scheme was being scrapped, intense debate in Alaska focused on another megaproject, Rampart Dam. Here the plan was to erect the largest dam in the world on the Yukon River near the village of Rampart. A vast marshy lowland known as the Yukon Flats would be submerged, and the 10,000-square-mile lake created would cover an area larger than Lake Erie. Again business and labor leaders, politicians, and the university president rallied behind Rampart. And again opposition arose from the usual quarters: principally, the Alaska Conservation Society allied with Alaska Native villagers. Boosters touted 5 million kilowatts generated annually, twice as much as produced by any other dam in the world. No one was more outspoken in support of Rampart than Alaska's U.S. senator and former governor, Ernest Gruening. Of the Yukon Flats, he said: "Scenically it is zero. In fact it is one of the few really ugly areas in a land prodigal with sensational beauty." Joining the ranks of developers who saw northern lands as desolate and even repulsive, Gruening told audiences that the region to be inundated was "nothing but a vast wasteland . . . notable chiefly for swarming clouds of mosquitoes."

Perhaps there *was* one man more outspoken than Gruening—his administrative assistant, former editor of the *Fairbanks Daily News-Miner,* George Sundborg. "Search the whole world over," said Sundborg, "and it would be difficult to find an equivalent area with so little to be lost through flooding. . . . In fact, those who know it best say the kindest and best thing one could do for the place is put it under 400 feet of water." Apparently, those who knew the area best were not the Native people who had occupied the territory for perhaps 10,000 years, but men more of Sundborg's stripe. Critics pointed out that seven Athabascan villages would be drowned by the project, but that didn't amount to much of a loss, according to Sundborg. After all, he said, the whole area contained "not more than ten flush toilets." It was a novel way to quantify the value of a culture, but not out of character for the comparatively uncivilized boosters. When Rampart Dam proponents paid any attention at all to the 1,200 people whose homeland and livelihood would disappear, they sometimes did so with a degree of ethnocentrism not readily distinguishable from insult. Gruening didn't mind saying, for instance, that these people lived in "an area as worthless from the standpoint of human habitation as any that can be found on earth."

Besides the impact to human beings, there were other serious ecological consequences of the proposed dam. The potential cost was summed up in plain language by the U.S. Fish and Wildlife Service: "Nowhere in the history of water development in North America have the fish and wildlife losses anticipated to result from a single project been so overwhelming." As usual, Sundborg was ready with a comeback: "Did you ever see a duck drown?" If the battle over Rampart had hinged on force of rhetoric alone, the dam probably would have been built. But conservationists drew on the lessons gleaned from the Chariot controversy and brought a new awareness of ecological relationships to the debate. Besides that, the fact that no market existed for the huge quantity of electricity did little to inspire federal support for the $1.3 billion construction-cost. Rampart was a dead issue by 1968, though an occasional letter to the editor in the Fairbanks paper will still call for its construction.

                ●   ●   ●

Another development scheme involving Alaska in the early sixties was promoted by the R. M. Parsons Company, a large southern California-based engineering firm. This scheme would reverse the flow of the Yukon and Tanana rivers. A system of dams, canals, and pumping stations—the canals excavated by AEC teams using nuclear explosives, the pumps powered by nuclear reactors—would send the Far North's glacier-fed water down the Rocky Mountain trench in western Canada, all the way to the arid American Southwest. Dubbed the North American Water and Power Alliance (NAWAPA), the dream was revived in the early 1980s and again in 1992 by right-wing political activist Lyndon LaRouche. In his 1992 presidential campaign literature, LaRouche noted with pleasure that Alaska governor Walter Hickel had proposed the construction of a plastic offshore pipeline to send Alaska water to California. "These initiatives," according to the LaRouche pamphlet, "show that once again, there is emerging a mood to solve problems, rather than to succumb to anti-growth propaganda and national economic devastation."

So far, escalating costs, if not environmental sensibilities, have kept NAWAPA merely on paper. Of course, the environmental consequences of radically altering these natural systems down the length of the continent would be staggering; but those who track water politics in America believe that NAWAPA's construction—albeit probably without the benefit of nuclear excavation—is a matter of "when," not "if."

Another megaproject touted at various times over the past 100 years by both American and Soviet technologists represents perhaps the ultimate contempt for the Arctic world: a plan to thaw it. Bering Strait would be dammed and huge pumps would force Arctic Ocean water into the north Pacific. In theory, the displaced cold water would allow the warmer Atlantic Gulf Stream to be drawn north, thawing the polar ice pack. If the scheme worked mechanically—which is doubtful—it could have the potential to alter climate on a global scale. Atomic Energy Commission chairman Glenn Seaborg thought such nuclear-aided proposals to alter weather and climate "intriguing," but risky.

                ●   ●   ●

Pierre Salinger, Kennedy's press secretary, recalled hearing through back channels between 1961 and 1963 that "a plan was proceeding in the Defense Department to construct some nuclear testing sites for the United States in Alaska." Salinger thought testing nuclear bombs in close proximity to the Soviet Union might have repercussions that Kennedy would not favor. He told the president, who had not heard about it, and "a call from the President to Secretary of Defense Robert McNamara (who had also not heard of the plan) brought the whole idea to a shuddering stop."

For the moment.

At one point during the Project Chariot controversy, Alaska's senator Ernest Gruening opined: "If they wanted to blow a hole in the ground they should have picked an uninhabited island where there would be no possibility of danger to anyone." The AEC seemed to take note because in 1964 the agency, along with the Department of Defense, set up operations on Amchitka Island in Alaska's Aleutian Chain, and declared it a temporary nuclear test site. The island was said to be uninhabited, the nearest village being more than 200 miles away. As Melvin L. Merritt, whom the AEC hired to write an environmental study, liked to say: "There's no people for several hundred miles for the simple reason that nobody would ever want to live on Amchitka." Of course, Merritt didn't mention that the U.S. military had thoroughly trashed a good bit of the island a few decades earlier during the World War II Aleutian campaign. From the air, an extensive network of ruts from military vehicles was still visible in the tundra twenty-five years later; a couple of thousand tumbled-in Quonset huts had been left behind to rot; an estimated 1 million telegraph poles jutted at odd angles or dangled, broken, from the wires; and under the constant rain and fog, numberless oil drums rusted and leached their toxic contents into the soil. As one writer observed when he toured the "subarctic junkyard" during the AEC's drilling operations, "an ecologist or naturalist . . . feels a revulsion, as an artist would standing before an ancient masterpiece defiled by vandals."

In spite of the military and the AEC's degradation of the landscape, wildlife abounded on Amchitka. The island's waters were home to pink, coho, and sockeye salmon; cod, halibut, and sole; and Stellar's sea lion, seals, and one of the world's largest single populations of sea otter. On shore, perhaps 100 pairs of American bald eagles nested on the rocky cliffs, as did twenty pairs of peregrine falcon. The cliffs were also home to dense colonies of seabirds: puffins, cormorants, murres, and guillemots. Emperor geese stopped at the island on their migration. Amchitka had been part of a national wildlife refuge for more than fifty years before the AEC and Department of Defense arrived. But the 1913 executive order that established the refuge permitted use of the land for "lighthouse, military or naval purposes." The document had been written decades before nuclear fission had even been imagined, but the AEC and the secretary of the interior found that testing atomic weapons at the wildlife refuge was "consistent with the spirit of the Executive Order."

While Gruening's comment may have revived the AEC's interest in Amchitka, the agency had had its eye on the island since early 1950. According to the AEC's official history, when bomb testers began a search for a continental test site, Amchitka was the first such location recommended by the AEC and the DOD for a nuclear test. An atmospheric and an underground test were scheduled for the fall of 1951. In late 1950, President Truman even approved the plans, called Operation Windstorm. But a search for a more convenient site continued, and toward the end of 1950 planners zeroed in on the Las Vegas bombing and gunnery range in Nevada. The Nevada Test Site accommodated all the AEC's testing until the 1960s, when the military wanted to test a warhead too large for safe detonation in Nevada. After what the AEC claimed was a "considerable search," the bomb testers went to Amchitka.

Three underground nuclear detonations occurred on Amchitka: Long Shot in 1965, Milrow in 1969, and the largest underground test in U.S. history, the five-megaton Cannikin shot in 1971. Long Shot helped military analysts discriminate between seismic signals generated by underground nuclear tests and earthquakes. These data were of use in verifying Soviet compliance with any test-ban treaties negotiated. With Milrow the AEC set off a one-megaton detonation as a "calibration" test to see what sort of geologic effects might be expected from Cannikin's five-megaton yield. In the protest against Cannikin, a new group of political activists joined the antiwar and disarmament protesters: environmentalists. The Sierra Club and Friends of the Earth focused on the possibility that Cannikin might trigger a massive earthquake and tidal wave. A dozen protesters from Vancouver, British Columbia, set off for Amchitka in an old wooden halibut packer. To celebrate the fusion of environmental and political activism, they called their voyage the Greenpeace, and launched a new international environmental activist group.

When fired, the Cannikin blast lifted the earth over ground zero with such violence that shorebirds standing on the beach above had their legs driven up into their bodies. Overpressures in the ocean caused the eyeballs of sea otters and seals to burst through their skulls. On the island, rocky bluffs and sea stacks crashed into the water, lakes drained, and at surface zero the land collapsed into a subsidence crater. What long-term radiological hazard will result from the Amchitka detonations is not known. Radioactive tritium, however, was detected in mud above the Long Shot ground zero within a few years of the detonation. AEC geologists had calculated that no such migration of radionuclides would occur for 400 years.

In 2000, Amchitka construction workers were added to the list of nuclear weapons workers declared by the Congress to be eligible for compensation for "the health effects of their labors," as one senator said. Workers with most forms of post-exposure leukemia, multiple myeloma, non-Hodgkin's Lymphoma, and any of a specified list of diseases associated with radiation exposure—or their heirs should they have died from these illnesses—became eligible for $150,000 payments. Of the 1,400 former Amchitka workers who registered with the program, a higher number reported cancers than would be expected from a group of the same size and age from the general population. Thirty-five leukemia cases, for example, were reported, when only 1.9 would have been expected. Ten times the expected number of pancreatic cancers were reported.

•   •   •

Walter Hickel became governor of Alaska in 1966 and tried to hitch his dream of "opening up" the Arctic for resource extraction to a rising nuclear star. He wrote AEC chairman Glenn Seaborg that one of his dreams was to "see economic development in Interior Alaska" by "extending the Alaska railroad into the heavily mineralized area of the North country." Hoping that the AEC might assist with that effort, Hickel offered Alaska's North Slope as a new testing ground for nuclear bombs: "I would appreciate your opinion of the potential and possible use of Arctic Alaska as an atomic testing site." And the AEC did conduct a "geological exploration" of an area north of the Brooks Range near Point Lay as an alternate site in case of unforeseen difficulties at Amchitka Island, which was the agency's preferred location. Hickel sent the AEC repeated inquiries about the Point Lay option, until the agency finally told him that "work near Point Lay would clearly be more difficult and costly than on Amchitka. We . . . do not plan to return to that area."

•   •   •

In the 1960s, when the U.S. military considered where in the world to test deadly nerve gas and germ-warfare agents, they chose Alaska. At the secret Gerstle River Test Site, part of the 1,200-square-mile Fort Greely Military Reserve in Interior

Alaska, the army experimented with some of the most deadly chemical agents known to man. Mustard gas and the lethal nerve gases known as VX and GB were packed into rockets and artillery shells and either launched or fired from howitzers into the spruce forests and marshes of the Gerstle River area. Of course, not every piece of ordnance detonates as it is supposed to do, and "the test area remains a no-man's-land" of unexploded ordnance, according to a military historian.

Sixty miles east of the Gerstle River testing grounds, the army selected a site near Delta Creek as a place to test bacterial disease agents in the open air. It was one of only two locations in the United States where germ-warfare organisms are acknowledged to have been released into the environment. In 1966 and 1967, the army's tests at Delta Creek sought to determine the effectiveness of the tularemia bacteria in subarctic conditions. Tularemia (after Tulare County, California, where it was first found) is sometimes called "rabbit fever," though it can occur in 100 mammals, in insects, birds, fish, and water. It is an acute infectious disease related to bubonic plague. Onset symptoms occur suddenly and include extreme weakness, headache, recurring chills, and drenching sweats from high fever. Untreated, death occurs in about 6 percent of cases.

In one incident uncovered by the Alaskan investigative reporter Richard Fineberg, the army lost hundreds of rockets laden with an aggregate ton of lethal nerve gas. The rockets, which were slated to be destroyed, were stacked on a frozen lake in the winter of 1965. But, for some reason, the soldiers failed to retrieve the rockets before the spring thaw, and they sank to the bottom of the lake, apparently forgotten. In a few years, with personnel turnover, the story of lethal nerve gas rockets lying at the bottom of one of the lakes in the military reserve slipped into local folklore. In 1969, however, a new commander at the test center followed up on the rumors, tracked the evidence to a lake about a mile from the Gerstle River facility, and ordered it pumped dry. On the muddy bottom, the crew found more than 200 nerve-gas rockets—one leaking. A small drop of the stuff on the skin can kill a person in minutes.

The military undertook a general "cleanup" of the Gerstle River Test Site in 1970, though perhaps it is more accurate to say the contaminants were consolidated. The army simply heaped up 4 million pounds of chemical munitions, gas masks, contaminated clothing, and equipment into two mounds and covered them with dirt. An attempt to transfer the "restored" land to the Bureau of Land Management resulted in BLM declining the offer. The army cannot certify that the land is decontaminated because, as one historian has written, "when the program terminated in the late 1960s, records of the testing inexplicably disappeared and remain missing, apparently destroyed. What files remain confirm sloppy record-keeping which failed to identify the type of weapons being tested or how and when they were disposed of."

＊   ＊   ＊

If the logistical difficulties and cost of transporting radioactive wastes from the Lower Forty-eight to Alaska were not so great, the Department of Energy would no doubt be eyeing the state as a terrestrial dump site (it did license Alaska offshore dumping, as mentioned elsewhere in this section). The Soviets, however, found it quite convenient to call on their northern regions to absorb the environmental burden of cold war militarization. In the Kara Sea off the Arctic island of Novaya Zemlya, they have scuttled ships with damaged nuclear reactors onboard, including no fewer than five nuclear submarines containing ten reactors loaded with nuclear fuel. At the western edge of the Asiatic steppes, a person who merely stands for one hour along the bank of Lake Karachay near the plutonium plant at Chelyabinsk will receive a lethal dose of radiation.

＊   ＊   ＊

"CHASE" was the name of a U.S. Navy program to scuttle aging Liberty Ships loaded with surplus explosives. The acronym stood for "cut holes and sink 'em." In 1967, the navy loaded the Liberty Ship *Robert Louis Stevenson* with more than 6 million pounds of unexpended munitions: TNT, rocket fuel, bombs, torpedoes, mines, etc., then towed her to a spot thirty-five miles off Amchitka. The ship had been fitted out with pressure-activated detonators designed to blow at a depth of 4,000 feet. After the scuttling valves were opened, however, the *Stevenson* didn't sink as quickly as expected. Wind and current had begun to carry her back toward Amchitka when she disappeared into a fog bank. Eventually, the navy found the vessel through the use of sonar devices. It had come to rest in 2,800 feet of water, seventeen miles from Amchitka. The ship sat—and still sits—half the intended distance from Amchitka, a 6-million-pound bomb in Alaskan fishing waters. Environmentalists fear that an earthquake might send the ship sliding down the steep slope upon which it rests, triggering a detonation equivalent to a two-kiloton nuclear explosion. In any event, they say, mercury, lead, and other toxins will find their way into marine organisms as the weapons casings corrode.

Alaska's Aleutian Islands also attracted the U.S. military when it needed to establish a dump site for chemical warfare agents. Between 1947 and 1949, the military dumped nearly 2 million pounds of poison gas canisters (mustard and arsenic blister agents) into the sea just twelve miles off Chichagof Harbor on the Aleutian Island of Attu.

There has been no change in a U.S. policy that apparently considers Alaska a wasteland suitable as a dumping ground, as a test site for dangerous technologies, and as a practice bombing range. The military has expanded its training activities in Alaska, including the use of vast new areas of the country for its "live fire" exercises.

●   ●   ●

Early on the morning of Good Friday, March 24, 1989, the tanker *Exxon Valdez* ran aground on Bligh Reef, spilling 11 million gallons of Alaskan crude oil into Prince William Sound in Southcentral Alaska. It was the largest and most disastrous oil spill in U.S. history. Despite repeated and categorical assurances from the oil companies that contingency plans were in place to handle any eventuality, the spill response was demonstrably inadequate. Also clear to heartsick observers around the world was the fact that Alaska remains vulnerable to the so-far empty promise that technology can manage development in wilderness.

In addition to the national missile defense silos installed at Fort Greely, another launch facility was built in Alaska on Kodiak Island. Vowing that all operations would be entirely commercial and non-military, Alaska's state legislature had begun the spaceport project as a state-owned corporation intended to stimulate economic development. But when private investors declined the opportunity to invest, millions in federal funding arrived through the Air Force, courtesy of Alaska's Senator Ted Stevens, then chairman of the Appropriations Committee. The complex soon became a military facility, or, as the mayor of Kodiak equivocated, "a privately-owned launch pad doing military launches." Missions at the Kodiak Launch Complex include missile defense testing operations.

The AEC

From 1954 to 1974, the Atomic Energy Commission operated as a unique agency mandated to both develop and regulate nuclear technologies. In effect, the agency policed itself. Finally in 1974, Congress split off the regulatory function, establishing the Nuclear Regulatory Commission (NRC) to oversee the private nuclear power industry and renaming the research divisions the Energy Research and Development Administration (ERDA). ERDA included some nondefense energy research, but was dominated by military applications. In 1977, Congress, acting at the behest of President Jimmy Carter, created a new cabinet department, the Department of Energy (DOE), to consolidate various energy-related agencies. The principal business at DOE is still related to nuclear weapons, and that effort is still under military control. The DOE continues to employ sophisticated public relations techniques in pursuit of a favorable public image. In fact, the U.S. General Accounting Office has reported that the DOE was spending "only 22 percent of its money on its mission, with the balance going to such things as public relations." The agency still proclaims that a vigorous nuclear testing program is necessary for national security. As one local DOE official said, "It's still the same; the only difference is the name over the door."

AFTERWORD

A writer of a work of this sort—historical investigative journalism, perhaps—sifts through thousands of bits of information, evaluating them from an evidentiary standpoint, then screening them again, choosing some as illustrative of important themes, discarding others as extraneous. It is the writer's choice, too, when it comes to the presentation of the selected facts. Some are set up as crucial information, others treated as background details. The point is that covering history is inherently subjective, a matter of judgment, an interpretive business. And because of that, I feel a responsibility to state explicitly what views inform my approach to this story, and to summarize what I take to be its meaning.

Project Chariot helps to illuminate a transitional period in environmental history and Alaska Native history. And, as a study of one small episode in the American nuclear experience, it animates many themes of the wider nuclear story. But perhaps more important, Project Chariot raises ethical issues that we still face and that touch many of our central institutions.

* * *

Although antipathy toward the federal government was an established trademark of Alaska journalism, in the case of Chariot, editors advised their readers simply to trust Uncle Sam to do right by Alaska. Editorial discussion was always predicated on the assumption that a desire for technological progress was the reality everyone shared. Of course, it was an ideology, and one that everyone did not share.

While they are private sector entities, newspapers enjoy special protection under the constitution. In turn, the public has come to expect a measure of public responsibility from the press: a healthy skepticism—if not a vigorous search for contradictions—in the official story; the willingness to look at a wide variety of sources and give voice to a wide range of perspectives; a commitment to objectivity, rather than promotion of an ideology. It is an expectation rarely met in or outside Alaska.

•   •   •

Religious groups perennially wrestle with the question of militant involvement in social and political issues. Is the Church just another pillar of the establishment, with a vested interest in a stable, hierarchically organized social structure? Should it confine itself only to the periphery of political conflict? It would seem reasonable, if the Church is to provide credible leadership in matters of moral philosophy, that it also provide leadership in applying that philosophy to social issues. While some church leaders in Alaska apparently suppressed criticism of this government project, the Reverends Richard Heacock and Keith Lawton saw their public anti-Chariot advocacy as part and parcel of their ministry. They recognized that it cannot be sufficient simply to defer to the judgment of government officials in matters of social responsibility. For to do so may be to tolerate ethnic cleansing, apartheid, genocide, slavery, and so on.

•   •   •

The reasons cited by the chambers of commerce and other business and labor leaders for embracing Project Chariot have a familiar ring to those who have followed the debates over economic development projects in rural areas: It will bring federal dollars to the area, create jobs for people, and put the area "on the map." Short-term economic contributions dominate the equation. Long-term, less visible, or noneconomic costs are seldom fully considered.

•   •   •

In many respects, science is big business, too. It is a fact of life that scientific research is expensive. And funding tends to come from large government agencies or large corporate entities, each of which can have strong, vested interests in the results. The Project Chariot case shows that it is not always possible to regard research opportunities from an apolitical, amoral, valueless vantage, especially when sponsors draw improper conclusions from the data. Les Viereck spoke to this issue in his letter of resignation from the University of Alaska: "A scientist's allegiance is first to truth and personal integrity and only secondarily to an organized group such as a university, a company, or a government." In a word, corporate and even scientific objectives must be subordinate to philosophical values. And a regard for human welfare must be the overarching canon of the scientific code of ethics.

●   ●   ●

The Western university is built on a concept of academic freedom—the opportunity to pursue any avenue of inquiry in an atmosphere of pure scholarship and freedom from political pressures. This special latitude allows a researcher to follow obstinately his or her own idea until its truth might ultimately be established. But throughout history, one ideology or another has infected the university: religious ideology in the Middle Ages, racist ideology in Germany, social ideology in communist Russia, and a mercantile ideology in America. During the 1950s and 1960s, the cold war came onto American campuses. It is not surprising that the University of Alaska lost sight of its mission and yielded to the cold war political ambience by dismissing Pruitt and Viereck. What *is* surprising—and the lesson here—is that some scholars were able to hold strong to the principle of academic freedom and were willing to pay the highest professional cost in defense of that ideal.

●   ●   ●

A reverence for such ideals as justice and truth is understood to be among the philosophic underpinnings of democratic governments. Yet it has become a matter of record (since the Freedom of Information Act became law and a number of whistleblowers within the agency have sounded off) that the U.S. Atomic Energy Commission and its successor agency, the Department of Energy, compiled a stunning record of willful manipulation of facts. "There is nothing comparable in our history," says former secretary of the interior Stewart Udall, "to the deceit and the lying that took place as a matter of official Government policy in order to protect [the nuclear arms] industry. Nothing was going to stop them and they were willing to kill our own people." At issue is the capacity and tendency of a government agency to circumvent the lawful administration of public affairs in order to advance its own agenda. Behind such institutional corruption may be a desire to save the country from a threat that, it is claimed, the citizenry does not fully appreciate. The fallacy, of course, is that in the process, the zealots trample the very institutions they rush to protect. Rationalizations that bypass the public in matters of public policy threaten democracy in the most basic way; they usurp what Jefferson called the "ultimate powers of society" from their only "safe repository . . . the people themselves." It is not too exaggerated to say, as Stewart Udall has done, that "the atomic weapons race and the secrecy surrounding it crushed American democracy."

*   *   *

The Project Chariot story is a tale of conflict—even scandal—involving passionate, radical, pioneering people. But it is more than that. Chariot illustrates why the most cherished institutions of a free society—a democratic government, a free press, the university, even the Church—cannot necessarily be accepted as seats of objectivity and candor. The lesson Project Chariot offers is that a free society must be a skeptical one, that rigorous questioning and dissent protect, rather than subvert, our freedoms.

Appendix

METHODOLOGY

The impertinent question is the glory
and the engine of human inquiry.

—Gary Trudeau,
Smith College commencement
address, 1987

This did not start out to be a book. In late 1987, I received a grant from the Alaska Humanities Forum, the state council that administers National Endowment for the Humanities money, to produce a film documentary on Project Chariot for public television. Three months into researching the project, my collaborators, the public television station affiliated with the University of Alaska Fairbanks, began to have second thoughts. At a meeting of the station's management, as one senior staff member told me, an administrator suggested that it might not be in the station's best interest to produce this particular documentary. The university budget had been radically cut in 1987, and the elimination of the station was briefly considered as a cost-saving measure. My script outline—though it had been vetted by the scholars on the Alaska Humanities Forum board—was judged unacceptable by the administrator. Among other things, he worried that the program would antagonize former university president Wood, who still had many allies on campus. In a time of severe budget constraints, it was reasoned, the station could ill afford to make enemies of senior university officials. When asked how the station could gracefully manage the unprecedented action of withdrawing from a grant proposal *after* it had been funded, the administrator proposed that I could simply give the money back. Apparently no one at the meeting rose to remind management just what the mission of public television might be, or to observe that the administrator's suggestion amounted to censorship.

In a short while, the station formally withdrew from the project, citing financial exigencies and the press of other work. Perhaps these realities had some influence on the station's decision, but it is hard to imagine that the self-preservation motive did not hold sway, as well.

Though I didn't think so at the time, it was a lucky break. Gary Holthaus, the director of the Alaska Humanities Forum, suggested that I keep the money and redesign the project. I decided to develop an archival collection on Project Chariot for the University of Alaska Fairbanks archives. I would conduct and transcribe oral history interviews, visit government repositories, assemble documents, photographs, films, other tape recordings, clippings, articles, and so forth. I would use this material as the basis of a book, but I would also leave the collection behind for other researchers, scholars, producers.

* * *

Tape-recorded oral history interviews seemed a good way to document this story for several reasons (beyond the significant advantage of being able to hear *how* the words are said). First, the people who were most deeply involved with Chariot tended to be busy people unlikely to take the time to write down their recollections if asked to do so. But they were likely to give an interview, even one of a couple hours duration, and the result might be fifty or more pages of transcript, a substantial primary source document. Second, the paper trail within the American nuclear establishment is difficult to access, which makes personal accounts the more valuable. Third, audio recordings (as well as photos and films) would be useful in producing future radio or television documentary programs. And finally, unburdened by a cameraman and bulky video gear, I could travel farther, interview more people, and still do it more cheaply.

In the spring of 1988, I traveled to eleven cities in the Lower Forty-eight states and Canada, stretching my grant dollar by sleeping at youth hostels, and even the Thirty-fourth Street YMCA in New York City. Besides the interviews, which included all the living members of the Chariot bioenvironmental committee, as well as Barry Commoner, Al Johnson, and others, I visited the Lawrence Livermore National Laboratory archives.

Throughout the summer of 1988, I recorded interviews in Fairbanks, Anchorage, and Juneau. I tried to interview every significant player in the thirty-year-old episode, but some key people were already deceased. In fact, included in this category were three people who would have been at the top of my list: Don Foote, John Wolfe, and Daniel Lisbourne. David Frankson did not die until the following year, but I understood from Point Hope people that he was in a nursing home and non compos mentis. Others who would have been important sources, but who died before I could interview them, include Howard Rock, Albro Gregory, and Kermit Larson.

In August of the same year, I traveled to Point Hope to hear firsthand, and to record, the views and remembrances of the Inupiat people. I even managed to take a 100-mile, all-night, Honda three-wheeler ride from Point Hope to the Chariot site and back. A word on this excursion might not be out of order in a discussion of research methodology, Alaska-style.

●   ●   ●

Reggie Oviok, whom I chanced to meet on the Point Hope spit one mild evening, offered to show me the way to Ogotoruk Creek. The next night, after he got off work, I rented his father's three-wheeler and we started off on our two bright red, big-tired machines. We zipped down the beach on the Tikigaq spit, past the tent camps that the people use in summer, past dozens of beached walrus carcasses and fifty-five-gallon drums (both rust-colored and decomposing), twenty-five miles to the Cape Thompson hills. Here, though we were only perhaps seven miles from Ogotoruk Creek, the cliffs forced us inland and we slowly negotiated miles of tussocks. In first gear, I carefully ascended each individual two-foot-high hump, only to crash down the other side or pitch sideways at the peak until I had to jump off into a foot of muck in order to keep the machine upright. The ride was unbelievably jolting.

Eventually, we ascended a promontory (known to the flatlander Point Hopers as Suicide Hill, but which struck me as a merely respectable knob) that opened up a grand view of the Tikigaq spit arcing out to sea like a causeway to the horizon. It was the very same view Beechey had noted in his ship's log, I remembered.

Reggie was a better driver than I, and his four-wheeler was more stable than my three-wheeled machine. The distance between us grew until, just as I got the vehicle good and stuck in a mud patch and called to him, he slipped over a ridge and out of sight. With each futile attempt to blast out, I sank deeper into the ooze. Finally, I discovered that the bike would rise and inch forward if I got off and operated the throttle while slogging alongside through the bog. Apparently, it was the buoyancy afforded by the fat tires that did the trick.

At the last ridge before the Ogotoruk Creek valley, Reggie was waiting for me. As we dropped down the steep grassy hillside to the creek, I noticed a sun-bleached skull and rack of a caribou not far from the scavenged carcass of a long-dead three-wheeler. We crossed and recrossed Ogotoruk, aiming from bar to bar, and here the tires saved me again. Several times, I picked crossings that were not as shallow as I had reckoned. With a very low-angle sun bouncing off the water, what I took to be riffles caused by shallows four or five inches deep sometimes turned out to be wind-caused ripples on the surface of water that was four or five *feet* deep. But instead of sinking to the bottom as I would have expected, I found that the tires' flotation permitted me actually to bob along upright, the muffler bubbling, the nubbins on the spinning tires paddle-wheeling me across to the other bank. Amazing machines, I thought.

By midnight, Reggie and I were eating *muktuk* at the mouth of Ogotoruk Creek. Ground zero was a picture of peace. The little creek flowed freely; light-orange clouds radiated from the horizon over a gently lapping, almost lavender sea. A small ghost town of derelict huts survived at the AEC camp, and here and there on little knolls, long immobile and rusting road graders and weasels stood a kind of lonely watch, like some species of faithful domestic animal, confused as to why so much had *almost* happened, but hadn't happened.

Because we were so far west in the time zone, solar midnight—and darkness—did not catch up with us until a few hours later when we were well into the hills. Reggie's headlight was a pinpoint of light three-quarters of a mile ahead as I jounced over the tussocks. I thought my teeth would rattle out of my head; my thumb had all but lost the strength to hold the accelerator down; my wrists and knees were shot.

An hour and a half into the Cape Thompson hills, Reggie's headlight seemed to stop moving. It turned out he'd pulled up where a grizzly bear had jumped out in front of him. The bear had a marmot in its mouth, thankfully, and ran off. But, since he had the only rifle, Reggie thought he'd wait for me. We made the beach at three-thirty in the morning and stopped for tea from the Thermos. Looking up, we saw a fog rolling in with amazing speed. The hills we'd just descended were socked in. It was good timing. Reggie said that if it had come two hours earlier, we would have gotten lost in the hills for sure. We loaded up, and I followed the receding beacon of

Reggie's red taillight through the damp mist, dodging the hazy outlines of walrus carcasses, and made Point Hope by 5:00 A.M. Eleven days later, my hands and wrists were still so numb that I could barely type up an account of the trip.

●   ●   ●

Other research activities were less taxing, physically.

Though my travel funds had all been spent, I knew that my list of interviewees did not sufficiently represent the Livermore/AEC viewpoint. Steve Wofford, the assistant archivist at Livermore, helpfully tracked down for me several former laboratory officials and other AEC personnel who had worked on Chariot, and in the summer of 1989, using my wife's frequent flyer miles, I again traveled outside for a round of interviews. I also visited again LLNL's archives, as well as the Department of Energy's Nevada Operations Office (NVOO) in Las Vegas. I was even able to arrange for a tour of the Nevada Test Site and to visit the enormous Sedan crater, the hole blasted in lieu of the Chariot experiment.

Altogether I obtained about eighty hours of recorded interviews with about forty people in the Lower Forty-eight and Canada and across Alaska. I also located eight other relevant recordings in the oral history collection at Rasmuson Library. Most of these recordings have been transcribed and total about 1,000 single-spaced pages. They are bound with an index in the two-volume work *Project Chariot: A Collection of Oral Histories*, 1989, D. O'Neill (comp.). Copies are available at Rasmuson Library, Lawrence Livermore National Laboratory, and at various repositories in Alaska.

●   ●   ●

I do not know exactly where the line between oral history and journalism might be drawn, particularly when the topic under investigation is controversial, and the interviewer is obliged at times to be persistent, even tough. I do know that interviewing Edward Teller is a very different proposition from sitting at the kitchen table of a kindly old-timer and getting a few yarns down on tape. I know that little news or much of interest to researchers and scholars will come of an interview like one Teller told me he'd enjoyed, wherein the interviewer, he said, "let me talk about anything I liked." But there are a few differences between journalism and oral history that do come to mind.

First of all, there are curatorial matters. I always obtained signed releases, which transferred copyright ownership of the recording to the University of Alaska library. The recordings have been indexed in various ways, stored in a public facility, and are available for anyone to use, copy, and quote. Secondly, I think my interview strategy might differ somewhat from a more journalistic approach. For instance, I tried to start the interview with questions related to the interviewee's personal background, sometimes extending back to his or her forebears. It has always astonished me how fruitful such seemingly pro forma questions can be.

Also, I tried to illicit the interviewee's ideas without first suggesting either what the salient issues were, or any particular analysis of them. This sort of interview contrasts, I think, with those generally conducted by reporters. I wanted to offer my informants the opportunity to define the issue. I hoped that, well in advance of being put on the defensive, my interviewees would raise for discussion a particular controversy, frame the analysis along the rhetorical lines of their choosing, and present their views in the most unruffled, articulate, and persuasive way. But after that was done, I saw it as my duty to call their attention to internal inconsistencies, or information from the historical record, or to the statements of writers and thinkers, especially when that information was at variance with my source's views. In a word, I considered it my job to ask the question that would logically occur to an informed and skeptical listener today, or fifty years from now. In this respect, I borrowed more from the journalists' stock-in-trade than from the typical practitioners of oral history.

Sometimes this approach got me in trouble, as the interviewees thought they perceived a direction to my questioning and ascribed to me a point of view critical of them. For example, I might first ask a newspaper publisher to outline what he saw as the responsibilities of the press when covering controversial public policy issues. Then I might ask a question like: "Working from that definition of professional standards, how do you assess the performance of the Alaska press during Project Chariot?" Finally, and especially if the editor did not raise the subject himself, I might summarize the published work of a mass media scholar who had criticized the Alaska press's coverage of Chariot and ask for the interviewee's reaction. This seemed to me a reasonably value-neutral way to gather information, but sometimes the interviewees would bristle, saying something like, "I see from the drift of your questions that you believe we behaved unethically." The truth was, I was as likely to agree as disagree with the notions to which I asked the interviewees to react. And sometimes I honestly had not yet formed an opinion at all.

In only two of the forty-odd interviews was my interlocutor genuinely hostile. Edward Teller, for example, did more than bristle. Eight minutes into our interview, Teller's irritation reached criticality. Calmly, he asked to take a break. When he was assured the recorder was shut off, Teller detonated. He cursed loudly and with great facility. He said my questions were politely phrased, but substantively rude. Several times, when I tried to explain that the questions were important for the historical record—difficult, perhaps, but not inappropriate or disrespectful—he cut me off, simply repeating his position in a louder voice. At one point, he accused me of asking questions that were inescapably incriminating, similar to, he said: "Have I stopped beating my wife? When did I stop beating my wife?" I was astounded. I knew this logical fallacy, sometimes called "presupposition," and I had not committed it. I was speaking with a person whose intelligence was indisputably colossal, and yet he was telling me something that he must have known to be untrue, that I knew to be untrue, and that right before us and eminently retrievable sat tape-recorded *proof* that it was untrue. A discussion of how we might proceed got nowhere, and eventually Teller asked for the return of the release he had signed. I gave it to him; he tore it up. Still, the interview was, for me, enormously illuminating and useful. I did not learn much from Teller about Project Chariot, per se. But I think I got a glimpse of the man: I beheld the reaction these topics could spark within him, and I saw his method activated.

In the other rancorous encounter, the interviewee also asked me to turn off the tape recorder for a moment. I had asked him, as a point of clarification, if he thought that the work of one of the Chariot investigators was insufficiently scientific—a charge that, it seemed to me, he had just, in essence, made. When I turned the recorder off, he too exploded. He said the person we had been speaking of was a dirty, filthy pig who lived amid human excrement in his cabin (a reference, I assumed, to the "honey buckets" used as a toilet facility in some parts of Alaska). Furthermore, he said, spitting out the words, the Eskimo people "hated his guts." "Now," he demanded, "is that the sort of statement you want to put in your archives?" Barely, I resisted a mulish inclination to reply in the affirmative. After all, the information—especially the part about the Eskimos' disposition toward the Chariot critic—was clearly material. (Later, I made a point of checking out this assertion, and all of the Eskimos in question with whom I spoke—perhaps a dozen—repudiated the claim.) Also, had the man's sentiments been recorded, it would have documented an attitude of extreme hostility on the part of a key Chariot proponent toward a key Chariot critic. And that would have been relevant.

Despite these disquieting experiences, the interview strategy outlined seems to me to be reasonable and necessary. An important question—however diplomatically phrased—may be taken as impertinent. The result may be unpleasant. The question still must be asked.

●   ●   ●

Besides the oral history interviews, other primary source documents came from federal agencies and archives. During one of my trips to the States, I was able to visit the Department of Energy's repository for unclassified documents in Las Vegas known as the Coordination and Information Center (CIC), which is run by Reynolds Electric under contract with DOE. CIC maintains a computerized data base in which documents can be searched by key word. In addition to personal visits to CIC and Livermore, I filed Freedom of Information Act (FOIA) requests with various agencies, including the DOE's History Division, the DOE's Nevada Operations Office (NVOO), the Federal Bureau of Investigation, Central Intelligence Agency, Defense Intelligence Agency, and the Department of the Air Force. At various times, I sought the aid of Alaska's U.S. senators in securing responses to my requests. In general, as any investigative journalist knows, the effort to obtain information from defense, intelligence, or nuclear agencies is an exercise in exasperation. A few examples are illustrative: When I first visited Livermore, I was told several times in categorical terms that nothing at Livermore pertaining to Chariot was classified. I was brought a large stack of unorganized documents and told emphatically that this comprised the entire Chariot collection. On looking through it, I discovered six yellow index cards, each containing numbers and dates. As a secretary made copies from the file for me, I asked her if the cards held the place of classified documents not included in the file. She said that was correct.

Later, I wrote to request classification review of the six documents. Very promptly, two documents were declassified in their entirety, two were declassified with deletions, and two were judged too sensitive for declassification, even in part. Regarding the two documents that I received whole, though they were very telling, nothing in them remotely suggested the need for their classification originally, let alone their restriction for thirty years. Regarding the two documents that I received with sections blacked out, I doubted from the context of the documents that the withheld sections contained only information that required restriction. I decided to appeal the decision.

In the case of one of these partial documents, a single paragraph had been deleted. After more than a year of correspondence between classification officers at Livermore and at the Pentagon (the Pentagon procrastinated), the paragraph was released to me along with the admission that the information had probably never been legitimately classified to begin with. (The banned information turned out to be a simple reference to the existence of Distant Early Warning sites in Alaska, a fact that had been common knowledge to everyone in Alaska since the 1950s.)

In the case of the second document that had been sent to me with deletions, several whole pages had been withheld, and here the Livermore official would not budge. He said the pages contained information that must remain classified and that there was no likelihood of my ever seeing them in the foreseeable future. But some months later I found a copy of the same document at another repository. All the pages had been declassified excepting only two very small blacked-out segments that could have involved no more than a few words or numbers. Again, there appeared to be no reason why the bulk of the document had ever been restricted, or why Livermore still withholds several whole pages from researchers.

As a second example, I would mention the Department of Energy's Division of History. If you call there and ask for so ordinary a document as the minutes of an AEC commission meeting, you will learn two things. First, the archivist can put her hands on the document in a matter of minutes, and second, if even one word of the document is classified, you will wait two years for someone to blacken it out and send you a copy.

Federal agencies are required by law to respond to Freedom of Information Act requests within ten working days, and to appeals within twenty working days. But requests at many agencies can take years to process, law or no law. It took the CIA nearly a year and a half to answer my FOIA and allow that they did, in fact, have a document relating to Don Foote. As mentioned earlier, two years and seven months transpired before the agency finally denied my requests for a copy of it.

To my FOIA request of the FBI in Anchorage, I received a photocopy of a scrap of paper upon which the following had been written in longhand: *Index: Operation Plowshare, Project Chariot, AEC, Cape Thompson.* It could be seen from the photocopy that the paper contained staple holes and the holes of a two-hole punch. It was apparently a cover sheet of a larger document upon which an agent had written relevant key words for indexing. The key words had been checked off, and a stamp block had been initialed, indicating that the document had been indexed and filed. But the FBI insisted that this sheet comprised the entire document, notwithstanding the staple holes, the two-hole punch holes, and the illogic of indexing the list of index words. Furthermore, the mere words *Operation Plowshare, Project Chariot, AEC, Cape Thompson* had struck FBI agents in Anchorage as so potentially sensitive that they insisted on clearing its release through FBI headquarters in Washington, a move that ensured a year and a half's delay in the document's release. After exhausting appeals, I have received only this meaningless page, which obviously, at some time, accompanied a more pertinent document.

Many of my FOIAs, however, resulted in thousands of pages of documents. In addition to these, I also obtained extremely valuable correspondence files from several key figures in the episode, including Al Johnson, Les Viereck, Doris Haddock, and

Elizabeth Foster. Nearly all of the material I collected is being deposited into the collections at Rasmuson Library, University of Alaska Fairbanks. The library's collections already hold a sizable amount of Chariot materials, including, most notably, Don Foote's papers, collections donated by Les Viereck, William Pruitt, and Brina Kessel, and the files of the university administration. All of these materials, which probably amount to the largest collection anywhere dealing with Project Chariot, are available for use by researchers.

●　　　●　　　●

For all the difficulties I encountered in obtaining government records during the 1980s and 1990s when I did the research, the job would be far more difficult today—perhaps impossibly so—in light of the unprecedented restrictions to federal records imposed by the administration of George W. Bush. The ability of scholars, researchers, journalists, producers, or ordinary citizens to check on the activities of their government has been systematically restricted since Bush first took office. As one scholar has written: "The Bush administration waged a campaign against openness even before the 9/11 attacks." At the time of the attacks, September 2001, Bush's attorney general John Ashcroft had already begun work on a memo he would release a month later that would so tighten the provisions of the Freedom of Information Act as to thwart its purpose. Reversing Clinton administration attorney general Janet Reno's policy to "err on the side of disclosure," Ashcroft encouraged federal agencies to err on the side of secrecy by resisting disclosure under FOIA wherever technically allowable, and by expanding criteria for withholding documents. For example, he "directed agencies to withhold from release a new category of documents—those that could be considered 'sensitive but unclassified,'" according to a report from the House Committee on Government Reform prepared for Rep. Henry Waxman. He sent a strong signal to the agencies that when they chose to suppress disclosure of government records, they would be defended by his attorneys: "When you carefully consider FOIA requests and decide to withhold records . . . you can be assured that the Department of Justice will defend your decisions."

Delaying actions also became commonplace at federal agencies, and FOIA backlogs reached a record high thirty-one percent in 2005. As University of Massachusetts professor Philip Melanson told the House committee: "The Bush administration has radically reduced the public right to know via executive orders, court cases, and policy memos, more so than any administration in modern history."

Classified documents are, of course, exempt from requests under the Freedom of Information Act. But many classified documents are old. While outdated information represents no threat to national security, it may be of great value to historians, for example. The Clinton administration accelerated declassification processes, making millions of documents of interest to researchers available for the first time. In 1997 alone, Clinton oversaw the declassification of 204 million pages of documents. In 2005, the Bush administration processed roughly one seventh that number. Meanwhile, between 1995 (Clinton's term) and 2005 (Bush's term), the annual number of new documents being classified quadrupled.

Since the Freedom of Information Act was enacted in 1966, it has been a paradigm of enlightened public policy that has been emulated around the world. Today more than seventy nations have freedom of information laws. In addition, the emergence of electronic document management and the World Wide Web has meant that records can be easily, economically, and instantly available. Despite these trends, according to the Committee on Government Reform report, "no administration in modern times has done more to conceal the workings of government from the people" as has the administration of George W. Bush.

ACKNOWLEDGMENTS

It is appropriate that I should offer the first word of thanks to the late Don Charles Foote. Though he died before I came to Alaska, his vision, intellect, and ethicality have inspired me. The archival collection he left at Rasmuson Library both hooked me on the story and was the single most important source of information for my research.

I would like to acknowledge and thank the Alaska Humanities Forum (1987) for funding my initial research, and the Alaska and Polar Regions Department, Elmer E. Rasmuson Library, University of Alaska Fairbanks for continuous additional support. Bill Schneider, Gary Holthaus, and Don Muller gave me particular assistance at the forum, and at Rasmuson Library, Schneider, Pauline Gunter, and David Hales supported my digging into a campus episode that was still, to say the least, politically charged. No one helped me negotiate more obstacles in the course of this work than did Bill Schneider, my boss, mentor, colleague, next-door neighbor, and good friend. Our department head, David Hales, was particularly fearless and supportive as the nature of my investigations and writing began to draw notice.

I thank especially the men and women who consented to often lengthy tape-recorded oral history interviews: Elijah and Doris Attungana, Robert Atwood, Jim and Bonnie Babb, Max Britton, James Brooks, Ernie Campbell, Peter Coates, Barry Commoner, Patrick Daley, Teddy Frankson, Wayne Hanson, Dick Heacock, Gary Higgins, Dave Hopkins, Celia Hunter, Beverly James, Al Johnson, Gerald Johnson, Brina Kessel, Kitty Kinneeveauk, Rudi Krejci, Art Lachenbruch, Keith and Jackie Lawton, William Pruitt, Robert Rausch, George Rogers, Allyn Seymour, Michael L. Smith, Tom Snapp, Bill Stern, George Sundborg, Gerry Swartz, Edward Teller, Les Viereck, Alice Weber, Norman Wilimovsky, Ginny Wood, and William R. Wood.

Joe Foote gave me crucial assistance at the earliest stages of the project, and photographs and helpful information over the years. Berit Foote corresponded with me and also gave me a batch of excellent photos. Elizabeth Foster sent me a detailed and beautifully written account of the anti-Chariot activities of "The Campaign"; correspondence with her has been one of the unexpected joys of this research. Mrs. Foster, Doris Haddock, and Keith and Jackie Lawton sent me a sheaf of useful documents. I thank the people of Point Hope who made me feel welcome, especially: Reggie Oviok, Francis Schaeffer, Rex and Piquk Tuzroyluk, and J. J. Lane.

My friend Peter Coates, a British historian whose dissertation research helped get me started, and with whom I corresponded for six years, offered reactions to my ideas that I tended to regard as the measure of their validity. Al Johnson corresponded with me for five years and always provided—besides historical details—a tough intellectual critique of my developing opinions. Of the other scholars who helped me think through the many issues upon which Chariot touches, I thank particularly Bill Schneider, Rudi Krejci, Terrence Cole, Andrea Helms, Mary Mangusso, Mary Kancewick, Jim Gladden, Pat Daley, Beverly James, and Dean Kohloff.

In addition to their citation in the notes and bibliography, I want to further recognize here several people whose writings were particularly useful to me: Stephen Hilgartner, Richard C. Bell and Rory O'Connor, A. Costandina Titus, Peter Coates, Joe Foote, David Bradley, Trevor Findlay, Harvey Wasserman and Norman Solomon, Patrick Daley and Beverly James. Richard Sylves shared his insight and his notes with me, Stephen Skartvedt kindly sent me his thesis.

It can hardly be overemphasized that the assistance archivists provide to researchers and scholars is a key ingredient in a process of national self-examination that keeps a democracy healthy. I received excellent cooperation from Steve Wofford of Lawrence Livermore National Laboratory. LLNL has asked me to state that the views expressed in this book are the views of the author only and may not be the views of Lawrence Livermore National Laboratory, the regents of the University of California, or the Department of Energy. I also want to acknowledge Jim Carothers of LLNL, Berni Maza and Yvonne Townsend of CIC in Las Vegas, and Roger Anders and Betsy Scroger of the DOE historian's office. Deserving of special thanks is Ernie Campbell of NVOO for sharing his near-total recall of events thirty years old (and for the chance to witness that his talent for information management extends to leaping up from the dinner table to rifle through his library for a particular passage from Twain). I thank Betty Campbell, also, for warm hospitality and her take on the Bioenvironmental Committee from the vantage of the "Chariot wives."

On the subject of excellent archival assistance, I thank Diana Kodiak, who, in addition to her regular duties, provided the dose of insanity that kept me sane in the subterranean environs of the University of Alaska archives. Others at Rasmuson Library who helped me—usually beyond the call of duty—include: Marge Naylor, Ron Inouye, Joanna Philips, Laurie Boon, Richard Veazey, Dixon Jones, Peggy Asbury, Mary Larsen, Gretchen Lake, Judy Triplehorn, Susan Fisher, and Paul McCarthy. I thank particularly the students who painstakingly transcribed many hours of recorded interviews: Tanya Grodidier, Annamarie Kuhn, Keli Hite, Suzette Holman, Erina McGeorge, and Jan Neimeyer. Karen Gash of the University of Nevada at Reno archives also gave me excellent help.

To my friends, whose interest (and tolerance) sustained me over the course of seven years, I owe a lot. Several people's enthusiasm seemed always to come when I most needed a boost: Bob Day, Linda Schandelmeier, Marjorie Cole, Terrence Cole, Carl Benson, June Weinstock, Jeannie O'Neill. And I thank Eli Ulvi for babysitting.

Richard Fineberg, Marjorie Cole, Terrence Cole, Joe Foote, Stafford Campbell, and M. L. Campbell read the whole manuscript at an early stage and offered valuable suggestions. John Morgan, Frank Soos, Linda Schandelmeier, Dave Stark, and Lisa Chavez read several early chapters and gave me good advice and needed encouragement.

I thank my original agent, Gail Ross, and St. Martin's Press editors Michael Sagalyn, and especially Eric Wybenga for his enthusiasm, support, and intelligent editing. Most of all I thank my wife, Sarah Campbell, for her years of concessions to this project, and my son Kyle, who endured limits on his access both to computer games and to his father.

For this edition, I thank my agents Anna Cottle and Mary Alice Kier of Cine/Lit Representation. At Basic Books I thank particularly John Sherer for seeing the merit in reprinting the book; Julie McCarroll, Carol Smith, Josephine Mariea, Chris Greenberg, and Amy Scheibe; and especially Maris Kreizman for outstanding editing and attention to all the details that guided this reissue into production.

NOTES

Abbreviations and Sources

OHI: Oral history interview, tape recording available at UAF Archives.
UAF Archives: Alaska and Polar Regions Department, University of Alaska
Fairbanks.
LLNL: Lawrence Livermore National Laboratory.
CIC: Coordination and Information Center, U.S. Dept. of Energy, Las Vegas, NV.
DOE Archives: History Division, Office of the Executive Secretariat,
Washington, D.C.
Foote Collection: Don Charles Foote Collection, UAF Archives.
USAEC: United States Atomic Energy Commission
USGS: United States Geological Survey
Les Viereck papers: The private papers of Leslie A. Viereck.
Al Johnson papers: The private papers of Albert W. Johnson.

1. The Sea People of Tikigaq

1   "No truly . . . Arctic": Larsen and Rainey (1948), p. 27.
2   "Sometimes the Tikiramiut . . . dominions": Burch (1981), pp. 11, 15.
2   "the aristocrats of the Arctic": Morgan (1988), p. 4.
2   "By the late 1700s . . . south": Burch (1981), pp. 11, 14.
2   "It was during this time . . . spit": Wilimovsky and Wolfe (1966), p. 1.
2   "Shortly after Cook's . . . population": Burch (1981), p. 14.
2   "Sailing by . . . Golovin": Wilimovsky and Wolfe (1966), p. 1; Burch
        (1981), p. 43.
3   Beechey's account: Beechey (1831), pp. 359–64.

3   "And, though their general . . . alcohol": Foote (1966), p. 1046; Fortuine (1989), p. 316.
4   "But, starting in 1849 . . . Strait": Foote (1966), p. 1045.
4   "In little more than . . . well": Burch (1981), p. 16; Foote (1966), p. 1046.
4   "Turning their technology . . . herds": Burch (1981), p. 16, 26.
4   "Anthropologists estimate . . . two-thirds": Foote (1966), p. 1046; Rainey (1947), p. 283.
4   "The Tikirarmiut . . . east": Burch (1981), p. 17.
4   "American whalers . . . nations": Larsen and Rainey (1948), p. 14.
4   "From November . . . zero": Larsen and Rainey (1948), pp. 26–27; Rainey (1947), p. 238.
4   "Strong surface winds . . . world": Allen and Weedfall in Wilimovsky and Wolfe (1966), p. 13.
5   "Men disappeared . . . sea": Larsen and Rainey (1948), pp. 26–27.
5   "There seems to be . . . sea people": Rainey (1947), p. 235.
5   Traditional whale hunting: Giddings (1967), pp. 231–32; Rainey (1947), pp. 259–62; Brower (1942), pp. 43–55; Burch (1981), p. 23; Sonnenfeld (1956), pp. 82–104; Pulu (undated), pp. 2–34; Lowenstein (1986), pp. 43–63.
5   "At home the women . . . whale": Rainey (1947), p. 259.
6   "When the whale . . . again": Brower (1943), p. 51.
6   "Then the lance . . . drowned": ibid., p. 52.
7   "the most ancient village site in the Arctic": Giddings (1967), p. 115.
7   "We now realize . . . observers": Burch (1981), p. 42
7   "During the 1800s . . . Thompson": ibid., p. 43.
7   "The mound . . . from the air": Giddings (1967), p. 108.
8   "But no . . . and try": ibid., p. 113.
8   "by far the most extensive . . . region": ibid., p. 117.
8   "It revealed that . . . pyramids": Writing in the North Slope Borough's publication, *Uiniq* (Fall 1991), Mayor Jeslie Kaleak notes "Point Hope residents will proudly tell you that theirs is the oldest continuously occupied settlement in North America." Journalists Brooks and J. Foote (1962), p. 60, write that the Point Hope settlement has "been in continuous existence for at least 5,000 years." Anthropologists are somewhat more cautious. Shinkwin (1978), p. 14, concludes that Eskimo residency on the Point Hope peninsula "begins some 2,000 years ago," Rainey (1947), p. 235, says the people have lived on the point "for many centuries," and according to Burch (1981), p. 11, "it must have been several thousand years ago" that people first arrived in the Point Hope region. The great pyramids were built about 4,500 years ago.
8   "Today, 99 percent . . . sea": Burch (1981), p. 11.

8   "cohesive society . . . years": Lowenstein (1986), p. 8.
9   "They readily took . . . government": K. Lawton OHI in O'Neill (1989), p. 710.
9   "the whaling captains directed . . . affairs": Rainey (1947), p. 282.
9   "By 1940 . . . territory": Foote (1966), p. 1053.
9   "of belonging . . . where they were": Lowenstein (1986), p. 9.

2. The Firecracker Boys

10   "It is not too much . . . meter": Quoted in Ford, (1986 ed.), p. 50.
10   "Writing in *Collier's* . . . airplanes": Hilgartner, et al. (1983), p. 18.
10–11   Quotations from the Langer article are from ibid.
11   "Oak Ridge enriched . . . itself": Sylves (1987), pp. 265–67.
11   "No one was permitted . . . pass": Hewlett and Duncan (1969), p. 36.
11   Security regulations: Jungk (1958), pp. 115–16; Titus (1986), p. 8.
12   "Suddenly, there was . . . toward one": quoted in Rhodes (1988), p. 672.
12   "A millisecond after detonation . . . mile": Lamont (1965), pp. 235–39.
13   "that we puny things . . . Almighty": quoted in Hilgartner (1983), pp. 31–32.
13   "Oppenheimer recalled . . . bitches": quoted in ibid., p. 33.
13   "I was looking . . . impressed": Langone (1984), p. 65.
13   Gouldsmit story: Jungk (1958), p. 171.
14   "During 1943 . . . countries": quoted in ibid., p. 178.
14   "Perhaps American . . . Soviets": Giovannitti (1965), p. 329.
14   "Perhaps by 1945 . . . leadership": ibid., p. 331; Jungk (1958), p. 208.
15   Hiroshima death toll: Rhodes (1988), pp. 733–34.
15   "Seventy-five hours . . . five years": ibid., pp. 740–42; Lamont (1965), pp. 265–66; Neel and Schull (1956), p. 29.
15   "the greatest . . . history": White House press release, August 6, 1945, *A Statement by the President of the United States*, text in Williams and Cantelon (1984), pp. 68–70.
15   "consider promptly . . . power": Williams and Cantelon (1984), p. 70.
15   "But the scientific . . . civilians": Titus (1986), pp. 24–25.
16   "fully 70 percent . . . job": ibid., pp. 27–28.
16   "The act permitted . . . years": ibid., pp. 26–27.
16   "No safety standards . . . completely": ibid., pp. 31–33.
17   "I hoped to rely . . . work": quoted in Coughlan (1954), p. 65.
17   "What Edward can't . . . bother with": ibid., p. 61.
17   "nine out of ten . . . genius": Hans Bethe (1954), *Comments on the History of the H-bomb*, text in Williams and Cantelon (1984), p. 137.
17   "When he was ten . . . suffrage": Rhodes (1988), p. 110.
18   "Violence broke out . . . lampposts": Blumberg and Panos (1990), p. 20.
18   "In day-to-day . . . work": quoted in ibid., p. 34.

19   "The atom bomb . . . on a point": McPhee (1974 ed.), p. 58; Rhodes (1986), p. 576.

19   "After the war . . . H-bomb": Hans Bethe (1954), *Comments on the History of the H-bomb,* text in Williams and Cantelon (1984), p. 134.

20   "The people assigned . . . suggesting": ibid., pp. 135–39.

20   "The University . . . York": Sylves (1987), pp. 259–60; Orlans (1967), p. 82.

20   "Teller alone was given personal veto . . . program": York (1987), p. 70.

21   Laurence article: quoted in Hilgartner, et al. (1983), p. 39.

21   "Heat will be so plentiful . . . oil": quoted in Ford (1986 ed.), pp. 30–31.

21   "will revolutionize . . . bases": quoted in Sylves (1987), pp. 136–37.

21   "Thirteen years and $1 billion . . . built": ibid., p. 139.

21   "most notable failures": Orlans (1967), p. 152.

21   *Nucleonics* editorial: quoted in Hilgartner, et al. (1983), p. 49.

22   Project Orion: McPhee (1976), pp. 123–32.

22   "this greatest of destructive . . . mankind": quoted in Sylves (1987), p. 189.

22   "deserts flourish . . . misery of the world": quoted in Titus (1986), p. 77.

22   "A massive public relations . . . Designed": ibid.

23   "Forging ahead . . . bombs": ibid., p. 79.

23   "In 1974 . . . program": Price (1982), p. 4; D. Albright and M. Hibbs "India's Silent Bomb," *Bulletin of the Atomic Scientists,* Sept. 1992, p. 28.

23   "Other nations thought . . . Australia": Seaborg and Corliss (1971), p. 319.

23   Lilienthal quote: Lilienthal (1963), p. 109.

24   "We looked at the . . . about": G. Johnson OHI in O'Neill (1989), p. 191.

24   "The Suez Crisis . . . at Livermore": G. W. Johnson in USAEC (1964a), p. 4.

24   "In 1949, the year . . . tundra": Blumberg and Owens (1976), p. 402.

24   "I wish that this . . . 1962": Teller and Brown (1962), p. 86.

25   "Shortly after the Suez . . . earth": *The New York Times,* 11 December 1961, p. 35 C.

25   "Our thought . . . pay for": G. Johnson OHI in O'Neill (1989), p. 197.

25   "The AEC gave . . . development": Hewlett and Holl (1989), p. 528.

25   "In November . . . Livermore": Internat. Atomic Energy Agency (1970), p. 35.

25   "Participants came . . . Laboratory": Zodner (1958), p. 3.

26   "some questions . . . that way": quoted in Hilgartner, et al. (1983), p. 51.

26   "One will probably . . . cause": Teller in Zodner (1958), p. 7.

26   "Symposium participants . . . canal": ibid., p. 3.

26   "Actually, by 1971 . . . imperative": Seaborg and Corliss (1971), p. 185.

26   "politically prohibitive": Vortman in Zodner (1958), p. 48.

26   Twenty-six devices, 16.7 MT: ibid., p. 55.

26   262 bombs, 270.9 MT: Defense Atomic Support Agency (1964), p. 20.

27   Harold Brown's concluding remarks: Brown in ibid., p. 96.

27   "Harold Brown . . . plowshares": quoted in Teller and Brown (1962), p. 82.

27   "The Atomic Energy . . . 1957": Internat. Atomic Energy Agency (1970), p. 35.

27   "Still leery . . . $100,000": Hewlett and Holl (1989), p. 529.

27   "highlight the peaceful . . . tests": quoted in ibid., p. 529.

27   "Project Plowshare was . . . 1957": Internat. Atomic Energy Agency (1970), p. 35.

28   "By 1959 . . . $6 million": Hewlett and Holl (1989), pp. 528–29.

28   "by 1964 . . . $12 million per year": USAEC (1964a), p. 6.

28   "At its height . . . tool": Liberatore (1982), p. 11.

28   "a slightly flawed planet": Seaborg and Corliss (1971), p. 174.

28   "To remedy nature's oversights . . . instant harbors": ibid., p. 188.

28   "They would slice . . . ranges": Teller, et al. (1968), pp. 239–44.

28   "Edward Teller spoke . . . suit us": Teller and Brown (1962), p. 84.

28   "stem the surging tide . . . landscapes": Sanders (1962), p. 194.

28   "Nuclear explosions would dam . . . zones": Seaborg and Corliss (1971), pp. 191–96.

28   "To rescue Arctic . . . coastal cities": ibid., p. 194.

28   "A classic suggestion . . . flooding": Hilgartner, et al. (1983), p. 50.

29   "On a more local scale . . . lanes": Teller, et al. (1968), p. 306.

29   "Mining would . . . released": Seaborg and Corliss (1971), pp. 175–83.

29   "And should the need . . . space": ibid., pp. 196–97.

29   Wendell Berry quote: Berry (1987), p. 138.

29   Konrad Lorenz quote: Lorenz (1987), p. 14.

29   "Here the Livermore . . . business": G. Johnson OHI in O'Neill (1989), p. 202.

29   "The Panama . . . level target": ibid.

29   Briefed Eisenhower: ibid., Ambrose (1984), p. 568.

29   "Exact scaling laws . . . experiment": G. Johnson OHI in O'Neill (1989), pp. 218–19.

30   "Plowshare should be . . . others": Teller and Brown (1962), p. 84.

30   "It is naive to . . . fail": Sanders (1962), p. 145.

30   Johnson quotes: G. Johnson OHI in O'Neill (1989), p. 219.

31   "They said, Oh . . . problems": Hopkins OHI in O'Neill (1989), p. 424.

31   "bright young guys . . . to do it": ibid.

31   "autonomous and irresponsible bureaucracy": ibid., p. 426.

31   "Once before, Hopkins . . . of my life": Hopkins OHI in O'Neill (1989), p. 421.

32   "But a short time later . . . Seppings": Pewe, et al. (1958), p. 8.

32   The description of Daniel Lisbourne's hunting trip is adapted from an account by Don Charles Foote, Foote Collection, UAF Archives.

3. "We Looked at the Whole World. . . ."

34   Kiste quote: Kiste (1974), p. 28.
35   "no one knew they were coming": *Fairbanks Daily News-Miner,* 15 July 1958.
35   "We looked at the whole . . . job": *Fairbanks Daily News-Miner,* 15 July 1958.
35   Postwar economic decline: Coates (1989), pp. 2–3.
35   "two-thirds in labor . . . here": *Fairbanks Daily News-Miner,* 16 July 1958, p. 1.
36   "the largest . . . said Teller": *Anchorage Daily Times,* 16 July 1958, p. 15.
36   Port Moller canal: ibid.
36   Norton Sound harbor: ibid.
36   Umiat harbor: ibid., 15 July 1958, p. 1.
36   "I'm delighted . . . looking for": *Fairbanks Daily News-Miner,* 15 July 1958.
36   "We came here to be . . . undertake": ibid.
36   "must stand on its own . . . justified": ibid.
36–37   George Rogers quotes: Rogers OHI conducted by R. K. Inoye, in O'Neill (1989), pp. 654–55.
37–38   Quotes from Longyear report: E. J. Longyear Co. (1958), pp. I–II, LLNL.
38   "LRL had advised . . . Seppings": G. Johnson OHI in O'Neill (1989a), p. 229; Pewe, et al. (1958), p. 421.
38    $176 million: Longyear Co. (1958), p. II.
38    $50 to $100 million: *Anchorage Daily Times,* 15 July 1958, p. 1.
38   "we have to be . . . in hand": ibid.
38   "While the costs . . . study": Longyear Co. (1958), p. III, LLNL.
38   "figures of 50 to . . . exaggerated": *Anchorage Daily Times,* 15 July 1958, p. 15.
38   "foolish to give figures now": ibid.
38   "more like $5 million": *Fairbanks Daily News-Miner,* 15 July 1958.
39   "contingent upon . . . facilities": Atomic Energy Commission, "Current status of the program for the peaceful uses of nuclear explosives," in *Proposed Revisions to the Program for Peaceful Uses of Nuclear Explosions,* 3 March 1959, p. 2, declassified with deletions, Doc. #0070391, CIC.
39   "a faculty member . . . thoughtfully": A. Johnson to D. Foote, 25 March 1961, p. 1, Al Johnson papers.
39   "Many of these . . . said": A. Johnson to L. Viereck, 21 February 1961, p. 1, Foote Collection.

39   "We have . . . cosmic rays": *Anchorage Daily Times,* 16 July 1958.

40   "Those scientists who were . . . damage": Pauling (1983), pp. 67–68.

40   "The statement . . . come of it": A. Johnson to D. Foote, 25 March 1961, p. 2, Al Johnson papers.

40   "you have the fewest . . . people": *Fairbanks Daily News-Miner,* 16 July 1958, pp. 1, 3.

40   "big state . . . people": A. Johnson to Viereck, 21 February 1961, p. 1, Foote Collection.

40   "dig a harbor . . . desired": *Fairbanks Daily News-Miner,* 17 July 1958, p. 3.

40   "Alaska Can Have . . . yesterday": ibid.

41   Written by Sundborg: G. Sundborg to D. O'Neill, 9 November 1988, O'Neill Collection, UAF Archives.

41   Quotes from editorial: *Fairbanks Daily News-Miner,* 24 July 1958, p. 4.

41   "If this project . . . undertake": ibid., 15 July 1958.

41   Studies of Port Moller and Katalla: E. Teller to A. Starbird, 15 August 1958, declassified, LLNL.

41   "two months earlier . . . Delaware": C. Bacigalupi to D. Kilgore, 20 May 1958, declassified, Col-58-107, LLNL.

41   "his associates landed at . . . placed": *See* Kachadoorian et al. (1958), pp. 8, 27, 29, 32.

42   "He showed the geologists . . . drilled": ibid.

42   "There is no analysis . . . figure": G. Rogers to D. Foote, 15 Feb. 1961, p. 2, Box 11, #21, Foote Collection.

42   "The report suggested . . . reserves elsewhere": ibid., pp. 1–8.

42   No commercial fish stocks: *See* Smith et al. and Alverson and Wilimovsky in Wilimovsky and Wolfe, eds. (1966).

43   "He could very well . . . not": G. Johnson OHI in O'Neill (1989a), p. 230.

43   "Writing to Gen. . . . Armstrong": E. Teller to A. Starbird, 15 August 1958, declassified, LLNL.

43   AEC meeting of May 22, 1959: *Atomic Energy Commission Meeting No. 1511,* May 22, 1959, declassified with deletions, Doc. #753, CIC.

44   "without endangering . . . size": E. Teller to H. Fidler, 7 October 1958, p. 3, declassified with deletions, Doc. #125233, CIC.

44   "In a second . . . radiation": E. Teller to E. Shute, 26 January 1959, in *Proposed Revisions to the Program for Peaceful Uses of Nuclear Explosions,* March 1959, declassified with deletions, AEC 811/37, Doc. #0070391, CIC. *Note:* LLNL also has the Teller letter, but refused a number of requests to declassify several pages of it. The document can be found at CIC with very minor deletions.

45   Harbor dimensions: *Fairbanks Daily News-Miner,* 17 July 1958, p. 3.

45   "four 100-kiloton . . . respectively": E. Teller to E. Shute, 26 January 1959, op. cit.

45   Convoy: Based on 20 feet (truck length) x 2.4 million = 9,090 miles.

45   40 percent of WWII firepower: Based on 6 MT figure in Pauling (1983), p. 5.

45   "The report of the USGS . . . feet": Kachadoorian et al. (1958), p. 29.

46   U.S. Weather Bureau report: *The Feasibility of Project Chariot from the Standpoint of Radioactive Fallout,* K. Nagler and F. Cluff (1958), U.S. Weather Bureau, Washington, D.C.

46   Livermore film: *Industrial Applications of Nuclear Explosives* (1958), Film A-81, LLNL.

46   Aerial snapshots: the photos can be seen at the Laboratory Archives, LLNL.

46   "Speaking at Point Hope . . . name": R. Ball in "Excerpts of a tape recording by Keith Lawton . . . on March 14, 1960," by D. Foote, Foote Collection.

46   Hundreds of billions: 70 million cubic yards excavated at 120 pounds per cubic foot, a reasonable average.

47   Quotes from film: *Industrial Applications of Nuclear Explosives,* 1958, Film A-81, LLNL.

47   "I don't see how . . . laboratory": G. Johnson OHI in O'Neill (1989), pp. 234, 237.

47   "could not have been safely . . . Nevada": ibid., pp. 224–225.

47   "I think the original . . . detonation": G. Higgins to D. O'Neill, 29 August 1989, O'Neill Collection, UAF Archives.

47   Higgins quotes: Higgins OHI, tape #H-90-27, UAF Archives.

49   "to the effects . . . nonsense": *Fairbanks Daily News-Miner,* 24 July 1958.

49   AEC authorized studies: USAEC (1964), pp. 1–7.

49   Scope of work order: ibid., p. 3-1.

50   Details of the storm are from: ibid., pp. 3-1, 3-3, D-4, D-6; J. Foote (1961), pp. 31–32.

4. A Bomb on the World Stage

51   "Teller and . . . explosions": Ambrose (1984), p. 568.

51   "By 1957 the nuclear . . . few years": Pauling (1983), pp. 112–23; *Consumer Reports,* March 1959, pp. 102–11.

51   "Albert Schweitzer . . . testing": Divine (1981), p. 124.

52   "would be perfectly . . . tests": ibid., p. 125.

52   "to modify the weather . . . air": quoted in Ambrose (1984), p. 399.

52   "He told the president . . . years": Broad (1992), p. 46.

52   "a crime against humanity": quoted in ibid., p. 403.

53   "he understood from AEC . . . few years": Divine (1981), p. 126.

53   "the most eminent . . . weapons": quoted in Ambrose (1984), p. 399.

53   "David Lilienthal . . . testing": Lilienthal (1969), p. 205.

53   Starbird's letter: Broad (1992), p. 46.

53   "In fact, no absolutely . . . radiation": ibid., p. 48.

53   "Nevertheless, by selling . . . race": Magraw (1988), pp. 32–33.

53   "The test ban issue . . . war": Divine (1981), p. 128.

54   "But impetus . . . adviser": ibid., pp. 128–29.

54   "But the AEC's Strauss . . . possible": Ambrose (1984), pp. 430–32.

54   "By mid-August . . . kilotons": Divine (1981), p. 129.

54   "Eisenhower set to . . . position": Ambrose (1984), p. 479.

55   "McCone made one last . . . say": ibid.

55   "to proceed promptly . . . from weapons tests": USAEC (1959), p. 180.

55   "But before Khrushchev . . . ahead": Ambrose (1984), pp. 479–80.

55   Thirty-seven bombs, twenty-five atmospheric: U.S. Dept. of Energy (1988), pp. 8–10. *Note:* Both Ambrose and Divine use the figure of nineteen U.S. bombs; perhaps prior to the publication I cite, the DOE had not "announced" all of the Hardtack II detonations.

56   "Eisenhower calmly . . . cheating": Ambrose (1984), p. 565.

56   "can be evaded . . . us": ibid., p. 568.

56   "He told a cabinet . . . testing": ibid.

57   "In a follow-up . . . dream": E. Teller to The President, 11 May 1960, Doc. #108639, CIC.

57   "Eisenhower gave permission . . . stage": ibid., pp. 565–68.

57   "Nearly half of the massive . . . transparencies": USAEC (1959), pp. 129, 137, 167.

57   Strauss's wand: ibid., pp. 144–47.

58   "Scientists from forty-eight . . . papers": ibid., p. 89.

58   Teller's talk: ibid., p. 103.

58   Johnson's talk: ibid., p. 108.

58   "On hearing Johnson's . . . effect": G. Johnson OHI in O'Neill (1989), pp. 254–55.

58   Military "add-on" experiments: Sylves (1987), p. 191.

58   "And Glenn Seaborg . . . negotiated": Seaborg and Loeb (1981), pp. 247–48.

58–59 "spreading life, happiness . . . years": quoted in G. Johnson (1960a), p. 2.

59   "Within days . . . Union": "Department of State, Division of Language Services (Translation)," (transmitted on 15 September 1958), in AEC 811/25, 3 October 1958, DOE Archives.

59   "The open hostility . . . conference": *The New York Times,* 11 Sept. 1958.

59   "Willard Libby . . . harbor": ibid.

59   "But in another . . . conference": USAEC (1959), p. 166.

60   Quotes from Carothers and Chase: Ralph Chase OHI by James Carothers, 1986, transcript, LLNL.

60   "Even the movie camera . . . frame": *Nuclear Excavation: A Status Report,* Film A85, Lawrence Radiation Laboratory, LLNL.

60   XW-47 Polaris missile warhead: C. Hansen, "Neptune test collapse a 115-ton accident," *Bulletin of the Atomic Scientists,* March 1990, p. 47.

60   Visual details of Neptune: *Nuclear Excavation: A Status Report.*

61   "When test officials . . . (.09 kilotons)": Ralph Chase OHI by James Carothers, 1986, transcript, LLNL; E. Teller to E. Schute, 26 Jan. 1959, in *Proposed Revisions in the Program for the Peaceful Use of Nuclear Explosions,* 3 March 1959, declassified with deletions, AEC 811/37, Doc. #0070391, CIC/LLNL.

61   "The Neptune Event . . . Experiment": Shelton, A.V., et al (1960).

61   "First, Neptune showed . . . error": G. Johnson (1960b), p. 7.

61   "it was possible . . . possible": A. Shelton, et al. (1960), p. 12.

61   "The second—and even . . . realized": ibid., p. 13.

61   "The significance . . . levels": E. Teller to E. Shute, 26 January 1959, in *Proposed Revisions in the Program for the Peaceful Use of Nuclear Explosions,* 3 March 1959, declassified with deletions, AEC 811/37, Doc. #0070391, CIC/LLNL.

62   "Livermore had . . . $1 million": University of California Radiation Laboratory, "Proceedings of the Second Annual Plowshare Symposium, May 13–15, 1959, San Francisco, CA, (UCRL-5675), p. 2.

63   "In a classified letter . . . viable": E. Teller to E. Shute, 26 Jan. 1959, in *Proposed Revisions in the Program for the Peaceful Use of Nuclear Explosions,* 3 March 1959, declassified with deletions, AEC 811/37, Doc. #0070391, CIC.

5. Fracas in Fairbanks

64   "To be fair . . . warning": A. Johnson to L. Viereck, 21 Feb. 1961, p. 3, Foote Collection.

65   "In 1901, a trader . . . Alaska": Cole (1984), p. 18.

65   "Barnette would get off . . . be": ibid., p. 21.

65   "anxious to buy . . . sale": quoted in ibid., p. 25.

66   "largest log-cabin town . . . capital": ibid., pp. 32, 53–59.

66   "In 1959, the University . . . horizon": See Patty (1969).

67   "moose browsing . . . Lake": Marks, David B., *Farthest North University,* Undated article clipped from unidentified journal, Alaska Collection, UAF Archives.

67   Dog teams on campus: A. Johnson (undated), *Science, Society and Academic Freedom,* paper presented to American Association of University Professors, p. 1, A. Johnson papers.

67   "northernmost star . . . firmament": ibid., p. 2.

68   "As the game . . . fights": Patty (1969), p. 22.

68   "There was no tenure . . . friendly": A. Johnson (undated), "Science, Society and Academic Freedom," paper presented to American Association of University Professors, p. 6.

68   "Johnson heard from a colleague . . . afternoon": A. Johnson to L. Viereck, 21 Feb. 1961, p. 2, Foote Collection.

69   "Newspaper and other . . . it": Since all of Keller and Shelton's meetings in Alaska seem to have covered the same ground, direct quotes come from newspaper accounts of one or another of the meetings held in Alaska between 6 and 10 Jan. 1959. See *Anchorage Daily News,* 7 Jan. and 27 Jan. 1959; and *Fairbanks Daily News-Miner,* 9, 10, 13, 19, and 27 Jan. 1959.

70   "blast will not be . . . feet": *Fairbanks Daily News-Miner,* 15 July 1958.

70   "We want not just a hole . . . used": ibid.

70   "Before the AEC . . . facilities": *Anchorage Daily Times,* 15 July 1958, p. 1, and 16 July 1958, p. 15.

70   "Now it seemed the AEC . . . jobs": *Fairbanks Daily News-Miner,* 9 Jan. 1959.

70   "mineral potential . . . large": *Anchorage Daily News,* 7 Jan. 1959, p. 5.

70   "The officials pointed . . . economics": *Fairbanks Daily News-Miner,* 10 Jan. 1959, p. 1.

70   "in discussing the . . . basis": E. Teller to A. Starbird, 15 Aug. 1958, declassified, LLNL.

70   "vast mineral reserves": *Fairbanks Daily News-Miner,* 10 Jan. 1959, p. 1.

71   "Shelton continued to . . . areas": ibid., 1 Jan. 1959.

71   "He told the gathered . . . probably so": A. Johnson to L. Viereck, 21 Feb. 1961, p. 2, Foote Collection.

71   "Why, Al Johnson asked . . . last fall": A. Johnson OHI in O'Neill (1989), p. 295.

71   "Keller and Shelton . . . read": A. Johnson to D. Foote, 25 March 1961, p. 1, Foote Collection.

71   Patty letter: E. Patty to AEC, 7 Oct. 1958, President's papers, Box 8, #117, UAF Archives.

71   "Gerry Johnson wrote back . . . form": G. Johnson to E. Patty, 29 Sept. 1958, President's papers, Box 8, #117, UAF Archives.

72   "What Patty told the AEC . . . here": E. Patty to AEC, 7 Oct. 1958, President's papers, Box 8, #117, UAF Archives.

72   "But Tom English . . . do so": A. Johnson to L. Viereck, 21 Feb. 1961, Foote Collections.

72   "Keller and Shelton trotted . . . X-rays": ibid.

73   "At one point . . . furious": Lee Salisbury, personal communication, 15 June 1989.

73   "We harassed . . . warning": A. Johnson to L. Viereck, 21 Feb. 1961, p. 3, Foote Collection.

73   "Finally, Keller . . . button": Lee Salisbury, personal communication, 15 June 1989; also A. Johnson to L. Viereck, 21 Feb. 1961, p. 3.

73   "We are told . . . design": An untitled copy of the statement drafted by T. English and read to Keller and Shelton dated 9 January 1959 can be found in T. English to G. Johnson, 21 Jan. 1959, President's papers, Box 8, #117, UAF Archives.

74   "We have been subjected . . . AEC": ibid.

74   "was obviously stumped . . . work": A. Johnson to L. Viereck, 21 Feb. 1961, p. 3, Foote Collection.

74   "the University is . . . Alaska": T. English to G. Johnson, 21 Jan. 1959, President's papers, UAF Archives.

74   "Five days later . . . problems": G. Johnson to T. English, 26 Jan. 1959, Project Chariot Collection, Box 3, #120, UAF Archives.

74   "I am personally . . . goods": Undated clipping, *Fairbanks Daily News-Miner*, Foote Collection, Box 11, #27.

74   Similar letter: T. English to President Anch. C. of C., 29 Feb. 1959, Project Chariot Collection, Box 3, #20, UAF Archives.

74   "He was at his witty . . . boys": A. Johnson to L. Viereck, 21 Feb. 1961, p. 3, Foote Collection.

75   Cernick's editorial: *Fairbanks Daily News-Miner*, 10 Jan. 1959, p. 4.

76   Atwood's editorial: *Anchorage Daily Times*, 10 Feb. 1959, p. 4; see also R. Atwood OHI in O'Neill (1989), p. 546.

76   "The Fairbanks . . . order": *Fairbanks Daily News-Miner*, 13 Jan. 1959, p. 3.

76   "Alaskans, who . . . Century": McIntyre to McCone, 11 April 1959, PC Correspondence, Viereck papers.

76   "Long have I waited . . . child": quoted in ibid.

77   "I'd been a bush . . . ridiculous": G. Wood OHI in O'Neill (1989), p. 627.

77   Ginny Wood's letter: *Fairbanks Daily News-Miner*, 19 Jan. 1959, p. 6.

78   Al Johnson's letter: Undated clipping from *Fairbanks Daily News-Miner* (letter dated 13 Jan. 1959), Foote Collection.

78   Irving Reed's letter: (dated 27 Jan. 1959), ibid.

79   Olaus Murie's letter: (dated 2 Feb. 1959), ibid.

79   Robert Needham's letter: (undated), ibid.

79   "The *News-Miner* . . . questions": *Fairbanks Daily News-Miner*, 31 Jan. 1959.

79   "Patty continued . . . appropriate": E. Patty to G. Johnson, 27 Jan. 1959, Box 8, #117, President's papers, UAF Archives.

79   "On January 20 . . . February": G. Johnson to E. Patty, 20 Jan. 1959, ibid.

79   "On February 18 . . . features": E. Patty to R. Wiegman, et al., 18 Feb. 1959, Box 3, #20, Project Chariot Collection, UAF Archives.

80   "Though they had . . . funding": A. Johnson to B. Bartlett, 21 Feb. 1959, Box 3, #20, Project Chariot Collection, UAF Archives.

80   "John McCone . . . Congress": Clipping dateline: 18 Feb. 1959, Foote
          Collection.

80-81   "The thing that . . . cared": G. Johnson OHI, tape #H-90-03-02 (portion
          quoted not transcribed), UAF Archives

6. Tent Camp at Ogotoruk Creek

82   "A good field . . . again": W. Pruitt OHI in O'Neill (1989), p. 348.

82   "The Inupiaq word . . . productivity": Orth (1971), p. 716.

83   "Ogotoruk creek heads . . . repeats": Ruben Kachadoorian, *Geographic
          Setting,* in Wilimovsky and Wolfe (1966), p. 45.

83   "In the face of AEC . . . studies": E. Campbell to W. Osburn, 3 Aug.
          1976, Doc. #132475, CIC.

84   "The first parties . . . program": ibid.

84   "Even though it is . . . demands": See USAEC (1962), pp. 1–2; G.
          Johnson in O'Neill (1989), p. 262.

84   "Before Gerry Johnson . . . officials": *Juneau Empire,* 26 Feb. 1959, p. 1;
          Anchorage Daily Times, 27 & 28 Feb. 1959; see also J. Anderegg to For
          the Record, 27 Feb. 1959 in Doc. #20931, CIC.

84   "One member of his . . . them": E. Campbell, interview, 11 July 1989,
          author's notes.

85   "95 percent more . . . yield": *Polar Star,* 6 March 1959, p. 3.

86   "Professor Al Johnson . . . Chariot": A. Johnson to E. Bartlett, 15 Feb.
          1959, Box 3, #20, Project Chariot Collection, UAF Archives.

86   "Bartlett replied . . . subject": E. Bartlett to A. Johnson, 18 Feb. 1959a, in
          ibid.

86   "Within a few . . . confusion": E. Bartlett to A. Johnson, 18 Feb. 1959b in
          ibid.

86   "Al Johnson answered . . . week": A. Johnson to E. Bartlett, 21 Feb.
          1959, in ibid.

86   "Bartlett's terse reply . . . Chariot": E. Bartlett to A. Johnson, 24 Feb.
          1959, in ibid.

86   "I had better tell . . . evaporated": E. Bartlett to G. Johnson, 24 Feb.
          1959, in ibid.

86   "I have now become . . . elsewhere": *Washington News Letter* (of
          Senator Bartlett), 20 Feb. 1959, in ibid.

87   "But Gerry Johnson . . . cleared up": A. Johnson to D. Foote, 25 March
          1961, p. 3, Al Johnson papers.

87   "everyone was getting . . . biologists": ibid.

87   "community opinion leaders": USAEC (1964), p. 5-1.

87   "effectively illustrated . . . endorsed": *Relating to the Atomic Energy
          Project at Cape Thompson,* House Joint Resolution #9, First Alaska
          Legislature, First Session.

87   "Dr. John N. Wolfe . . . Washington": See M. Britton OHI in O'Neill
        (1989), pp. 2–5.

87   "In March of 1959 . . . detonation": USAEC (1964), p. 1-6.

88   "It [was] scarcely . . . point of view": J. Wolfe, "The Ecological Aspects of
        Project Chariot," *Proceedings of the Second Plowshare Symposium,* May
        13–15, 1959, San Francisco, CA, Part II, p. 62, Lawrence Radiation
        Laboratory and AEC San Francisco Operations Office, UCRL 5676.

88   "brilliant": *Fairbanks Daily News-Miner,* 27 Aug. 1959.

89   "eminent": ibid., 18 Aug. 1960.

89   "misinformed": ibid.

89   "A team composed . . . today": ibid., 13 March 1959.

89   "Wolfe said that environmental . . . off": ibid.

89   "Nevertheless, John . . . immediately": A. Johnson to D. Foote, 25 March
        1961, p. 3.

89   "essentially signed off . . . '59": A. Johnson OHI in O'Neill (1989), p.
        299.

90   "There [were] a number . . . again": W. Pruitt OHI in O'Neill (1989), p.
        348.

90   "that we were the tame . . . bought": ibid., p. 347.

90   "We sincerely hope . . . project": Quoted in *Jessen's Weekly,* 26 March
        1959.

90   Pruitt's biography: W. Pruitt OHI in O'Neill (1989), p. 327.

91   "demanded we hire . . . since": interview of B. Kessel, 31 March 1989,
        author's notes.

92   "We should understand . . . more": Univ. of Alaska, *An Ecological Study
        of the Flora and Fauna of the Cape Thompson-Ogotoruk Creek Region,
        Alaska,* March 1959, Box 1, #7, Project Chariot Collection, UAF
        Archives.

92   "By May 22 . . . work": J. Wolfe to E. Patty, 26 May 1959, Presidential
        papers, UAF Archives; also *Press Release from University of Alaska,* 19
        June 1959, Box 1, #7, Project Chariot Collection, UAF Archives.

92   "Besides the University . . . radioecology": Wilimovsky and Wolfe (1966),
        xiii–xvi; USAEC (1964), pp. 2-4 to 2-5.

92   "They didn't literally . . . fingers": B. Kessel OHI in O'Neill (1989), p.
        270.

94   "This looks . . . wildlife": L. Viereck OHI, 8 April 1986, UAF Archives.

94   "about 85% . . . research": A. Johnson to L. Viereck, 24 March 1959, Box
        4, #28, Foote Collection.

95   "Thanks to Al . . . began": "Progress Reports," 29 Sept. 1959, Box 9,
        #82, Project Chariot Collection; L. Viereck to A. Johnson, 16 June 1959,
        Box 4, #28, Project Chariot Collection, UAF Archives.

95   "Nearly half a million . . . sea": G. Swartz, "Sea Cliff Birds," in
        Wilimovsky and Wolfe (1966), p. 611.

95   "Viereck and his assistants . . . Ogotoruk": L. Viereck to A. Johnson, 3 July
        1959, Box 4, #28, Project Chariot Collection, UAF Archives.

7. Drop Us a Card

96   "If your mountain . . . card": *Anchorage Daily Times,* 26 June 1959, p.
        11.
96   "in honor . . . nation": E. Patty to J. Hoover, 1 Feb. 1960, President's
        papers, Box 2, #33, UAF Archives.
96   "the head . . . police": I. F. Stone, 1963, *The Haunted Fifties,* p. 257,
        Random House.
96   "Nothing derogatory . . . FBI": "Note" appended to J. Hoover to E. Patty,
        10 Feb. 1960, #100-373-606-20, FBI.
97   "Citing pressing official . . . doctorate": J. Hoover to E. Patty, 10 Feb.
        1960, President's papers, Box 2, #33, UAF Archives.
97   Von Braun: E. Patty to W. Von Braun, 10 Dec. 1959, in ibid.; L. Hunt,
        Secret Agenda, 1991, pp. 17, 240, St. Martin's Press.
97   Rickover: E. Patty to H. Rickover, 14 Dec. 1959, President's papers, Box
        2, #33, UAF Archives.
97   Bruckner: E. Patty to W. Bruckner, 30 Dec. 1959, in ibid.
97   Teller's commencement speech: OH recording #U-45, UAF Archives; see
        also *Fairbanks Daily News-Miner,* 18 May 1959, p. 1.
97   Fort Greely reactor: "Ft. Greely Nuclear Power Plant," press release, U.S.
        Corps of Engineers, Alaska, Huber Collection, Box 68, #0, UAF Archives.
98   Commoner quotes: B. Commoner OHI in O'Neill (1989), pp. 279–80.
99   "Once there it can . . . leukemia": Pauling (1983 ed.), pp. 114–15.
99   "in recognition . . . tyranny": from a photograph of the citation, University
        Relations Collection, UAF Archives.
99   Brooks quotes: J. Brooks OHI in O'Neill (1989), pp. 487–89.
100   "That is like a little girl . . . card": *Anchorage Daily Times,* 26 June 1959,
        p. 11.
100   Teller's frontier rhetoric: *See* Peter A. Coates, *The Trans-Alaska Pipeline
        Controversy: Technology, Conservation and the Frontier,* Lehigh
        University Press, 1991, p. 111.
100   "Anything new . . . states": Teller's 1959 commencement speech, OH
        recording #U-45, UAF Archives. Teller also made the remark in 1958 (see
        A. Johnson to L. Viereck, 21 Feb. 1961, Les Viereck papers).
100   "the world's foremost . . . scientists": *Nome Nugget,* 26 June 1959, p. 1.
101   "He continued to refer . . . coal": ibid., 29 June 1959, p. 1.
101   "Bacigalupi also seems . . . favorable": ibid.
101   "The AEC preferred . . . yield": D. Foote, *Project Chariot and the Eskimo
        People of Point Hope, Alaska,* 1961, p. 2; G. Higgins to D. Foote, 4
        November 1960, Foote Collection.

101   "he flew in a twin . . . Beechcraft": L. Viereck to D. Foote, 14 Dec. 1960, Viereck papers.

102   "From his camp . . . contract": USAEC (1964), pp. 3-10 to 3-11.

102   "Several hours after . . . oil": D. Foote, "History Notes—B," Foote Collection.

102   "all hell broke loose": L. Viereck to D. Foote, 14 Dec. 1960, Les Viereck papers.

102   "High winds . . . stopped": ibid. and G. Keto to R. Ball, 10 Aug. 1959, p. 1, declassified, Doc. #26498, CIC.

102   "Two pilots flying . . . Kotzebue": L. Viereck to D. O'Neill, 3 Feb. 1992; T. Schultz, pers. com., August 1992.

103   "Men were accumulating . . . researchers": L. Viereck OHI in O'Neill (1989), p. 399.

103   "In the race to unload . . . contents": D. Foote, "History Notes—B," Foote Collection.

103   "By July 7 . . . researchers": ibid.

103   "Not until July . . . begun": USAEC (1964), p. 3-11.

103   "permanent camp included . . . latrines": USAEC (1964), pp. 3-5, 3-13.

103   Sixty-six men: D. Foote, "History Notes—B," Foote Collection.

104   "Former Wien Airlines . . . measuring": personal communication, Tony Schultz, 21 Aug. 1992.

104   Kessel quotes: B. Kessel, OHI in O'Neill (1989), p. 266.

105   "Since the 800 . . . Hope": G. Keto to R. Ball, 10 Aug. 1959, declassified, Doc. #26494, CIC; USAEC (1964), pp. 3-6, 3-7.

105   "Boyles Bros. . . . lost": ibid., pp. 3-7, 3-11, 3-12.

106   "acutely aware": G. Keto to R. Ball, 10 Aug. 1959, declassified, #26498, CIC.

106   "No one from the AEC . . . why": D. Foote to E. Campbell, 26 Nov. 1959, p. 2, Foote Collection.

8. Polarbasillen

107   "We like him . . . yeah": K. Kinneeveauk OHI in O'Neill (1989), p. 815.

107   "He was my . . . us": E. Attungana OHI in ibid., p. 824.

107   "He lived closely . . . us": T. Frankson OHI in ibid., p. 843.

108   "Foote was born . . . New Hampshire": Interview with Joe Foote, 11 June 1993.

108   D. Foote's secondary schooling: Interview with Joe Foote, 24 April 1989.

109   "From there it was . . . objective": ibid.

109   "regional studies . . . world": D. Foote to G. Johnson, 24 March 1959, Foote Collection.

109   Western Union telegram: J. Wolfe to D. Foote, 5 May 1959, ibid.

109   "involves a complete . . . industry": D. Foote to J. Wolfe, 6 May 1959, ibid.

110   Quotes from Don Foote's journal: D. Foote, "Diaries," Box 27, Foote Collection.

112   "It is my personal . . . carried out": D. Foote to E. Campbell, 26 Nov. 1959, Foote Collection.

113   Foote's rented house at Point Hope was owned by Charles Adams, D. Foote to M. Johnson, 7 Nov. 1959, Foote Collection.

113   "without insulation . . . leaks": D. Foote to E. Campbell, 29 Sept. 1959, Foote Collection.

113   "Georgia O'Keeffe . . . them": Fejes (1967), pp. 74–75.

113   "with those few materials . . . cardboard": D. Foote to E. Campbell, 29 Sept. 1959, Foote Collection.

113   "House far too cold . . . uncomfortable": D. Foote, "Diaries," Box 27, Foote Collection.

113   "By November . . . solid": D. Foote to AEC, 1 Nov. 1960, Foote Collection.

114   "He did everything . . . in": Evelyn Tuzroyluk Higbee, personal communication, 15 December 1988.

114   "both Tom [Stone] and I . . . system": D. Foote to E. Campbell, 29 Sept. 1959, Foote Collection.

114   "There is little . . . etc.": D. Foote to A. Cooke, 6 December 1959, Foote Collection.

114   "She has her new . . . feast": D. Foote to W. Pruitt, 29 Nov. 1959, Foote Collection.

114   "She was just like . . . do": Francis Schaeffer, personal communication, Aug. 1988.

114   "She started to dress . . . circle": J. Lawton OHI in O'Neill (1989), pp. 695–98.

114   "If Don was otherwise . . . do": Joe Foote, personal communication, 24 April 1988.

115   "They weren't too happy . . . anyway": J. Lawton OHI in O'Neill (1989), pp. 695–96.

115   "As I write . . . day": D. Foote to J. Foote, 23 Nov. 1959, Foote Collection.

115   "I wrote Les . . . fourteenth": D. Foote to W. Pruitt, 17 May 1961, Foote Collection.

115   "allay the . . . natives": E. Campbell to D. Foote, 17 Nov. 1959, Foote Collection.

115   "be honest . . . knowledge": D. Foote to E. Campbell, 26 Nov. 1959, Foote Collection.

115   "most Point Hopers . . . zero": ibid.

116   "well read and . . . area": ibid.

116   "The council discuss . . . explosion": D. Foote to E. Campbell, 29 Nov. 1959, Foote Collection, see also Doc. #16878, CIC, for original.

116   "We the undersigned . . . time": Point Hope Village Council to AEC, 30 Nov. 1959, in Doc. #16878, CIC.

116   "In mid-October . . . eyes": D. Foote to A. Cooke, 6 Dec. 1959.

116-117 "Foote had hired . . . team": Foote's guides were Bernard Nash out of Point Hope and Jimmy Hawley out of Kivalina.

117   "Eventually, Foote . . . dogsled": D. Foote to M. Smith, 17 April 1960, Box 25, Foote Collection.

117   "finally thrown open . . . newsmen": *Fairbanks Daily News-Miner,* 21 Aug. 1959.

117   "within three weeks": ibid., 20 Aug. 1959.

117   "scientists would be . . . blast": ibid., 27 Aug. 1959.

117   "will be . . . 1961": ibid., 20 Aug. 1959; see also 21 Aug. and 25 Aug. 1959.

117   "The scientists now roaming . . . time": ibid., 25 Aug. 1959.

118   "which most scientists . . . interest": ibid., 25 Aug. 1959.

118   "ready to accept . . . Arctic slope": ibid., 24 Aug. 1959.

118   "these Natives, living . . . effects": ibid., 21 Aug. 1959.

118   "only a few Natives pass by": ibid., 20 Aug. 1959.

118   "In the vicinity . . . herds": C. Bacigalupi to Distribution, "Summary of Meeting with Alaska District Corps of Engineers," 27 May 1958, COL-58-112, LLNL.

118   "would apparently . . . fired": *Fairbanks Daily News-Miner,* 27 Aug. 1959.

118   "the explosion . . . ground": ibid., 20 Aug. 1959.

119   "it is a matter . . . weapon": ibid.

119   "That same summer . . . fallout": *Nome Nugget,* 12 June 1959, AP story by Jack Stillman.

119   "Any nuclear . . . medium": Defense Atomic Support Agency, "Peaceful Uses of Nuclear Explosives: Project Plowshare," 1 September 1964, p. 32, LLNL.

119   "As fall descended . . . remained": USAEC (1964), p. 3-18. Details and entries from Spencer's journal: "Journal of Harry Spencer, meteorologist, Project Chariot, Winter 1959–60," p. 2, Foote Collection, UAF Archives.

121   "no significant questions": USAEC (1964), pp. B-1 to B-12.

9. The AEC Meets the Eskimos

122   "I'm pretty . . . Hope": K. Kinneeveauk in "Excerpts from a tape recording made by Rev. Keith Lawton of the Episcopal Mission in Point Hope Alaska during a public information meeting given by members of

the USAEC and Environmental Committee for Project Chariot on March 14, 1960," by D. Foote, Foote Collection.

122   "an Eskimo fad": *Tundra Times,* 1 Oct. 1962.

122   "spliced tape . . . catalogs": ibid.

123   Downscaled to 280 kiloton: USAEC press release, "Project Chariot—1960," 4 March 1960. *Note:* two quite different versions of this press release exist.

123   "Internal documents show . . . sufficient": G. Johnson to G. Keto, 6 Dec. 1960, p. 4, Doc. #64411, CIC.

124   Weaver's report: C. Weaver to Office of Test Operations Files, "Report of Trip to Alaska During Period March 2 to March 11, 1960," 17 March 1960, p. 2, Doc. #78632, CIC.

124   "the laboratory was confident . . . atmosphere": ibid.

124   "Jim Brooks said . . . chains": R. Southwick to E. Shute, "Trip Report on Visit to Alaska—Project Chariot," 21 March 1960, in USAEC (1964), p. B-4.

124   "this donnybrook . . . Anchorage": ibid., p. B-5.

125   "certain questions as to the effects . . . Eskimos": ibid., pp. B-5, B-6.

125   "I discovered that he . . . Wolfe": ibid., p. B-6.

126   "I think I was taken . . . Hope": R. Rausch OHI in O'Neill (1989), p. 50.

126   "Rausch found himself in the awkward . . . hazards": ibid., pp. 51–60.

126   "Atom Scientists . . . project": *Nome Nugget,* 14 March 1960.

126   "The only hint . . . opposition": R. Southwick to E. Shute, "Trip Report on Visit to Alaska—Project Chariot," 21 March 1960, in USAEC (1964), p. B-10.

127   "since February 1957": Defense Atomic Support Agency, "Peaceful Uses of Nuclear Explosives: Project Plowshare," 1 Sept. 1964, p. 32.

127   "second-hand and warped": D. Foote to E. Campbell, 26 Nov. 1959, Foote Collection.

127   "Bartlett had responded . . . ineffectual": See E. Campbell to J. Philip, 25 April 1962, p. 12, LLNL; P. Brooks and J. Foote, "The Disturbing Story of Project Chariot," *Harper's,* April 1962, p. 60.

127   "presented a prepared . . . contents": D. Foote to E. Campbell, 26 Nov. 1959, Foote Collection.

127   "a mixture of hearsay": D. Foote (1961), p. 9.

127   "thus in several . . . north": D. Foote to E. Campbell, 26 Nov. 1959, Foote Collection.

127   Radios: D. Foote to R. Ball, 16 March 1960, Foote Collection.

128   Quotes from the meeting: "Excerpts from a tape recording made by Rev. Keith Lawton of the Episcopal Mission in Point Hope Alaska during a public information meeting given by members of the USAEC and Environmental Committee for Project Chariot on March 14, 1960," by D. Foote, Foote Collection; tape recording made by Keith Lawton of

March 14, 1960, AEC meeting at Point Hope, Alaska, UAF Oral History Collection, UAF Archives.

128   Elvis: K. Lawton OHI in O'Neill (1989), p. 688.

129   Quotes and details of the film: LLNL, *Industrial Applications of Nuclear Explosives,* 1958, Film A-81.

130   "he had been cutting meat . . . blade": K. Lawton, personnel communication, 28 Sept. 1989.

131   "little boys . . . pathological glee": R. Rausch, personnel communication, 6 May 1988, author's notes.

131   "a very unusual . . . information": R. Rausch, in "Excerpts from a tape recording made by Rev. Keith Lawton . . . on March 14, 1960," by D. Foote, Foote Collection.

131   "the statements of the AEC's John Wolfe": See Foote's notes of the Anchorage meeting, p. 6, Box 25, Foote Collection.

132   "namely Allyn Seymour": ibid., pp. 4–6.

132   "relying on information . . . Weaver": ibid.

134   Kitty Kinneeveauk quotes on her testimony at public hearings: K. Kinneeveauk OHI in O'Neill (1989), pp. 794–807.

138   November 30, 1959, petition: Point Hope Village Council to USAEC-San Francisco Operations Office, 30 Nov. 1959, in Doc. #16878, CIC.

138   "The January 18 . . . Chariot": E. Bullock to H. Keller, 18 Jan. 1960, Foote Collection.

138   "Bullock's company had landed . . . $200,000": USAEC (1964), pp. 3-3, 3-10, 3-21, 3-30, and 3-34.

139   "Thousands, many thousand . . . all": The Foote transcript notes "end side one" after the word "times," and the rest of the phrase is not transcribed. The complete phrase to the word *all* is recorded on the Lawton tape.

140   "The only thing . . . published": Teller also made the claim several times: ". . . the whole Plowshare subject is now declassified with the exception of the black box that contains the nuclear explosive itself. . . ." (*Popular Mechanics,* March 1960, p. 100).

141   "the verbatim . . . Seymour": D. Foote to J. Foote, 26 March 1961, Foote Collection. *See also* Foote's notes on Seymour and Weaver's presentations at the Anchorage meeting, Box 25, Foote Collection, and Albert Johnson's notes of the same meeting, Al Johnson papers. Note: corroboration of this point can be found in Albro Gregory, "Cape Thompson's Atomic Project Described As Feasible One," *Jessen's Weekly,* 18 Aug. 1960. Gregory writes: "Not one fish was killed by radiation in the Pacific blasts, said Dr. Allyn Seymour. And down in Nevada, where a herd of cattle has been grazing for several years near ground zero, not one suffered ill effects. On being slaughtered they have been used for food by humans. Marines were

in trenches close to the Nevada blasts and they suffered no ill effects from radiation and fallout."

141   "rather aggressive": R. Rausch OHI in O'Neill (1989), p. 51.

10. Bikini, Nevada, Tikigaq: A Trail of Empty Words

142   *Radioactive Eskimo:* from "Peter La Farge on the Warpath," Folkways Records (FN 2535), 1965. Note: Folkways is now owned by Smithsonian Folkways Records, lyrics quoted with permission.

143   "less than one month's advance notice": Kiste (1974), p. 34.

143   "The military governor . . . Land": quoted in ibid., p. 27.

143   "for the good . . . wars": quoted in ibid., p. 28.

143   "If the United . . . elsewhere": quoted in ibid.

143   "near starvation": ibid., p. 3.

143   "poorly conceived . . . executed": O'Keefe (1983), p. 135.

143   "Furthermore, contrary to the statements . . . thing for man": Bradley (1983 ed.), p. 125.

144   "What is true . . . radioactive": ibid., pp. 125–26.

144   "all showed considerable radioactivity": ibid., p. 130.

144   "I believe that . . . radioactivity": ibid., p. 126.

145   "it was safe to eat the fish": all quotes from the AEC meeting at Point Hope on March 14, 1960, are from: "Excerpts from a tape recording made by Rev. Keith Lawton of the Episcopal Mission in Point Hope Alaska during a public information meeting given by members of the USAEC and Environmental Committee for Project Chariot on March 14, 1960," transcript by D. Foote, Foote Collection.

145   "dead sea life . . . millions": quoted in Wasserman and Solomon (1982), p. 85.

145   "nearly a million pounds . . . destroyed": Hilgartner et al. (1983 ed.), p. 99.

145   "the only contaminated . . . trawler": quoted in Titus (1986), p. 50.

145   "due to the chemical . . . radioactivity": quoted in Hilgartner et al. (1983 ed.), p. 99.

146   "well within the danger zone": quoted in Titus (1986), p. 49.

146   "all the evidence available . . . contrary": Hewlett and Holl (1989), p. 177.

146   "eighty to ninety miles . . . ships": ibid., p. 175; Williams and Cantelon (1984), p. 178.

146   "six-megaton shot": Hewlett and Holl (1989), p. 173.

146   "The resulting fallout . . . measures": Glasstone (1957), p. 423.

146   "Subsequent editions . . . miles": Glasstone (1964 ed.), pp. 640–61.

146   "atolls . . . outside exclusion zone": Hewlett and Holl (1989), p. 172.

146   "tragic error": Teller and Brown (1962), p. 171.

147   "at best an educated guess": ibid., p. 172.
147   "Seeing the gauges . . . building": Wasserman and Solomon (1982), p. 85; O'Keefe (1983), p. 196.
147   "ten miles up island . . . would have died": Pauling (1983 ed.), p. 136.
147   "And had they been . . . dead": Glasstone (1964 ed.), pp. 462–63; Hilgartner et al. (1983), p. 99; Titus (1986), p. 47, O'Keefe (1983), pp. 196–97.
147   "Utirik Atoll . . . three days after the blast": Titus (1986), p. 47.
147   "We had meters . . . take": O'Keefe (1983), p. 196.
147   "given no instructions . . . dose": Hilgartner, et al. (1983 ed.), p. 99.
147   "they suffered . . . diarrhea": Titus (1986), p. 48.
147   "severely burned by beta . . . skin": Bradley (1983 ed.), p. 177.
148   "AEC issued a press release . . . homes": quoted in Wasserman and Solomon (1982), p. 86.
148   "None of the 28 . . . contracted": quoted in Titus (1986), p. 47.
148   "and nine years before . . . children": Hilgartner et al. (1983 ed.), p. 100.
148   "Nineteen of twenty-two . . . died": Wasserman and Solomon (1982), p. 86.
149   "*Life,* including Kitty Kinneeveauk": O'Neill (1989), p. 804.
149   "Many people considered Bikini . . . twenty more years": Bradley (1983 ed.), pp. 180–81.
149   "And missiles fired . . . Kwajalein": ibid., p. 178.
149   "The Marshallese . . . missile range": T. Hamilton, "Beyond Bravo," *The Bulletin of the Atomic Scientists,* March 1991, p. 45.
149   Amchitka: Hewlett and Duncan (1969), p. 535.
150   "On December 18 . . . Frenchman Flat": Titus (1986), pp. 55–58.
150   "100 aboveground detonations": USDOE (1988), p. V.
150   "in any other environment . . . state": Williams and Cantelon (1984), p. 203.
150   "When, in 1979 . . . weapons": Titus (1986), pp. 131–33.
150   "all evidence suggesting . . . protection": quoted in Hilgartner et al. (1983 ed.), p. 84.
150   "Nevertheless the AEC . . . assessments": ibid., p. 85; Titus (1986), p. 132.
151   "Additionally, the Eckhardt . . . fallout": ibid.
151   "By 1979 . . . expected": Hilgartner et al. (1983 ed.), p. 85.
151   "Sheep injuries . . . deaths": quoted in ibid., p. 91.
151   "negligently and wrongfully . . . duty": B. Jenkins quoted in Ball (1986), p. 159.
151   "a factual portrait . . . program": Ball (1986), p. 169.
151   "Dr. Harold Knapp . . . Branch": Wasserman and Solomon (1982), p. 77.
151   "critical omissions . . . sheep": quoted in Hilgartner et al. (1983 ed.), p. 94.
152   "normal in size . . . birth weight": quoted in ibid.
152   "But Knapp located . . . matter of days": ibid.

152   "The Nevada cattle study . . . on these animals": USAEC press release, "The Nevada Test Site Animal Investigation Project," 6 April 1960, p. 2.

153   "findings and conclusions": ibid.

153   "The latency period . . . cancer": Gofman (1981), p. 660.

153   "Different fallout . . . half-lives": Glasstone (1964 ed.), pp. 473–74.

153   "Isotopes of the gases . . . dirt": Sanders (1962), p. 76.

153   "The very fine . . . food": Glasstone (1964 ed.), pp. 473–74.

153   "hot spots . . . bombing range": Wasserman and Solomon (1982), p. 76.

153   Feed and water hauled to herd: USAEC press release, "The Nevada Test Site Animal Investigation Project," 6 April 1960, p. 5.

154   "Large doses . . . cancers": Teller and Latter (1958), p. 118.

154   "The first definite . . . years": Glasstone (1957), p. 481.

154   "There are a number . . . exposure": ibid., p. 480.

155   "a marked increase . . . retardation": ibid., p. 482.

155   "Linus Pauling wrote . . . radiation": Pauling (1983 ed.), p. 97.

155   "Pauling also noted . . . beings": ibid., p. 102.

155   "agreed with Pauling . . . expectancy": ibid., p. 108.

155   "Within a year . . . cancer": ibid., p. 94.

155   "The tragic case . . . anemias": ibid., pp. 94–95; Hilgartner et al. (1983 ed.), pp. 9–11.

156   "Also common knowledge . . . cancer": Pauling (1983 ed.), p. 95; Wasserman et al. (1982), p. 147.

156   "Hiroshima and . . . noted": ibid., pp. 95–96.

156   "mutations can . . . always be there": Teller and Latter (1958), p. 129.

156   "The Inca empire . . . Tibet": ibid., p. 130.

157   "an estimated 15-percent . . . children in Tibet": Pauling (1983 ed.), p. 135.

157   "potentially fateful . . . fallout": J. Arnold, "Effects of Recent Bomb Tests on Human Beings," *The Bulletin of the Atomic Scientists,* Nov. 1954, p. 347.

157   "no exposure is so tiny . . . certainty": H. J. Muller, "The Genetic Damage Produced by Radiation," *Bulletin of the Atomic Scientists,* June 1955, p. 210.

157   "Any radiation . . . general harmful": quoted in Pauling (1983 ed.), p. 68.

158   Popular press: "The Milk All of Us Drink," *Consumer Reports,* March 1959, p. 103.

158   Table figures: ibid., p. 104.

158   "Looking back on those . . . stupid": R. Rausch OHI in O'Neill (1989), p. 60.

159   "negligible, undetectable, or possibly nonexistent": USAEC (1960), p. 55.

159   Curies released at TMI and Chernobyl: Press release of U.S. Senator Frank Murkowski, 13 May 1992, pp. 2–3.

159   1.5 billion curies: G. Johnson to J. Philip, 31 March 1961, declassified, Doc. #5855, CIC. *Note:* an unclassified version of this letter exists that omits the quantity of vented radioactivity; it is Doc. #16869, CIC. See also: A. M. Piper, "Potential Effects of Project Chariot on Local Water Supplies, Northwestern Alaska," U.S. Geological Survey Professional Paper 539, 1966, pp. 1–2.

159   seventeen to thirty-eight Chernobyls: 1,500 MC/86 to 1,500 MC/40.

159   "Another government estimate . . . Chariot": U.S. Dept. of Commerce, Weather Bureau, Special Projects Section, Weather Bureau Research Station, Las Vegas, "Surface Radiation Estimate for Project Chariot," Nov. 1960. DOE-NV.

160   "This document questioned . . . detonation": ibid., pp. 2–3.

160   "314 to 675 Chernobyl disasters": 27,000 MC/86 MC to 27,000 MC/40 MC.

160   "The magnitude . . . indoors": Titus (1986), p. 113.

160   "fallout . . . 100,000 curies": from Piper (1966), pocket map: "Map Showing Expectable Fallout and Sampling Points for Drainage Basins in the Chariot-site Area, Northwestern Alaska."

160   "At this distance . . . that": R. Ball in "Excerpts from a tape recording made by Rev. Keith Lawton of the Episcopal Mission in Point Hope Alaska during a public information meeting given by members of the USAEC and Environmental Committee for Project Chariot on March 14, 1960," transcribed by D. Foote, Foote Collection; tape recording of March 14, 1960, AEC meeting at Point Hope, Alaska, Oral History Collection, UAF Archives.

161   "twenty kilotons each . . . period": Glasstone (1964 ed.), p. 196.

161   "seismic shock . . . eighty miles": A. Shelton to J. Reeves, 26 March 1959, p. 2, Doc. #78234, CIC, predicts that earth shock from a 460-kiloton blast would be felt at eighty miles. Another AEC document published a few months after the Point Hope meeting concludes that seismic shock from a 280-kiloton blast would be felt at Point Hope: "Public Safety Plan—Project Chariot—Plowshare Program," USAEC Office of Test Operations, Albuquerque, N.M., November 1960, p. 13, Doc. #3807, CIC. Other documents also confirm the eighty-mile seismic effect for the 280-kiloton configuration: General Manager, USAEC to Sen. E. Bartlett, 18 Jan. 1961, p. 6, Bob Bartlett Collection, UAF Archives; USAEC (1964), p. 4-10.

161   "With strontium 90 . . . centuries": Gofman and Tamplin (1971), p. 56.

161   "the whole Plowshare . . . itself": E. Teller, "We're Going to Work Miracles," *Popular Mechanics,* March 1960, p. 100.

161   "classification restraints . . . fallout": USAEC (1964), p. 5-2.

162   Rausch quote: R. Rausch in O'Neill (1989), p. 62.

162   "a quite high-up . . . Chariot": R. Rausch interviewed by author, 6 April 1989, author's notes.

11. The Committee and the Gang of Four

163   "If mankind is going . . . ignorant": N. Wilimovsky OHI in O'Neill (1989), p. 111.

163   "The majesty of technology . . . environment": J. Wolfe in Wilimovsky and Wolfe (1966), p. vii. *Note:* the statement is signed by the full committee, but members say Wolfe wrote it.

164   "filled with excitement . . . group": M. Britton OHI in O'Neill (1989), p. 3.

164   "Wolfe, says Britton . . . Commission": ibid., p. 5.

164   "There were no holds . . . burned": N. Wilimovsky OHI in O'Neill (1989), p. 120.

164   "I'm too blunt . . . no": ibid., p. 109.

164   "a kind of respect . . . sense": ibid., p. 101.

164   "Another of Britton's . . . things": M. Britton OHI in O'Neill (1989), p. 7.

165   Rectal temperatures: E. Campbell to J. Wolfe, 1 Dec. 1972, Doc. #131349, CIC.

165   "The last of the ones . . . could": M. Britton OHI in O'Neill (1989), p. 8.

166   "a very quiet chap . . . speak": ibid., p. 9.

166   "a very quiet, unassuming . . . scientist": ibid.

166   "an eager beaver . . . interminably": ibid., p. 10.

166   "only the chairman . . . staff": USAEC press release, "Plowshare Program Fact Sheet: Project Chariot," Revised Aug. 1962, p. 4.

167   "hard-headed . . . reason to": M. Britton OHI in O'Neill (1989), pp. 15, 26.

167   "If people tried . . . do it": A. Lachenbruch OHI in ibid., p. 73.

167   "I can guarantee . . . people": A. Seymour OHI in ibid., p. 168.

167   "I can remember . . . hell": N. Wilimovsky OHI in ibid., p. 119.

167   "many of us . . . Pacific": A. Seymour OHI in ibid., pp. 168–9.

167   "If it happened . . . game": N. Wilimovsky OHI in ibid., p. 119.

167   "we might have . . . that": A. Lachenbruch OHI in ibid., p. 73.

167   "but no incidents . . . mind": ibid., pp. 73–74.

167   "with impunity": M. Britton OHI in ibid., p. 15.

167   "I do not recall . . . anything": ibid., p. 14.

168   "over half a million dollars a year": USAEC (1964), pp. 6-2 to 6-5.

168   "Documents in . . . community": Joseph Foote to Paul Brooks, "Background materials on Project Chariot," 28 Oct. 1961, p. 9, Foote Collection.

168   "I absolutely remember . . . [laughs]": M. Britton OHI in O'Neill (1989), p. 18.

168   "I remember the day . . . statement": R. Rausch OHI in O'Neill (1989), pp. 48, 50.
168   "It is my recollection . . . thing": ibid., p. 49.
169   "I recall quite distinctly . . . by that": Author's notes from interview, 6 May 1988.
170   "It was his perception . . . did not": ibid.
170   "From the very beginning . . . standpoint": *Fairbanks Daily News-Miner,* 13 March 1959.
170   "the Committee's . . . if any": Wilimovsky and Wolfe (1966), p. x.
170   "the studies are not . . . standpoint": Minutes of AEC meeting 1511, 22 May 1959, p. 348, Doc. #753, CIC.
171   "prove the safety . . . experiment": USAEC press release, "Project Chariot," 24 June 1959, USAEC (1964), p. C-7.
171   "For instance . . . me": A. Seymour to J. Wolfe, 20 Sept. 1960, Doc. #50431, CIC.
171   "You'd have to kind . . . ahead": A. Seymour OHI in O'Neill (1989), p. 153.
171   "Whether the shot . . . responsibility": N. Wilimovsky OHI in O'Neill (1989), p. 109.
171   "assure that the . . . protected": USAEC press release, "Project Chariot," 6 March 1959, USAEC (1964), p. C-6.
172   Tracer experiment: A. M. Piper, *Potential Effects of Project Chariot on Local Water Supplies Northwestern Alaska,* U.S. Geological Survey, Professional Paper 539, U.S. Govt. Printing Office. *See also* USAEC (1964), p. x, where the tracer experiment is said to be "not part of Chariot."
172   "Ogotoruk wind . . . extent": Piper (1966), p. 31.
173   "where most coconut . . . scrub": Kiste (1974), p. 175.
173   "outside sponsorship . . . forthcoming": AEC staff paper 811/37, p. 2, DOE Archives.
173   "by January 1, 1960 . . . suspended": Minutes of AEC meeting 1511, 22 May 1959, p. 350, Doc. #753, CIC; see also AEC staff paper 811/37, 3 March 1959, p. 2, Doc. #70391, CIC.
174   "Preliminary Evaluation . . . year": R. Ball to Distribution, 2 Nov. 1959, Box 25, Foote Collection.
174   "hammered out . . . AEC": D. Foote to J. Foote, 23 Nov. 1959, Foote Collection.
174   Committee statement: USAEC press release, "Statement of Committee on Environmental Studies for Project Chariot," 7 January 1960, in USAEC (1964), pp. C-11 to C-12.
174   "Few birds . . . studies": ibid.
175   "based on the results . . . March 2, 1960": Minutes of AEC meeting 1597, 2 March 1960, p. 166, DOE Archives.

175   "Scientists from both . . . it": ibid., p. B-4; Johnson et al. to Committee on Environmental Studies, Project Chariot, undated, Project Chariot Collection, Box 1, UAF Archives.

175   "Notwithstanding . . . press": See ibid. and D. Foote to J. Wolfe, 17 June 1960, p. 3, AEC Corresp., Foote Collection.

175   "based on data . . . committee]": USAEC press release, "Statement of Committee on Environmental Studies for Project Chariot," 7 January 1960, in USAEC (1964), pp. C-11 to C-12.

176   "the Chariot detonation . . . delay": AEC Staff Paper 811/51, 17 Dec. 1959, p. 9, Doc. #70389, CIC.

176   Quotes from the Johnson, et al., letter: A. Johnson, W. Pruitt, L. Swartz, and L. Viereck to Committee on Environmental Studies, Project Chariot, "Subject: Committee Statement of January 7, 1960," Project Chariot Collection, Box 1, #1, UAF Archives. *Note:* the document is undated, though one copy bears a penned-in date: 8 June 1960. In J. Wolfe to A. Johnson et al., 11 July 1960, Al Johnson papers, Wolfe acknowledges receiving the letter on 9 June.

176   "deposited on the ice . . . breakup": USAEC press release, "Statement of Committee on Environmental Studies for Project Chariot," 7 January 1960, in USAEC (1964), pp. C-11 to C-12.

177   Quotes from the Brooks letter: J. Brooks to W. Egan, 6 April 1960, from a tape recording of the letter made by D. Foote, Box 58, #10, Foote Collection.

178   "a decline in marine . . . land": D. Foote to M. Johnson, 5 May 1960, Foote Collection.

178   "take advantage of these . . . Thompson": D. Foote, *The Economic Base and Seasonal Activities of Some Northwest Alaskan Villages,* November 1959, p. 38.

179   "proved, without doubt, the validity": D. Foote to J. Wolfe, 17 June 1960, Foote Collection.

179   "despite my protest . . . *Times*": D. Foote to J. Wolfe, 17 June 1960, Foote Collection.

179   "I will not allow . . . Chariot": ibid.

179   "regular, thrice-daily . . . site": J. Wolfe to A. Johnson et al., 11 July 1960, Al Johnson papers.

179   "It is clear . . . was": L. Viereck to W. Wood, 29 Dec. 1960, Viereck papers.

179   "But Foote had traced . . . satisfaction": D. Foote to J. Wolfe, 17 June 1960, Foote Collection.

179   "Wolfe went on to say . . . data": ibid.; J. Wolfe to A. Johnson et al., 11 July 1960, Al Johnson papers.

179   "But no such revision . . . scientists": L. Viereck to W. Wood, 29 Dec. 1960, Viereck papers.

180   "nor was it ever given . . . shows": USAEC (1964), pp. C-1 to C-37.

12. Wolfe, Meat, and Wood

181   "There are no . . . investigations": quoted in *The New York Times,* 17
         August 1960.

181   The account of caribou hunting is from D. Foote, *The Eskimo Hunters of
         Point Hope, Alaska: September, 1959, to May, 1960,* submitted to the
         USAEC, June 1960.

182   100,000 pounds: D. Foote and H. Williamson, "A Human Geographical
         Study," in Wilimovsky and Wolfe (1966), p. 1066.

183   "environmental scientists are . . . Chariot": J. Wolfe to A. Johnson et al.,
         11 July 1960, Al Johnson papers.

183   airplane flight: A. Gregory, *Jessen's Weekly,* 18 Aug. 1960.

183   Quotes from *NYT* article: *The New York Times,* 17 Aug. 1960.

184   "John Wolfe knew . . . firm conclusions": See J. Philip to G. Higgins, 28
         Feb. 1962, LLNL; J. Wolfe to J. Philip, 5 March 1962, LLNL.

184   "Don Foote had sent him a . . . Creek": D. Foote, *The Eskimo Hunters of
         Point Hope, Alaska: September, 1959, to May, 1960,* submitted to the
         USAEC, June 1960.

184   "fallout contours . . . grounds": See Piper (1966), Plate 1. *Note:* the
         contours are drawn from data provided by LRL in March 1961.

185   "Foote concluded that based . . . whale": D. Foote, *Project Chariot and
         the Eskimo People of Point Hope, Alaska,* submitted to the USAEC,
         March 1961.

185   "Wilimovsky once despaired . . . society": N. Wilimovsky OHI in O'Neill
         (1989), p. 111.

186   "I'm just proud . . . place": A. Gregory OHI in Jean Lester, *Faces of
         Alaska: A Glimpse of Alaska Through Paintings, Photographs and Oral
         Histories,* Tanana-Yukon Historical Society.

186   Albro Gregory article: *Jessen's Weekly,* 18 Aug. 1960.

186   "contradicts the facts . . . studies": D. Foote to A. Gregory, 23 Sept. 1960,
         Foote Collection.

187   "Gregory wrote back . . . people": A. Gregory to D. Foote, 7 Nov. 1960,
         Foote Collection.

187   "windowless hut . . . fur": Dorothy Walworth Crowell quoted in
         Fairbanks Daily News-Miner, 4 Dec. 1988, p. H-4.

187   "He allowed that he knew . . . before": from D. Foote's notes of interview
         with Gregory, 14 Feb. 1961, Foote Collection, Box 11, "History Notes 'C.'"

188   Rainey and Vanstone: See D. Foote, *Project Chariot and the Eskimo
         People of Point Hope, Alaska,* submitted to the USAEC, March 1961, p.
         10.

188   2,000 years: D. Foote to L. Viereck, 20 Jan. 1961, Viereck papers.

188   "Wolfe had read . . . had": W. Pruitt to D. Foote, 5 Sept. 1960, Foote Collection.

188   "A native from Point Hope . . . week": H. Spencer, *Journal of Harry Spencer, Meteorologist, Project Chariot, Winter 1959–60*, p. 9, Foote Collection.

188   "Two Eskimos staying . . . word": ibid., p. 11.

188   "The AEC arranged . . . visit": L. Viereck to D. Foote, 27 Aug. 1960, Les Viereck papers.

189   "Viereck wrote in a letter . . . Alaska": L. Viereck to A. Johnson, 20 Aug. 1960, Al Johnson papers.

189   "Stout, brought him . . . Reno": Hulse (1974), p. 59.

189   Details of Stout affair: ibid., pp. 52–59.

189   "You are hired . . . campus": quoted in ibid., p. 54.

190   "a small, dissatisfied minority group": ibid.

190   "quasi-military . . . community": ibid., p. 58.

190   "William R. Wood took over . . . failed": ibid., pp. 58–59.

190   *Who's Who:* W. Wood to The A. N. Marquis Company, 2 Nov. 1956, Wood Collection, UAF Archives.

190   "Wood included the listing . . . résumé": see, for example, W. Wood, undated, "Biographical Information," Wood Collection, UAF Archives.

190   "I think I should say . . . Nevada": R. Stevenson to W. Wood, 14 Dec. 1959, Wood Collection, UAF Archives.

190   "major factor in . . . plant": R. Wiegman to C. Sargent, 11 Dec. 1959, President's papers, Box 1, #10, UAF Archives.

191   "Desert Research Institute . . . program": W. Wood to R. Wiegman, 9 Dec. 1959, President's papers, Box 9, #142, UAF Archives.

191   "more in the promotion . . . fields": W. Wood to R. Wiegman, 16 Feb. 1960, President's papers, Box 9, #142, UAF Archives.

191   "sponsored scientific research . . . 1960s": Hulse (1974), pp. 89, 142.

191   "professional interest . . . war": *See* Michael J. Carey, "The Schools and Civil Defense: The Fifties Revisited," *Teacher's College Record,* Fall 1982, p. 115.

191   "He had attended . . . technologies": W. Wood OHI in O'Neill (1989), p. 430.

191   "in connection with work . . . Secret": W. Wood, "Security Investigation Data for Sensitive Position," Wood Collection, UAF Archives.

191   "a threat to all . . . America": W. Wood, "Renew Partnership to Advance Knowledge," *Fairbanks Daily News-Miner,* 2 Dec. 1987, p. 4.

191   "consummate interest . . . nurture": W. Wood to J. Wolfe, 3 Aug. 1961, President's papers, UAF Archives.

192   "completely down on the AEC": B. Kessel to A. Johnson, 25 Sept. 1960, Al Johnson papers.

192   "Viereck told friends . . . project": L. Viereck to D. Foote, 27 Aug. 1960, Viereck papers.

192   "extremely misleading . . . be": B. Kessel to A. Johnson, 25 Sept. 1960, Al
       Johnson papers.
192   "list of 'gripes' . . . data, etc.": B. Kessel to A. Johnson, 5 and 6 Nov.
       1960, Al Johnson papers.
192   "[Wood] has his own . . . use": ibid.
192   Wood had attended Teller speeches: ibid.
192   "some of the more radical . . . misleading": ibid.
193   "National Science Foundation . . . platters": B. Kessel to A. Johnson, 10
       Oct. 1960, Al Johnson papers.
193   Quotes from Kessel's account of her 26 Oct. 1960 meeting with Wood: B.
       Kessel to A. Johnson, 5 and 6 Nov. 1960, Al Johnson papers.
193   For more on Wood's concept of the university, see W. Wood OHI in
       O'Neill (1989), pp. 441–42.
195   "After Brina Kessel's . . . Viereck": ibid.
195   "it seemed of little . . . part": L. Viereck to A. Johnson, 18 Nov. 1960, Al
       Johnson papers.
195   "Still, he was . . . fault": L. Viereck to D. Foote, undated, Les Viereck papers.
195   "The one thing . . . commerce": ibid.
196   "Toward the conclusion . . . stipulations": C. Hunter to R. Rausch, 19
       Nov. 1960.
196   "Rausch claims . . . way": L. Viereck to D. Foote, undated, Les Viereck
       papers.
196   "I just can't . . . do": L. Viereck to A. Johnson, 18 Nov. 1960, Al Johnson
       papers.
197   "Les wants to cut . . . stands": B. Kessel to A. Johnson, 5 and 6 Nov.
       1960, Al Johnson papers.
197   "admired Les . . . now": ibid.
197   "working from a tape . . . prepared": L. Viereck to D. Foote, 14 Jan.
       1961, Les Viereck papers.
197   "the situation . . . project": L. Viereck to W. Wood, 29 Dec. 1960, Foote
       Collection.
197   "It has often been . . . government": ibid.
197   "could continue . . . runs": A. Johnson to L. Viereck, 24 March 1959,
       Project Chariot Collection, Box 28, UAF Archives.
198   "Her rationale went like . . . both": notes from interview of B. Kessel, 31
       March 1989.
198   "really quite pleased . . . member": B. Kessel to A. Johnson, 3 Dec. 1960,
       Al Johnson papers.
198   Herbarium position: B. Kessel to A. Johnson, 2 Jan 1961, Al Johnson papers.
198   "Wood signed off on all hiring": notes from interview of B. Kessel, 31
       March 1989, author's notes.
198   "Viereck spent about . . . with": L. Viereck to A. Johnson, 2 Feb. 1961, Al
       Johnson papers.

198   "As for keeping Viereck . . . money": L. Viereck to A. Johnson, 15 June 1961, Al Johnson papers.

198   "We have plenty . . . up": B. Kessel to A. Johnson, 12 May 1961, Al Johnson papers.

198   "Ordinarily Les would . . . now": B. Kessel to A. Johnson, 27 April 1961, Al Johnson papers.

199   "64 percent": L. Viereck to W. Wood, 16 June 1961, Les Viereck papers.

199   "When it came time . . . as well": L. Viereck to W. Wood, 16 June 1961, Viereck papers; also L. Viereck to A. Johnson, 15 June 1961, Al Johnson papers.

199   "with the bitter thought . . . Chariot": L. Viereck to A. Johnson, 15 June 1961, Al Johnson papers.

13. Going Public

200   "An imprudent . . . it": quoted in L. Salisbury to A. Johnson, 14 Oct. 1963, Al Johnson papers. Salisbury says, "this was written by an anonymous colleague in psychology who wants to stay on the staff for at least a year, so I am respecting his wishes."

200   "All the guys . . . women": G. Wood OHI in O'Neill (1989), pp. 651–52.

201   Details of Hunter and Hill's trip north: C. Hunter, pers. com., 8 Jan. 1993.

201   "would have been a snap . . . signals": G. Wood OHI in O'Neill (1989), p. 651.

201   "By the time . . . home": ibid., p. 652.

201   "somewhere along in the . . . Range": C. Hunter OHI in O'Neill (1989), p. 611.

202   "It's when environmental . . . typewriter": G. Wood OHI in O'Neill (1989), p. 627.

202   "We all feel . . . 1962": L. Viereck to J. Haddock, 2 March 1961, Les Viereck papers.

202   "something fairly drastic . . . done": L. Viereck to A. Johnson, 4 Sept. 1960, Al Johnson papers.

202   "In February, the ACS . . . area": ACS press release, 13 February 1961, Al Johnson papers.

203   "independent study . . . smoking": ibid.

203   "extremely mild . . . section": L. Viereck to A. Johnson, 2 March 1960, Al Johnson papers.

204   "really being grand . . . situation": B. Kessel to A. Johnson, 3 Dec. 1960, Al Johnson papers.

204   First Summary Report: USAEC, *Bioenvironmental Features of the Ogotoruk Creek Area, Cape Thompson, Alaska: A First Summary by the*

Committee on Environmental Studies for Project Chariot, December 1960 (TID 12439).

204   "the research 'to date'": USAEC (December 1960), p. 55

204   "Non-representative . . . reports": W. Pruitt to L. Viereck, L. Swartz, B. Kessel, 15 March 1961, Project Chariot Collection, UAF Archives.

204   "so completely at variance . . . results": ibid.

205   "We have emphasized . . . environment": W. Pruitt to L. Viereck, L. Swartz, B. Kessel, 15 March 1961, Project Chariot Collection, UAF Archives.

205   "It would appear . . . excavation": USAEC (Dec. 1960), p. 55.

205   "there are absolutely . . . throwout": W. Pruitt to L. Viereck, L. Swartz, B. Kessel, 15 March 1961, Project Chariot Collection, UAF Archives.

205   "His own references . . . report": quoted in B. Kessel to J. Wolfe, "Comments on 'Environmental Studies of Project Chariot' by University of Alaska Investigators," 7 April 1961, Project Chariot Collection, UAF Archives.

205   "there is a definite . . . botanists": ibid.

205   "The summary had said . . . provided": USAEC (December 1960), p. 53.

205   "It seems fairly . . . substitute": quoted in ibid.

205   "In general all . . . Chariot": B. Kessel to J. Wolfe, 7 April 1961, Project Chariot Collection.

206   "Your commentaries . . . full": J. Wolfe to B. Kessel, 12 April 1961 Project Chariot Collection.

206   Foote's reply: D. Foote to J. Wolfe, 14 March 1961, Foote Collection.

206   "Disruption of . . . geography": USAEC (December 1960), p. 53.

206   "The statement . . . did": D. Foote, "Comments on *Bioenvironmental Features of the Ogotoruk Creek Area, Cape Thompson Alaska: A Summary by The Committee on Environmental Studies for Project Chariot,* Plowshare Program, USAEC, December 20, 1960," p. 3, in D. Foote to J. Wolfe, 16 March 1961, Foote Collection.

207   "the chance . . . remote": USAEC (December 1960), p. 55.

507   "negligible . . . excavation": ibid.

207   "One reason I . . . abandoned": D. Foote to J. Wolfe, 16 March 1961, Foote Collection.

208   "Biologists in particular . . . soon": W. Pruitt OHI in O'Neill (1989), p. 352.

208   "A very prestigious . . . family": B. Kessel OHI with D. O'Neill, 10 Dec. 1985, #H85-324 A and B, UAF Archives.

208   James McKeen Cattell: see N. Reingold, *Dictionary of Scientific Biography,* 1971 (vol. III), pp. 130–31.

208   "In both his . . . presidents": ibid.

208   "didn't care . . . kept": W. Pruitt OHI in O'Neill (1989), p. 353.

209   "Pruitt had 18 . . . started": B. Kessel to A. Johnson, 25 Sept. 1960, Al Johnson papers.

209   open window: Interview with B. Kessel, 31 March 1989, author's notes.

209   "only one of a . . . difficult": W. Pruitt to D. Foote, 8 Dec. 1960, Foote Collection.

209   "We made trips . . . it": W. Pruitt OHI in O'Neill (1989), p. 357.

209   "In February of . . . alive": W. Pruitt, "Terrestrial Mammals Investigation, Ogotoruk Creek—Cape Thompson and Vicinity. Part A," January 1962, rough draft of final report to AEC, copy provided by Brina Kessel.

210   "During this trip . . . door": ibid.

210   "On this trip . . . layers of snow": ibid.; W. Pruitt, 1966, *Ecology of Terrestrial Mammals,* in Wilimovsky and Wolfe (1966), p. 519.

210   "I, of course . . . doing": B. Kessel to A. Johnson, 2 April 1961, Al Johnson papers.

211   "Al Johnson, who was inclined . . . agree": A. Johnson to B. Kessel, 9 April 1961, Al Johnson papers.

211   "permission to publish . . . investigators": J. Wolfe to "Those listed below," 3 Dec. 1959, Project Chariot Collection, UAF Archives.

212   "all hell broke . . . said)": W. Pruitt to D. Foote, 10 May 1961, Foote Collection.

212   "got fired up . . . all": W. Pruitt to D. Foote, 10 May 1961, Foote Collection.

212   "we really put our hearts into": W. Pruitt to D. Foote, 10 May 1961, Foote Collection.

212   "Viereck even headed . . . university": L. Viereck to R. Heacock, 24 Feb. 1961, Les Viereck papers.

212   "But there was something . . . idea": B. Kessel to A. Johnson, 5 and 6 Nov. 1960; B. Kessel to A. Johnson, 27 April 1961; A. Johnson to B. Kessel, 5 May 1961; A. Johnson to J. Marr, 12 May 1961; *See also* Albert W. Johnson, *Science, Society and Academic Freedom,* undated paper presented to the American Association of University Professors, all in Al Johnson papers.

212   "In June, Pruitt . . . 1962": "Fixed-Term Contract of Employment," dated 9 June 1961, Vice-President for Research Collection, UAF Archives; W. Pruitt to J. Foote, 23 Oct. 1961, Foote Collection.

213   "She had made numerous . . . censorship": W. Pruitt, "Statement to Project Chariot Environmental Committee and All Concerned," 25 April 1962, Vice-President for Research Collection, UAF Archives.

213   "Kessel had expected . . . approval": B. Kessel to J. Wolfe, 11 April 1962, President's papers, UAF Archives.

213   "had not had the opportunity . . . section": B. Kessel to W. Pruitt, 15 Feb. 1962, Vice-President for Research Collection, UAF Archives.

213   "that a major revision . . . investigations": ibid.

214   "I have received . . . altered": W. Pruitt to B. Kessel, 18 Feb. 1962, Vice-President for Research Collection, UAF Archives.

214   "ramifications . . . institutions": W. Pruitt to W. Wood, 2 March 1962, Vice-President for Research Collection, UAF Archives.

214   "You are hereby given . . . literature": B. Kessel to W. Pruitt, 29 March 1962, Vice-President for Research Collection, UAF Archives.

215   "Your puzzling statement . . . and how": W. Pruitt to B. Kessel, 30 March 1962, Vice-President for Research Collection, UAF Archives.

216   "Kessel also toned down . . . Pruitt": W. Pruitt, "Statement to Project Chariot Environmental Committee and All Concerned," 25 April 1962, Vice-President for Research Collection, UAF Archives.

216   "Pruitt had written . . . evaluated": W. Pruitt, "Terrestrial Mammals Investigation, Ogotoruk Creek—Cape Thompson and Vicinity. Part A," January 1962, rough draft of final report to AEC, copy provided by Brina Kessel.

216   "The deletions form . . . 'developers'": W. Pruitt, "Statement to Project Chariot Environmental Committee and All Concerned," 25 April 1962, Vice-President for Research Collection, UAF Archives.

217   "on April 11 . . . Investigations": ibid.; B. Kessel to J. Wolfe, 11 April 1962, President's papers, UAF Archives; W. Pruitt, *Project Chariot—Final Report, Terrestrial Mammals Investigations, Part A, General Studies and Small Mammal Biology,* Project Chariot Collection, UAF Archives.

217   "Kessel noted . . . investigations": B. Kessel to J. Wolfe, 11 April 1962, President's papers, UAF Archives; and B. Kessel to Environmental Committee Project Chariot, 13 April 1962, President's Papers, UAF Archives.

217   "The deleted statements . . . web": W. Pruitt, "Statement to Project Chariot Environmental Committee and All Concerned," 25 April 1962, Vice-President for Research Collection, UAF Archives.

217   "I was careful . . . humans": W. Pruitt OHI in O'Neill (1989), p. 366.

217   "While Pruitt did not . . . claimed": W. Pruitt, "Terrestrial Mammals Investigation, Ogotoruk Creek—Cape Thompson and Vicinity. Part A," January 1962, rough draft of final report to AEC, copy provided by Brina Kessel.

218   "I talked to her about it . . . satisfaction": A. Johnson to L. Hacker, 11 Oct. 1963, Al Johnson papers.

218   "It was Johnson's view . . . AEC": A. Johnson to L. Hacker, ibid.

218   "Gerry Swartz's . . . Brina": L. Swartz, interviewed by D. O'Neill, 18 Oct. 1988, Oral History Collection, UAF Archives.

219   "He reminded Wood . . . impressed": W. Pruitt to W. Wood, 15 April 1962, Vice-President for Research Collection, UAF Archives.

219   "unable to vouch . . . report": W. Pruitt, "Statement to Project Chariot Environmental Committee and All Concerned," 25 April 1962, Vice-President for Research Collection, UAF Archives.

219   "Instead of writing . . . problem": C. Elvey to W. Pruitt, 3 May 1962, Vice-President for Research Collection, UAF Archives.

219   "He was nasty . . . second best": W. Pruitt to D. Foote, 3 June 1962, Foote Collection.

219   "I really definitely . . . fired": W. Pruitt OHI in O'Neill (1989), pp. 371–72.

220   "But one document . . . the matter": J. Wolfe to B. Kessel, 17 Nov. 1961; W. Wood to C. Elvey, 25 Nov. 1961, both in President's papers, UAF Archives.

220   "When the American . . . agencies": L. Hacker to W. Wood, 8 Oct. 1963, Les Viereck papers.

220   "President Wood apparently . . . ACLU": W. Wood OHI in O'Neill (1989), p. 438.

220   "his contract came . . . background": quoted in W. Stern, "ACLU Probes Pruitt Case," *Polar Star,* 18 Oct. 1963.

220   "money for a permanent . . . administration": A. Johnson to L. Hacker, 11 Oct. 1963, Al Johnson papers.

221   "From my own . . . to me": L. Viereck to L. Hacker, 18 Oct. 1963, Les Viereck papers.

221   "Brina Kessel, too . . . level": interview of Brina Kessel, 31 March 1989, author's notes.

221   "And even Wood . . . scientists": W. Wood OHI in O'Neill (1989), p. 441.

221   "I recall vaguely . . . of it": ibid., p. 440.

221   "The Alaska Chariot . . . York": D. Foote to J. Foote, 9 Feb. 1961, Foote Collection.

222   "When in Kotzebue . . . him": D. Foote to A. Cooke, 9 Feb. 1961, Foote Collection.

222   "He is agent. Letter follows": A. Cooke to D. Foote, telegram 13 Feb. 1961, Foote Collection.

222   "I feel certain . . . office": A. Cooke to D. Foote, 13 Feb. 1961, Foote Collection.

223   "Don Foote had noticed . . . so": J. Foote to D. Foote, 15 Feb. 1961, Foote Collection.

223   FOIA request denial by CIA: F. Ruocco to D. O'Neill, 8 May 1992.

14. A National Protest

224   "More than likely . . . generation": C. Snow, 1961, address to the American Association for the Advancement of Science, quoted in *News Bulletin,* Alaska Conservation Society, March 1961.

224   "After World War . . . decisions": B. Commoner OHI in O'Neill (1989), pp. 272–74.

224   "None of the several associations . . . its will": AAAS Committee on Science in the Promotion of Human Welfare, "Science and Human Welfare: The AAAS Committee on Science in the Promotion of Human Welfare States the Issues and Calls for Action," *Science,* 8 July 1960, pp. 68–73.

225   "Accordingly, in 1958 . . . age": *Nuclear Information,* June 1961, end paper.

226   "Having become a major . . . itself": AAAS Committee on Science in the Promotion of Human Welfare, "Science and Human Welfare: The AAAS Committee on Science in the Promotion of Human Welfare States the Issues and Calls for Action," *Science,* 8 July 1960, pp. 68–73.

226   "Without mentioning . . . clouded": ibid.

226   "Wolfe felt . . . individually": J. Wolfe to J. Miller, 30 March 1960, L. Swartz papers.

226   "This CNI did . . . Plowshare": *Nuclear Information,* June 1960, and July 1960.

226   "stimulated such . . . material": K. Lark to L. Swartz, 17 Nov. 1960, L. Swartz papers.

226   Wolfe's letter: J. Wolfe to J. Miller, 30 March 1960, L. Swartz papers.

227   "I have been very . . . push": W. Pruitt to D. Foote, 10 Nov. 1960, Foote Collection.

227   "the flow of inquiries . . . CNI": J. Miller to J. Haddock, 22 Feb. 1961, Les Viereck papers.

227   "Air currents . . . regions": Commoner (1971), p. 54.

227   "But for some unknown . . . animals": B. Commoner, "Biological Risks from Project Chariot," *Nuclear Information,* June 1961, p. 9.

227   "And the stomach . . . units": ibid., p. 10.

227   "Furthermore, the caribou . . . world": ibid., p. 12.

228   "Unlike ordinary plants . . . it": ibid., p. 10.

228   "This arrangement . . . fallout": Pruitt (1962), p. 25.

228   100,000 pounds: D. Foote and H. Williamson, "A Human Geographical Study," in Wilimovsky and Wolfe (1966), p. 1066.

228   "It was my introduction . . . environmentalism": B. Commoner OHI in O'Neill (1989), p. 283.

229   "eyes start to light up": W. Pruitt OHI in O'Neill (1989), p. 336.

229   Hvinden and Gorham, as cited by Pruitt: Hvinden, T., 1958, "Radioaktivt ned fall, Norge," *Teknisk Ukeblad* 38–39, 3–18; Gorham, E., 1959, "A Comparison of Lower and Higher Plants as Accumulators of Radioactive Fallout," *Can. Journ. Botany* 37:327–329.

229   "The Canadian . . . region": as described in Alaska Conservation Society *News Bulletin,* March 1961, p. 23.

229   "not overrich in lichens . . . soil": see A. Betts to A. Luedecke, 2 August 1961, Attachment 3, Doc. #75525, CIC.

229   sedges as ranking next to lichen in radioactivity: Pruitt (1962), p. 25.

229   "ecologically obvious": ibid.

229   "vegetation-caribou-radiation . . . deleted": W. Pruitt to J. Wolfe, 8 Jan. 1962, Les Viereck papers.

230   "In his progress . . . document": B. Kessel to J. Wolfe, "Comments on 'Environmental Studies of Project Chariot' by University of Alaska Investigators," 7 April 1961, p. 4, Project Chariot Collection.

230   "somewhat out of . . . them": J. Wolfe to W. Pruitt, 19 Dec. 1961, Les Viereck papers.

230   "Once Wolfe found . . . thing": A. Johnson OHI in O'Neill (1989), p. 307.

230   "Special mention . . . reindeer": USAEC (1964), p. 2-7.

231   "There is almost no . . . Alaska": B. Commoner, "Biological Risks from Project Chariot," *Nuclear Information,* June 1961, p. 9.

231   "The grassland ecology . . . particles": W. Pruitt OHI in O'Neill (1989), p. 337.

231   "Fallout, which sifts . . . Sr90": B. Commoner, "Biological Risks from Project Chariot," *Nuclear Information,* June 1961, p. 10; *see also* E. Campbell to J. Philip, 25 April 1962, "Comments on Project Chariot Article in April 1962 Issue of *Harper's* magazine," p. 9, #34, LLNL. Here the AEC's Campbell states: "However, it is a valid point that nutrients absorbed through the root system of vascular plants are subject to additional step of discrimination in soil before assimilated [*sic*] within the cellular parts of the plants." The environmental committee's Allyn Seymour, however, seems to disagree that discrimination against the radionuclide occurs at the root wall (*see* A. Seymour OHI in O'Neill [1989], pp. 163–64).

231   "A great deal of . . . zones": B. Commoner, "Biological Risks from Project Chariot," *Nuclear Information,* June 1961, pp. 10–11.

232   "Washington University . . . Chariot": M. Friedlander, "Predictions of Fallout from Project Chariot," *Nuclear Information,* June 1961, pp. 5–8.

232   "Two of the shots . . . Blanca": from ibid.

232   "The AEC drew a curve . . . 25%": ibid., p. 8.

233   "With both publications . . . explosion": H. Margolis, "Project Chariot: Two Groups of Scientists Issue 'Objective' but Conflicting Reports," *Science,* 23 June 1961, pp. 2000–2001.

233   "Several of Margolis's . . . agency": A. Betts to A. Luedecke, 2 August 1961, Attachments 2 & 3, Doc. #75525, CIC.

233   "died down quickly": J. Wolfe to W. Wood, 26 July 1961, President's papers, UAF Archives.

233   "very cleverly managed": W. Wood to J. Wolfe, 3 August 1961, President's papers, UAF Archives.

233   "But the issue . . . about it": B. Commoner, "Project Chariot," *Science,* 18 Aug. 1961.

234   "*Science News Letter* also . . . Chariot": *Science News Letter,* 17 June 1961, p. 375.

235   "commends and supports . . . wildlife": *Sierra Club Bulletin,* May 1961, p. 4.

235   "And Gov. Bill . . . place": see *ACS News Bulletin,* March 1961, pp. 8–9.

235   "I should appreciate . . . based": W. Egan to E. Wayburn, 20 June 1961, in AEC 811/78, 10 July 1961, DOE Archives.

235   "I found myself . . . Range": E. Bartlett to W. Egan, 26 June 1961, in ibid.

235   "would unalterably . . . region": Wilderness Society Resolution in L. Crisler to O. Murie, H. Broome, and G. Marshall, 20 July 1960, Les Viereck papers.

235   "would be a . . . needs": quoted in ibid.

235   "distrusted . . . beings": L. Crisler to O. Murie, H. Broome, and G. Marshall, 20 July 1960, Les Viereck papers.

236   "part of our national . . . there": O. Murie to M. Thompson, 16 Sept. 1960, Les Viereck papers.

236   "thinking people . . . detonations": *National Wildlands News,* Feb. 1961, as quoted in ACS *News Bulletin,* March 1961, p. 10.

236   "The former published . . . Chariot": Coates (1991), p. 125.

236   "steeped . . . talk": P. Twyne to The President, 7 July 1961, in AEC 811/79, 14 July 1961, DOE Archives.

236   "The National Parks Association . . . conservationists": A. Smith to G. Seaborg, 9 June 1961, in AEC 811/74, 17 June 1961, DOE Archives.

236   "But 'Chariot' . . . cause": For additional discussion of conservationist groups' reticence in opposing Chariot, see ibid. pp. 127–31.

237   "The Massachusetts Audubon . . . few": E. Foster, "Project Chariot," undated ms., p. 11.

237   "We must be careful . . . in": quoted in Coates (1991), p. 129.

237   "mystery-cloaking . . . condemned": L. Crisler to O. Murie, H. Broome, and G. Marshall, 20 July 1960, Les Viereck papers. Crisler is responding to a letter from Richard Leonard, Sierra Club secretary.

237   "Not only sportsmen . . . plentiful": *Outdoor Life,* January 1961, pp. 10–11, 37.

238   "It was agreed . . . testing": R. Swann to L. Viereck, 20 March 1961, Les Viereck papers.

238   "In a 1964 review . . . fate": USAEC (1964), pp. 5-1, 5-2.

239   "Questioning began . . . opposition": ibid., p. B-9.

239   "If anybody at . . . outsider": K. Lawton OHI in O'Neill (1989), p. 681.

239   "I'm glad they found . . . be": ibid., p. 684.

240   "ridiculous": ibid., p. 686.

240   "Well the Eskimo people . . . on": ibid., pp. 687–88.

240   "As Lawton remembers . . . conscience": K. Lawton, personal communication, 7 Jan. 1993, author's notes.

240   "I must point out . . . project": W. Gordon to Mr. and Mrs. J. Haddock, 9 March 1961, Haddock-Foster papers.

241   "The truth is . . . hands": R. Heacock to C. Hunter, 20 Feb. 1961, Alaska Conservation Society Collection, UAF Archives.

241   "pretty scary . . . courage": R. Heacock OHI in O'Neill (1989), p. 731.

241   "Heacock asked . . . assembly": R. Heacock to C. Hunter, 20 Feb. 1961, Alaska Conservation Society Collection, UAF Archives.

242   "not much . . . light": W. Gordon, telephone interview, 12 Dec. 1993, author's notes.

242   "We continue . . . plans": Alaska Mission of the Methodist Church, "1960 Official Journal," 17–20 May 1960, p. 108.

242   "It has been described . . . organization": R. Heacock OHI in O'Neill (1989), p. 724.

242   "in response to . . . assured": T. Stevens to R. Heacock, 25 April 1961, Richard Heacock papers.

242   Quotes from 1961 recording at Point Hope: Aligah Attungana, Joseph Frankson, Teddy Frankson, Mrs. Frankson, John Oktollik, Jimmy Kilivook, Dan Lisbourne, and David Frankson, interviewed by David Frankson, 14 March 1961, Oral History Collection, UAF Archives; *see also* O'Neill (1989), p. 736, for transcript.

243   "And I'm sure . . . Heacock": R. Heacock OHI in O'Neill (1989), p. 723.

243   "In Norwich . . . Congress": A. Cooke to L. and T. Viereck, 24 May 1961, Les Viereck papers.

243   "in the face . . . against it": A. Cooke to C. Anderson, 5 May 1961, Doc. #100095, CIC.

243   "I imagine myself . . . ball": J. Haddock to L. Viereck, 21 Feb. 1961, Les Viereck papers.

244   Radio and television interviews: see E. Foster, "Project Chariot," undated ms., p. 2. A partial recording of an interview of Lawton on WGBH-TV in Boston exists on the Lawton tape of the March 14, 1960, AEC meeting at Point Hope, UAF Archives.

244   "can call up . . . the President": E. Foster, undated ms., "Project Chariot," p. 3.

245   Haddock's six-page paper: J. Haddock, D. Haddock, and V. Foster, "The Threat of Project Chariot to the People of Point Hope, Alaska, and Their Way of Life," Feb. 1961, Les Viereck papers.

245   "It would be instantly . . . risk": ibid.

245   "To this memorandum . . . possible": J. Haddock to L. Viereck, 21 Feb. 1961, Les Viereck papers.

245   "from a close personal . . . headmaster": ibid.

246   "Dean Acheson . . . Project Chariot": E. Foster, undated ms., "Project
          Chariot," p. 13; J. Haddock to L. Viereck, 10 April 1961, Les Viereck
          papers.

246   "McGeorge Bundy . . . Foster's": J. and D. Haddock to K. and J. Lawton,
          10 April 1961, Haddock-Foster papers.

246   "Bundy too . . . misgivings about Chariot": J. Haddock to L. Viereck, 10
          April 1961, Les Viereck papers; E. Foster, "Project Chariot," undated ms.,
          p. 13.

246   "Ted Sherburne . . . Commission": J. Haddock to L. Viereck, 7 March
          1961, Les Viereck papers.

246   "Max Foster also . . . story": J. Haddock to L. Viereck, 10 April 1961, Les
          Viereck papers.

246   "Haddock wrote the Lawtons . . . wrote": J. and D. Haddock to K. and J.
          Lawton, 10 April 1961, Haddock-Foster papers.

246   "he told a tale . . . reply": quoted in E. Foster, "Project Chariot," undated
          ms., p. 14.

246   "By May, the AEC's . . . expand": A. Betts to Commissioner Seaborg et al.
          15 May 1961, DOE Archives. *See also:* D. Ink to C. Anderson, 23 May
          1961, Doc. #100094, CIC.

247   "make available . . . tribes": *Indian Affairs: Newsletter of the Association
          on American Indian Affairs,* July 1961.

247   "assess the facts . . . interests": ibid.

15. Drumbeats on the Tundra

248   "These people . . . question": as quoted in Morgan (1988), p. 181.

248   "What happened in . . . revolution": Berry (1975), pp. 138–39.

249   "All the four seasons . . . explosion": Health Council of Point Hope to J.
          Kennedy, 3 March 1961, in Doc. #16872, CIC.

250   "make my peace . . . die": Morgan (1988), p. 158.

250   "responding to . . . requests": *Indian Affairs,* July 1961, p. 1.

250   "specific request for help": ibid., p. 2.

250   "But Viereck had . . . AAIA": Viereck does not remember writing AAIA
          (personal communication 17 Feb. 1993), and his extensive
          correspondence files contain no copy.

250   Frankson letter: quoted in ibid., pp. 1–2.

251   Organic Act: quoted in Arnold (1978), p. 69.

251   "While the law . . . action": ibid., pp. 68–69, 91.

251   "And by 1961 . . . rights": T. Hetzel, "Indian Rights and Wrongs in
          Alaska," *Indian Truth,* Oct. 1961.

251   "The Eskimos had never . . . village": ibid.; Morgan (1988), 173.

251   "the only one . . . later": H. Rock OHI, 25 March 1976, UAF Archives.

251   "Rock also mentioned . . . coalition": ibid.; Morgan (1988), pp. 173, 195.

252   "At this point . . . threatened": Members of the Village Council, Point Hope, Alaska, to S. Udall, 25 July 1961, in USAEC (1964), pp. B-25 to B-26.

252   "Four days after . . . allayed by the presentation": J. Kelly to For the Record, "Report on the Trip of July 24–August 2 to Livermore, California and Alaska," 7 Aug. 1961, Doc. #75527, CIC. *See also: Fairbanks Daily News-Miner,* 29 July 1961, p. 1.

253   "there was still . . . expressed": ibid., p. B–18.

253   "did not appear . . . information": J. Kelly to For the Record, "Report on the Trip of July 24–August 2 to Livermore, California and Alaska," 7 Aug. 1961, Doc. #75527, CIC, p. 3.

253   "As Southwick correctly . . . standpoint": R. Southwick, "Report of Alaska Trip: Project Chariot," in USAEC (1964), p. B-20.

253   Barrow eider duck controversy: *Indian Affairs,* July 1961; Blackman (1989), pp. 180–84.

254   Petition to Kennedy: Blackman (1989), p. 183.

254   "As it is now . . . hunt": quoted in ibid., p. 181.

254   "We did not know . . . everything": *Indian Affairs,* July 1961, p. 3.

255   Barrow Conference on Native Rights: the most descriptive account of the meeting can be found in Morgan (1988), pp. 180–87.

255   "The Eskimo leaders . . . language": *Indian Affairs,* December 1961.

255   "We the Inupiat . . . rights": from "A Statement of Policy and Recommendations adopted by the Point Barrow Conference on Native Rights, Barrow, Alaska, November 17, 1961," reprinted in *Indian Affairs,* Dec. 1961.

255   "We were told . . . rights": ibid.

256   "*Brainwashed* . . . villages": L. Madigan, "Liberation From Lies," in ibid.

256   "because we never . . . other": ibid.

256   "The delegates sorted . . . home": ibid.

256   "two problems . . . Hope": ibid.

256   "We deny . . . village": ibid.

256   "Previously, the *News-Miner's* . . . English": T. Snapp OHI with D. O'Neill, 20 June 1988, Oral History Collection, UAF Archives.

257   "The *News-Miner* . . . conference": ibid.

257   "I could hardly wait . . . funding": ibid.

258   "The two would-be publishers . . . their man": ibid.

258   "I sat down . . . *Miner*": ibid.

258   "forty-seven-page": Daley and O'Neill (1993), p. 268.

258   "$35,000 . . . one on one": T. Snapp OHI with D. O'Neill, 20 June 1988, Oral History Collection, UAF Archives.

258   "Snapp had already . . . journalism": ibid.

259   "Lael Morgan . . . team": Morgan (1988), p. 197.

259   "cited the settlement . . . today": *Tundra Times,* 1 Oct. 1962, p. 1.

259   "the problem of settling . . . face": ibid.
259   five reporters: H. Rock to H. Forbes, 30 Sept. 1962, quoted in Morgan (1988), p. 199.
259   "no mention whatever . . . Alaska": *Fairbanks Daily News-Miner,* 27 Sept. 1962, p. 1.
259   "The value of an . . . backers": H. Rock to H. Forbes, 30 Sept. 1962, quoted in Morgan (1988), p. 199.

16. Spiking the Wheels of Chariot

260   "It is, therefore, important . . . Chariot": J. Kelly to J. Foster, 18 May 1962 (received), LLNL.
260   "There is a distinct . . . *press*": D. Foote to L. Viereck, 23 Jan. 1961, Les Viereck papers.
260   "privately, most people . . . state": J. Foote to P. Brooks, "Background materials on Project Chariot," 28 Oct. 1961, Foote Collection.
261   "In February . . . safety": *Fairbanks Daily News-Miner,* 15 Feb. 1961, p. 3.
261   "running into heavy . . . explosion": *Anchorage Daily Times,* 31 March 1961, p. 4.
261   "And the *Times* editorial . . . mile": ibid., 29 June 1961, p. 4.
261   "CARIBOU'S FONDNESS . . . BLASTS": *The New York Times,* 4 June 1961.
261   "the first dash . . . *Times*": E. Foster, "Project Chariot," undated ms., p. 15.
261   "in the process . . . Eskimos": *The New York Times,* 4 June 1961.
261   "a strong . . . program": *Christian Science Monitor,* 21 August 1961.
261   "showed a desire . . . treatment": J. Foote to P. Brooks, 28 Oct. 1961, p. 13, Foote Collection.
261   "The four-part series . . . statements": *Fairbanks Daily News-Miner,* 19, 21, 22, 23 Aug. 1961.
261   "ALASKAN ESKIMOS BUCK AEC": *Christian Science Monitor,* 16 Dec. 1961.
261   "*Bulletin of the Atomic Scientists*": *Bulletin of the Atomic Scientists,* Dec. 1961, pp. 426–27.
261   "Here's the big . . . aired": W. Pruitt to D. Foote, 18 Sept. 1961, Foote Collection.
262   "David Brower, authorized . . . Alaska": D. Brower, personal communication, 6 March 1993.
262   "the answer to prayer": E. Foster (undated), p. 19.
262   "Is the plan . . . things": Brooks and Foote (1962), p. 61.
263   "morass of scientific . . . uncertainty": ibid., p. 66.
263   "Few of us . . . use": ibid., p. 67.

263   "We still prefer . . . article": E. Campbell to J. Philip, "Comments on Project Chariot Article in April 1962 Issue of *Harper's* Magazine," 25 April 1962, in J. Philip to J. Kelly, 27 April 1962, LLNL.

263   "As Gerald Johnson . . . lost": G. Johnson OHI with D. O'Neill, tape H90-03-02, UAF Archives.

264   "The decision . . . 1958": E. Campbell to J. Philip, "Comments on Project Chariot Article in April 1962 Issue of *Harper's* Magazine," 25 April 1962, in J. Philip to J. Kelly, 27 April 1962, LLNL.

264   "In fact . . . cites": C. Bacigalupi to D. Kilgore, 20 May 1958, classification canceled, LLNL.

264   "Campbell wrote that . . . files": E. Campbell to J. Philip, "Comments on Project Chariot Article in April 1962 Issue of *Harper's* Magazine," 25 April 1962, in J. Philip to J. Kelly, 27 April 1962, LLNL.

264   "But had he done . . . viability": see especially *Anchorage Daily Times,* 15 July 1958, and *Fairbanks Daily News-Miner,* 24 July 1958.

264   "had little sympathy for Chariot": J. Kelly to G. Seaborg, 7 May 1962, in AEC 811/101, 17 May 1962, DOE Archives.

264   "earnest and detailed . . . supply": O. La Farge to S. Udall, 28 March 1961, Doc. #24661, CIC.

265   "I have the complete . . . Berry": S. Francis to L. Viereck, 14 Aug. 1961, Les Viereck papers.

265   "involved exploration . . . Chariot": ibid.

265   "Howard Rock's Council": Point Hope Village Council to S. Udall, 25 July 1961, in USAEC (1964), pp. B-25 to B-26.

265   "heartwarming . . . profoundly": S. Francis to L. Viereck, 14 Aug. 1961, Les Viereck papers.

265   "Right now the . . . elsewhere": ibid.

265   "containing approximately . . . miles": USAEC press release, "Plowshare Program: Fact Sheet on Project Chariot," 8 April 1961, p. 5.

266   "On January 30 . . . participation": S. Francis to D. Foote, 1 Feb. 1962, Foote Collection; S. Francis to D. Foote, 8 May 1962, Foote Collection; J. Kelly to G. Seaborg, 7 May 1962, in AEC 811/101, 17 May 1962, DOE Archives.

266   "In our analysis . . . care": S. Francis to D. Foote, 1 Feb. 1962, Foote Collection.

266   "It was apparent . . . Chariot": J. Kelly to G. Seaborg, 7 May 1962, in AEC 811/101, 17 May 1962, DOE Archives.

266   "the AEC could neither . . . experiment": ibid.

266   "It would take . . . felt": ibid.

266   "The influence of 'hawk' . . . Sputnik": Schooler (1971), p. 107.

266   "As an official . . . Strauss": Hewlett and Holl (1989), p. 464.

267   "A-BLAST TO DIG . . . grounds": *The New York Times,* 13 May 1962.

267   "no biological objections": *The New York Times,* 17 Aug. 1960.

267   "educated guess . . . Chariot": *The New York Times,* 13 May 1962.

267   "no self-respecting . . . bucks": as quoted in E. Foster (undated), p. 20.

268   "I have become deeply . . . land": J. Reeves to E. Shute, 2 June 1961, Doc. #18760, CIC.

268   "a primary reason . . . environment": J. Foster to J. Kelly, 27 April 1962 (probably a draft version of the April 30 letter), LLNL; USAEC press release, "Project Chariot—1960," 4 March 1960, p. 2; USAEC (Dec. 1960), p. 5.

268   "First, the AEC required . . . April": J. Reeves to O. Roehlk, 8 June 1961, Doc. #16865, CIC.

269   "And John Wolfe . . . standpoint": *Nome Nugget,* 29 April 1960, p. 4.

269   "In a 1961 memo . . . publicly": W. Hess, G. Higgins, M. Nordyke, V. Shelton, and R. Wallstedt to G. Johnson, "Possible revision of Chariot to 1–200 kiloton shot," 5 January 1961, LLNL.

269   "a meaningful radioactivity experiment": J. Foster to J. Kelly, 27 April 1962 (probably a draft version of the April 30 letter), LLNL.

269   "signal importance . . . excavation": J. Philip to G. Higgins, 14 Aug. 1961, LLNL.

269   "At a meeting held . . . lost": USAEC, "Seventeenth Meeting of the Advisory Committee for Biology and Medicine, USAEC," 17–18 Oct. 1958, Doc. #105525, CIC.

269   "The human environmental . . . sensitivity": ibid., p. 5.

269   "All our efforts . . . basis": J. Reeves to E. Shute, 2 June 1961, Doc. #18760, CIC.

269   "there is any possibility . . . weather": ibid.

270   "saturated muck . . . slopes": J. Foster to J. Kelly, 27 April 1962 (probably a draft version of the April 30 letter), LLNL.

270   "The internal view . . . value": ibid.

270   "in a stand-by status": AEC 811/97, 7 Feb. 1962, p.3, DOE Archives

271   "orderly series . . . data": ibid.

271   "frustration . . . Johnson": G. Johnson OHI in O'Neill (1989), pp. 222–23.

271   "to specifically investigate . . . documents": J. Foster to J. Kelly, 27 April 1962 (probably a draft version of the April 30 letter), LLNL.

271   "intervening international . . . area": AEC meeting 1821, 14 Feb. 1962, Doc. #107610, CIC.

272   "President Kennedy had . . . technology": Findlay (1990), p. 49.

272   "presidential authorization . . . been sought": J. Foster to J. Kelly, 27 April 1962 (probably a draft version of the April 30 letter), LLNL.

272   "in view of the . . . future": AEC Meeting 1821, 14 Feb. 1962, Minutes dated 27 Feb. 1962, Doc. #107610, CIC.

272   "the need to establish . . . closed out": ibid.

272   "continue to hold . . . time": AEC 811/97, 7 Feb. 1962, p. 3, DOE Archives.

272   "Teller had resigned . . . weapons": Broad (1992), p. 50.

272   "we recommend . . . canceled": J. Foster to J. Kelly, 30 April 1962, LLNL.

272   "Foster outlined . . . Chariot": ibid.; J. Foster to J. Kelly, 27 April 1962 (probably a draft version of the April 30 letter), LLNL.

272   "looks like another . . . demands": W. Hess et al. to G. Johnson, 5 Jan. 1961, LLNL.

272   "Such an action . . . canal": J. Foster to J. Kelly, 30 April 1962, LLNL.

273   "prepare plans . . . Chariot": J. Kelly to J. Foster, 18 May 1962 (received), LLNL.

273   "By July 1962, . . . excavation": AEC staff paper 811/104, 10 July 1962, Doc. #75574, CIC.

273   "But Kelly's arguments *against* . . . 'uncertainties'": ibid.

274   "Kelly concluded . . . canceled": ibid., p. 4.

274   "Information Plan . . . no longer needed": ibid., p. 5.

274   "largely obviated . . . Chariot": ibid., p. 7.

274   "Within three seconds . . . miles": Defense Nuclear Agency (1983), p. 70.

274   "An hour later . . . site": USAEC press release, "The following statement was issued at Las Vegas, Nevada, at 11:00 A.M. PDT, July 6, 1962," NVOO.

274   "Twenty-one years . . . predicted": Defense Nuclear Agency (1983), p. 73.

275   "it seems not . . . five times": M. Friedlander, "Predictions of Fallout from Project Chariot," in *Nuclear Information,* June 1961, p. 7.

275   "twelve miles an hour": ibid.

275   "At 2:45 P.M. . . . visibility": USAEC press release, "The following statement was issued at Las Vegas, Nevada, at 2:45 P.M. PDT, July 6, 1962," NVOO.

275   "The dust was so thick . . . afternoon": Skartvedt (1992), p. 130.

275   "Five days later . . . Dakotas": USAEC press release, "The following statement was issued at Las Vegas, Nevada, at 2:45 P.M. PDT, July 6, 1962," NVOO.

275   "tanker trucks . . . ditches": Defense Nuclear Agency (1983), p. 104.

275   "Two years later . . . Canada": Findlay (1990), p. 27.

276   "The bomb had been buried . . . deep": Defense Nuclear Agency (1983), pp. 70–73.

276   "a significant contribution . . . technology": USAEC press release, "The following statement was issued at Las Vegas, Nevada, at 3:40 P.M. PDT, July 7, 1962," NVOO.

276   "publicly convincing . . . Chariot": AEC staff paper 811/104, 10 July 1962, Appendix A, p. 5, Doc. #75574, CIC.

276   "The next month . . . experiment": USAEC press release, "Project Chariot Decision Held In Abeyance," 24 Aug. 1962, NVOO.

276   "interestingly, when *The New York* . . . abeyance": *The New York Times,* 13 May 1962, p. 71.

276   "If Sedan and future . . . Australia": see Findlay (1990), p. 30; J. Foster to J. Kelly, 30 April 1962, LLNL.

277   "In 1962, however . . . demise": Findlay (1990), p. 30.

277   "small but very vocal groups": AEC staff paper 811/104, 10 July 1962, p. 3, Doc. #75574, CIC.

277   "the self-righteousness . . . canceled": S. Francis to D. Foote, 26 Aug. 1962, Foote Collection.

277   "Those involved with the project . . . obsolete": ibid.

277   "PROJECT CHARIOT CALLED OFF": *Fairbanks Daily News-Miner,* 24 Aug. 1962, p. 1.

278   "Alaskan Eskimos won . . . end": ibid.; *Anchorage Daily Times,* 24 Aug. 1962, p. 1.

278   "Crowley passed along . . . decision": ibid.

278   "human geographers . . . life": ibid.

278   "Another, somewhat more . . . energy": *Anchorage Daily Times,* 28 Aug. 1962, p. 6.

279   "The greatest champion . . . rights": *Tundra Times,* 1 Oct. 1962.

279   "an average man . . . before": J. Haddock to L. Viereck, 21 Feb. 1961, Les Viereck papers.

279   Max Foster's toast: quoted in E. Foster (undated), p. 21.

17. Blacklisting

281   "If you can't . . . speak it": A. Johnson, "Science, Society and Academic Freedom," undated paper presented to American Association of University Professors, p. 4, Al Johnson papers.

281   "In February 1963 . . . impressed": E. Pfeiffer, L. Graves, M. Chessin, O. Stein, in "Information Activity Hit: Zoologist Denied Job," *SSRS Newsletter,* August 1963, Society for Social Responsibility in Science. *See also* E. Pfeiffer to "Those who wrote to Dr. P. L. Wright, Chairman, Zoology Department, Montana State University in support of Dr. W. O. Pruitt's application for a temporary position in the MSU Zoology Department," 9 July 1963, Les Viereck papers.

281   "and suggested he . . . leave": W. Pruitt to D. Foote, 27 May 1963, Foote Collection.

281   "The faculty . . . New York": E. Pfeiffer, L. Graves, M. Chessin, O. Stein, in "Information Activity Hit: Zoologist Denied Job," *SSRS Newsletter,* August 1963, Society for Social Responsibility in Science.

281   "The two . . . years": W. Wood to M. Warden, 15 March 1959, Wood Collection, UAF Archives.

281   "Wood advised Newburn . . . Pruitt": E. Pfeiffer, L. Graves, M. Chessin, O. Stein, in "Information Activity Hit: Zoologist Denied Job," *SSRS Newsletter,* August 1963, Society for Social Responsibility in Science.

281   "Wolfe said, according to Abbot . . . him": quoted and paraphrased in ibid.

282   "Wolfe also suggested . . . application": ibid.

282   "Pruitt had indeed . . . Union": see L. Hacker to L. Viereck, 8 Oct. 1963, Les Viereck papers.

282   "the American Association . . . Welfare": B. Commoner to Les Viereck, 27 March 1961, Les Viereck papers.

282   "Society for Social . . . Science": see "Information Activity Hit: Zoologist Denied Job," *SSRS Newsletter,* August 1963, Society for Social Responsibility in Science.

282   "*Time* magazine": see *Time* 13 Sept. 1963, p. 63.

282   "general supervisor of this . . . findings": quoted in W. Pruitt to H. Forbes, 29 May 1963, Foote Collection.

282   "The statement . . . floored": ibid.

282   "might properly edit . . . action": quoted in ibid.

282   "Several members of the . . . AEC": E. Pfeiffer in "Information Activity Hit: Zoologist Denied Job," *SSRS Newsletter,* August 1963, Society for Social Responsibility in Scienc

283   "I frankly think . . . support": W. Pruitt to D. Foote, 23 April 1963, Foote Collection.

283   "But he didn't say . . . going": G. Wood OHI in O'Neill (1989), p. 642.

283   "Sorry, if I told . . . support": "Courageous Scientists," *Tundra Times* editorial (undated clipping), Doc. #163772, CIC.

283   "As one friend remembered . . . arrived": W. Fuller to Committee to Nominate Honorary Degree Recipients, 17 June 1992, University Relations Office, University of Alaska Fairbanks.

283   "Pruitt thinks he got . . . secretary": W. Pruitt OHI in O'Neill (1989), p. 378.

283   "but neither the eighty-six . . . 1993": G. Cross, telephone interview, 26 April 1993; Ada Arnold, telephone interview, 26 April 1993.

283   Alaska Methodist University: ibid., p. 380.

283   "his aged mother . . . FBI": ibid., p. 377.

283   "A particularly disturbing . . . community": Metzger (1972), pp. 256–57.

284   "President Patty . . . speak it": A. Johnson, "Science, Society and Academic Freedom," undated paper presented to American Association of University Professors, p. 4, Al Johnson papers.

284   "We thought you might . . . members": J. Philip to W. Wood, 30 June 1961, President's papers, UAF Archives.

284   "A former employee . . . things": J. Wolfe to W. Wood, 14 Aug. 1961, President's papers, UAF Archives.

284   "It is suggested . . . executive": J. Wolfe to Board of U.S. Civil Service Examiners (cc: W. Wood), 25 Jan. 1962, President's papers, UAF Archives.

285   "He reports that Alaska . . . as well": L. Viereck, personal communication, 21 June 1993. See also L. Viereck to D. and B. Foote, 13 April 1962, Les Viereck papers; C. Hunter to D. Foote, 9 April 1962, Foote Collection.

285   "Fairbanks legislator Warren . . . legislature": E. Beistline to W. Taylor, 1 March 1961, President's papers, UAF Archives.

285   "Almost immediately . . . ISEGR": see W. Dickson to D. Foote, 24 Nov. 1964, Les Viereck papers.

285   "he remembers Wood . . . body": V. Fisher, personal communication, 12 Dec. 1987.

286   "account they gave the police": Case #F-31962, Records and Identification, Dept. of Public Safety, State of Alaska.

286   "Heading back the four . . . car": *Fairbanks Daily News-Miner,* 27 Feb. 1969.

286   "Road conditions . . . slippery": ibid., 3 March 1969; Case #F-31962, Records and Identification, Dept. of Public Safety, State of Alaska.

286   "The trooper . . . percent": Case #F-31962, Records and Identification, Dept. of Public Safety, State of Alaska.

286   "several of Foote's friends": W. Pruitt OHI in O'Neill (1989), p. 379; K. Lawton OHI in ibid., pp. 719–21.

286   "It was Bonnie Babb . . . know": Jim and Bonnie Babb OHI by D. O'Neill, 1 Oct. 1988, Oral History Collection, UAF Archives.

287   "The cause of death . . . injury": Certificate of Death for Don C. Foote, Recorder's No. V-69-382, Fairbanks Recording District, Alaska.

287   "Just a few months . . . document": F. Ruocco to D. O'Neill, 8 May 1992, O'Neill Collection.

287   "The research halt . . . dissolved": *Fairbanks Daily News-Miner,* 4 March 1969, p. 1.

287   Viereck's eulogy: L. Viereck, "Don Foote," undated paper, Les Viereck papers.

Epilogue

289   Clarence Darrow quote: from speech to jury, Chicago, Illinois, 1920, in A. Weinberg, *Attorney for the Damned,* 1957, quoted in S. Donadio et al., *The New York Public Library Book of 20th-Century American Quotations,* 1992.

289   "the father of . . . ecology": Press release, Ministry of Indian and Northern Affairs, Canada, 8 Nov. 1989.

290   "In 1992, thirty years . . . biologists": see D. O'Neill to Committee to
       Nominate Honorary Degree Recipients, 23 May 1992, O'Neill Collection.
290   "led by two senior professors": they were Drs. Rudi Krejci and Jack
       Distad.
291   Legislative citation: "The Alaska Legislature, Honoring Leslie Viereck and
       William Pruitt," Eighteenth Alaska State Legislature, passed 8 May 1993,
       sponsored by Rep. John Davies (D), Fairbanks, and others.
291   "Many ideas, theories . . . today": P. Usher to Committee to Nominate
       Honorary Degree Recipients, 2 Sept. 1992, University Relations Office,
       University of Alaska Fairbanks.
291   "Dr. Wood has had back hoes . . . campus": R. Atwood, speech at the
       Alaskan of the Year ceremony, 30 March 1973.
291   "One of these projects . . . reactor": W. Wood OHI in O'Neill (1989), p.
       437.
291   "During his thirteen . . . statewide": "Dr. Wood's 13-year tenure seen as
       period of unprecedented growth and development for university,"
       undated and unattributed paper, Wood File, University Relations Office,
       University of Alaska Fairbanks.
291   "four separate incidents . . . subject": R. Fineberg, "UAF Journalism
       Department Dismisses Professor Students Call 'Outstanding,'" *Interior/
       North Alaska Newsletter Media Project,* 31 March 1980, p. 10.
291   "As one philosophy . . . understood": R. Krejci quoted in ibid.
291   "Wood climbed aboard . . . dam": Coates (1991), p. 143.
292   "anti-God, anti-man, anti-mind": quoted in ibid., p. 203.
293   "Looking back . . . movement": B. Commoner OHI in O'Neill (1989), p.
       284.
293   *Time* magazine cover and cover story: *Time,* 2 Feb. 1970.
293   "the dean of . . . movement": *The Earth Times,* 21 Oct. 1995.
293   "the father of . . . environmentalism": P. Montague, "Barry Commoner:
       The Father of Grass-Roots Environmentalism," in Kriebel (2002), p. 5.
293   Alaskan biologists had demanded study: see E. Campbell to W. Osburn, 3
       Aug. 1976, Doc. #132475, CIC.
293   "most comprehensive ever done": Sanders (1962), p. 79; USAEC press
       release, "Plowshare Program Fact Sheet—Project Chariot," revised Aug.
       1962, p. 4.
293   "Nowhere in the free . . . environments": J. Wolfe, "Comments on
       Biological Aspects of Report by St. Louis Committee for Nuclear
       Information," attachment to Gen. A. Betts to A. Luedecke, 2 Aug. 1961,
       in AEC staff paper 811/82, 15 Aug. 1961, Doc. #75525, CIC.
293   "Begun ten years . . . 1969": see Coates (1991), p. 205; J. Reed,
       "Ecological Investigations in the Arctic," *Science,* 21 Oct. 1966; E.
       Campbell to J. Kirkwood, 5 March 1970, Doc. #130821, CIC; Merritt
       and Fuller (1977), p. iv.

293   "In the summer of 1962 . . . issues": *Fairbanks Daily News-Miner,* 23, 25, 26, 27 June 1962.

294   "The 40 million . . . history": Arnold (1978), pp. 147–48.

294   "$18 million . . . $4 million": Findlay (1990), pp. 164–67.

294   "Richard Nixon . . . achievement": R. Nixon, "Remarks by the Vice President of the United States Prepared for Delivery Before Meeting of Sigma Delta Chi, Toledo, Ohio, October 26, 1960," Nixon-Lodge Campaign press release dated 27 October 1960, President's papers 1960–61, Box 4, UAF Archives.

294   "But according to Plowshare . . . $15 million": Findlay (1990), p. 168.

294   "In addition to the assorted . . . itself": Congressman Craig Hosmer, quoted in ibid., p. 170.

294   "The truth was . . . planet": ibid., pp. 170–75; Seaborg (1971), p. 174.

294   "When the Atlantic . . . feasibility": Findlay (1990), p. 174.

295   "The most incredible . . . not": ibid., p. 176.

295   "Plowshare could no longer . . . policy-making": Sylves (1987), pp. 208–09.

295   "The story of Project Plowshare . . . imminent": ibid., p. 212.

295   "In the Soviet Union . . . excavating": ibid., pp. 83–90.

295   "widespread nuclear contamination . . . 80s": Sen. Frank Murkowski press release, "Murkowski Calls for Arctic Radiation Research," 10 Feb. 1992.

295   "Four days later . . . Agency": Issue Brief: Office of Senator Frank Murkowski (Alaska), "The Problem of Russian Arctic Pollution," updated July 3, 1992.

295   "credence to his initial fears": Sen. Frank Murkowski press release, "Murkowski Calls for U.S.-Russian Radiation Testing," 28 Feb. 1992.

295   "a billion curies . . . explosions": ibid.

295   "In August . . . contamination": Hearing Before the Select Committee on Intelligence, 102nd Congress, "Radioactive and Other Environmental Threats to the United States and the Arctic from Past Soviet Activities," 15 Aug. 1992, U.S. Govt. Printing Office, 1993, p. 142.

296   "should have been done . . . not": audio tape recording of Teller's SDI speech to Commonwealth North at Captain Cook Hotel, 9 June 1987, made by Chris Toal, SANE Alaska; video tape recording made by author and KUAC-TV, Fairbanks, both in UAF Archives.

296   "In the future . . . safe": E. Teller on "Alaska News Nightly," Alaska Public Radio Network, 6 and 7 Aug. 1992.

296   "As the Soviet economy . . . cash": W. Broad, "A Soviet Company Offers Nuclear Blasts for Sale to Anyone With the Cash," *The New York Times,* 6 Nov. 1991.

296   "We are willing to entertain . . . used": D. Wolfson, quoted in W. Broad, "A Soviet Company Offers Nuclear Blasts to Anyone With the Cash," *The New York Times,* 6 Nov. 1991.

296   "If your mountain . . . card": E. Teller quoted in *Anchorage Daily Times,*
        26 June 1959.

296   "yard sale . . . history": R. Seitz, quoted in T. Rauf, "Cleaning Up with a
        Bang," *Bulletin of the Atomic Scientists,* Jan./Feb. 1992, p. 9.

296   "Chetek appears. . . enterprises": Nordyke (2000), p. 60.

297   "leading the laboratory . . . incomplete": Hans Bethe (1954), *Comments
        on the History of the H-bomb,* text in Williams and Cantelon (1984), p.
        134.

297   "well on its way to success": quoted in Magraw (1988), p. 33.

297   "During the moratorium . . . test": Liberatore (1982), p. 8.

297   "to turn away from nuclear weapons": ibid.

297   "the most expensive military project in the history": L. Rothstein, "No
        Party for Star Wars," *Bulletin of the Atomic Scientists,* June 1992, pp.
        3–4.

297   "a fraud . . . accomplish": quoted in ibid.

298   "Time will reveal . . . interests": Stone (1963), p. 330.

298   Quotes from Teller's speech: audio tape recording of Teller's SDI speech
        to Commonwealth North at Captain Cook Hotel, 9 June 1987, made by
        Chris Toal, SANE Alaska; video tape recording made by author and
        KUAC-TV, Fairbanks, both in UAF Archives. *See also: Fairbanks Daily
        News-Miner,* 10 June 1987, p. 6; *Anchorage Times,* 10 June 1987, p.
        B-1.

299   "But this 'terminal' . . . enough": Johnson (2006), p. 222.

300   "vigorous research . . . system": FitzGerald (2000), p. 119.

300   "But the arms manufacturers . . . space": Johnson (2006), p. 210,
        231–232.

300   "It is the policy . . . defense": See "National Policy on Ballistic Missile
        Defense Fact Sheet," in The White House web site,
        http://www.whitehouse.gov/news/releases/2003/05/20030520-
        15.html

300   "After three years . . . space": Johnson (2006), p. 212–214.

300   "What military analysis . . . failure": FitzGerald (2000), p. 495.

301   Artificialities of the missile defense testing program: Union of Concerned
        Scientists, "Limitations and Artificialities of the Testing Program,"
        December 2005, UCS web site:
        http://www.ucsusa.org/global_security/missile_defense/limitations-and-
        artificialities-of-the-testing-program.html

301   "very high quality . . . information": From "The Value of Fort Greely to
        the Intercept Test Program and a Block 2004 Contingency Defense, and
        the Use of Classification to Prevent Public Scrutiny;" testimony of Dr.
        Lisbeth Gronlund (Senior Staff Scientist, Union of Concerned Scientists
        and Research Fellow, Security Studies Program, Massachusetts Institute of
        Technology) at a Special Investigations Briefing on "Rushing to Failure in

2004: Is Missile Defense Testing Adequate?" held by John F. Tierney, U.S. House of Representatives Committee on Government Reform, Subcommittee on National Security, Veterans Affairs, and International Relations, 11 June 2002. See: http://www.ucsusa.org/global_security/missile_defense/the-value-of-the-fort-greely-test-bed.html

301 Impounding test data: Johnson (2006), p. 222.

302 "The very next month . . . water": *Fairbanks Daily News-Miner,* 30 March 2007, http://newsminer.com/wp-content/themes/fdnm/single-print.php?post=6218

302 "When Secretary Rumsfeld . . . existed": Media Availability with Secretary Rumsfeld and Lt. Gen. Obering on Missile Defense, Ft. Greely, Alaska. See: http://www.defenselink.mil/transcripts/transcript.aspx?transcriptid=3705

302 Approiximately $100 billion: Johnson (2006), p. 230.

302 "Fire, aim, ready": Gronlund, L. "Fire, aim, ready," in *Bulletin of the Atomic Scientists,* Sept./Oct. 2005, pp. 67–68.

302 "Something is better than nothing": Boese, W. "Top Military Brass Insists Missile Defense Ready to be Deployed," in *Arms Control Today,* April 2004. See: http://www.armscontrol.org/act/2004_04/MD.asp

302 "NUCLEAR WASTE DUMP . . . HAZARD": *Anchorage Daily News,* 6 Sept. 1992, p. 1.

302 "the scientists illegally . . . regulations": H. Book to E. Campbell, 2 Nov. 1962, in Doc. #16850, CIC.

302 "According to the documents . . . Plowshares": V. McKelvey to J. Reeves, 27 Feb. 1961, USGS, Reston, VA; J. Devine to Deputy Assistant Secretary for Water and Science, Assistant to the Secretary for Alaska, 14 Sept. 1992, USGS, Reston, VA.

302 "Just before the Sedan . . . zero": Piper (1966), p. 19.

303 "These three sand mixtures . . . curie": A. Baker to E. Price, 28 Feb. 1963, Doc. #16849, CIC.

303 "However, as much . . . granted": R. Woolsey to E. Price, 23 Oct. 1962, "U.S. Geological Survey, Water Quality Branch, Denver, Colorado— License 5-1399-3," Nuclear Regulatory Commission, Washington, D.C.

303 "Five curies would represent . . . mishap": Fifteen-curie figure for Three Mile Island comes from Office of Senator Frank H. Murkowski (Alaska), "The Problem of Russian Arctic Pollution," 3 July 1992.

304 "It was apparently not funded . . . generally": V. McKelvey to J. Reeves, 27 Feb. 1961, USGS Reston, VA; J. Devine to Deputy Assistant Secretary for Water and Science, Assistant to the Secretary for Alaska, 14 Sept. 1992, USGS, Reston, VA.

304 "At the twelfth site . . . downstream": A. Piper and D. Eberline to J. Philip, 9 Oct. 1962, p. 5, LLNL.

304   "To decontaminate the plots . . . pile": A. Baker to E. Price, 28 Feb. 1963, Doc. #16849, CIC.

304   "Other scientists working . . . fact": A. Johnson to D. O'Neill, 29 Sept. 1992. Johnson was at the camp and says, ". . . no one let on that anything like that was being done."

304   "No one monitored . . . review": North Slope Borough Science Advisory Committee, "Draft Review of the 'Project Chariot: 1962 Tracer Study Remedial Action Plan,'" U.S. Department of Energy, Feb. 1993, NSB-SAC-OR-125, March 1993.

305   "exceeded one thousand . . . burial": H. Book to E. Campbell, 2 Nov. 1962, in Doc. #16850, CIC.

305   "Nothing in this permit . . . lands": J. Carver to G. Seaborg, 8 June 1961, in AEC staff paper 811/75, 20 June 1961, DOE Archives.

305   "NORTH SLOPE DECRIES . . . broke": *Anchorage Daily News,* 10 Sept. 1992, p. 1.

305   "ATOMIC ARROGANCE . . . HOPE": *Tundra Times,* 15 Oct. 1992, p. 1.

305   "ESKIMOS FURIOUS . . . DUMP": *San Francisco Examiner,* 13 Sept. 1992.

305   "ESKIMOS UNCOVER . . . NEARBY": *Fresno Bee,* 15 Dec. 1992.

305   "ESKIMOS FEEL . . . EXPERIMENTS": *Seattle Times,* 6 Dec. 1992.

306   "The investigators . . . and left": Ronan Short, telephone interview, 27 April 2007. (Short was a member of the DEC team to visit the site.)

306   "I'm scared now . . . them": Ray Koonuk, Jr., quoted in *Anchorage Daily News,* 13 Sept. 1992, p. 1.

306   "desperately seeking . . . village": *Anchorage Daily News,* 13 Sept. 1992, p. 1.

306   "But three paper studies . . . pollutants": J. Magdanz, "The Fear at Point Hope," *We Alaskans* (Sunday magazine of the *Anchorage Daily News*), 29 Nov. 1992.

306   "Even though the cancer . . . 600": ibid.

306   "have never believed that": *Anchorage Daily News,* 13 Sept. 1992.

306   "I have a personal . . . Chariot": J. Kaleak quoted in J. Magdanz, "The Fear at Point Hope," *We Alaskans* (Sunday magazine of the *Anchorage Daily News*), 29 Nov. 1992.

306   "My government . . . village": E. Patkotak, "Radiation lies ensnared this health worker," *Anchorage Daily News,* 13 Sept. 1992, p. H-2.

307   "meaningful radioactivity experiment": J. Foster to J. Kelly, 27 April 1962 (probably a draft of Foster to Kelly, 30 April 1962), LLNL.

307   "A billion . . . vented": G. Johnson to J. Philip, 31 March 1961, declassified, Doc. #5855, CIC.

307   Thyroid experiment on Alaska Natives: see K. Rodahl and G. Bang, "Thyroid Activity in Men Exposed to Cold," Technical Report 57-36, Arctic Aeromedical Laboratory, Ladd Air Force Base, Alaska; K. Rodahl,

"Human Acclimatization to Cold," Technical Report 57-21, Arctic Aeromedical Laboratory, Ladd Air Force Base, Alaska.

307   "We had with us . . . thing": K. Rodahl, interviewed on "Eskimo Guinea Pigs," *CNN Special Reports,* produced by K. Rickenbaker, Cable News Network, the program aired on CNN in May 1993.

308   "Surviving participants . . . health": "Eskimo Guinea Pigs," ibid. Anaktuvuk Pass residents interviewed in the CNN segment who denied having been fully informed of the nature of the I-131 experiment were Bob Ahgook, Dora Hugo, and Justus Mekiana.

308   "You have poisoned our land": *The New York Times,* 6 Dec. 1992, p. 16.

308   "Murkowski met . . . priority": *Anchorage Daily News,* 14 Sept. 1992.

308    $3 million: *Anchorage Daily News,* 10 Feb. 1993.

308   GAO study: *Fairbanks Daily News-Miner,* 10 March 1993, p. B-2.

308   University of Alaska disposals: H. Book to G. Smith, 9 Feb. 1976; D. Holleman to Nuclear Materials Safety Section, U.S. Nuclear Regulatory Commission, 1 May 1985, p. 46, 48, NRC, Washington, D.C.

308   GAO final report: U.S. GAO (1994).

309    $150,000: In *Report from the Mayor,* North Slope Borough, Dec. 1992, p. 2.

309   "The Department of Energy . . . people": *Anchorage Daily News,* 10 Feb. 1993.

309    $5 million: *Anchorage Daily News,* 24 May 1993.

309    $6 million, fifty-one-man camp: D. Dasher, Alaska Dept. of Environmental Conservation, personal communication 30 June 1993.

309   "it could easily go down . . . mankind": D. Fisher of Battelle Pacific Northwest Laboratories on "Alaska News Nightly," Alaska Public Radio Network, 12 Nov. 1992.

310   "70,000 curies . . . Delta Junction": 70,000 curies left at Fort Greely: see W. Johnson, "Testing Nuclear power in Alaska: The Reactor at Fort Greely," M.A. thesis, University of Alaska Fairbanks, 1993; 2.5 million times: based on 70,000 curies at Greely and .026 curies buried at Chariot site.

310   Point Barrow five-megaton shot: G. Johnson to A. Starbird, 7 Feb. 1958, Doc. #108000, CIC.

310   Spiridon Lake project: W. Egan to G. Johnson, 2 May 1964, Plowshare Collection, LLNL.

310   Dredging Bering Strait: Seaborg (1981), pp. 193–94.

310   Fort Greely reactor: see W. Johnson, "Testing Nuclear Power in Alaska: The Reactor at Fort Greely," M.A. thesis, University of Alaska Fairbanks, 1993; "Alaska's Engineering Heritage," Alaska Section, American Society of Civil Engineers; "Fort Greely Nuclear Power Plant," press release U.S. Army Engineer District, Alaska, 18 Dec. 1957; "Fort Greely Power

Plant," *Alaska Construction & Oil,* Aug. 1979, p. 4; C. Kern, "Alaska's Atomic Power Project," *Alaska Construction News,* Summer 1960.

311   "5 million kilowatts . . . world": W. Hunt, *Alaska: A Bicentennial History.*

311   "Scenically it is . . . beauty": quoted in Coates (1991), pp. 150–51.

311   "nothing but a vast . . . mosquitoes": quoted in ibid., p. 151.

311   "Search the whole . . . water": quoted in ibid., 142.

311   "not more than ten flush toilets": quoted in Brooks (1971), p. 83.

311   "an area as worthless . . . earth": quoted in ibid., pp. 93–94.

311   "Nowhere in the history . . . overwhelming": quoted in ibid., p. 91.

311   "Did you ever see a duck drown?": quoted in ibid., p. 90.

312   "Another development scheme . . . Southwest": Seaborg (1971), pp. 111–15.

312   Lyndon LaRouche: see agendum for "A National Conference on Water from Alaska," 27 Feb. 1982, presented by the National Democratic Policy Committee, whose board chairman was Lyndon LaRouche; see also "America is Running Dry: Build Great Water Projects Now!" distributed by Democrats for Economic Recovery/LaRouche in '92.

312   "These initiatives . . . devastation": "America is Running Dry: Build Great Water Projects Now!" distributed by Democrats for Economic Recovery/LaRouche in '92.

312   "'when,' not 'if'": M. Reisner, *Cadillac Desert* (1986), p. 14.

312   "Another megaproject . . . risky": Seaborg (1981), pp. 193–95; see also D. Foote, "Geographical Engineering," *The Northern Engineer,* Spring 1969; M. Dunbar, "On the Bering Strait Scheme," *Polar Notes,* Occasional Publication of the Stefansson Collection, Dartmouth College, November 1960.

313   "Pierre Salinger . . . stop": Salinger (1966), p. 137.

313   "If they wanted . . . anyone": Coates (1991), p. 132.

313   "As Melvin L. . . . . . Amchitka": T. O'Toole, "Cannikin: The Last of the Big Blasts," *Washington Post,* 31 Oct. 1971.

313   "From the air, an extensive . . . soil": ibid.

313   "subarctic junkyard . . . vandals": Laycock (1971), pp. 117, 123.

314   "lighthouse, military or naval . . . Order": Merritt and Fuller (1977), p. iii.

314   "the agency had had its eye on . . . 1950": Laycock (1971), p. 126.

314   "According to the AEC's official . . . test": Hewlett and Duncan (1969), p. 535.

314   "An atmospheric . . . Windstorm": Miller (1986), p. 130.

314   "But a search . . . Nevada": Hewlett and Duncan (1969), p. 535.

314   "too large for safe testing in Nevada": Merritt and Fuller (1977), p. vi.

314   "considerable search": quoted in Laycock (1971), p. vi.

314   "In protest against Cannikin . . . group": Coates (1991), p. 132; R. Fineberg to D. O'Neill, 11 June 1993 (Fineberg sailed on the Greenpeace voyage).

314   "shorebirds . . . skulls": R. Rausch, personal communication, 6 May 1988, author's notes.

314   "rocky bluffs . . . crater": S. McCarthy, "Project Chariot, Long Shot, Milrow, and Cannikin. . . ." (no date), unpublished paper.

314   "Radioactive tritium . . . detonation": ibid., p. 632; Laycock (1971), pp. 128–29.

315   "AEC geologists . . . 400 years": Laycock (1971), pp. 128–29.

315   "In 2000 . . . payments": Press release from the office of U.S. Sen. Frank Murkowski, "Murkowski: President signs amchitka nuclear test workers to gain compensation for occupational illnesses," 31 Oct. 2000.

315   "Of the 1,400 . . . reported": Hunter, Don, "Data shows cancer spurt at Amchitka nuclear test site: Workers contracted several types at higher than average rates," *Anchorage Daily News,* 10 Aug. 2005.

315   "'opening up' the Arctic . . . site": W. Hickel to G. Seaborg, 28 Sept. 1967, Alaska State Library, Juneau, AK.

315   "geological exploration . . . that area": G. Seaborg to W. Hickel, 19 Oct. 1967 (received), Alaska State Library.

316   "The test area . . . historian": Nielson (1988), p. 210.

316   "In 1966 and 1967 . . . conditions": Richard A. Fineberg, "The Chemical War Tests in Alaska," *Anchorage Daily News,* 6 June 1970, p. 1, 7 June 1970, p. 4. See also: Fineberg (1971), p. 99.

316   Tularemia: Only about 200 cases of tularemia are diagnosed in the United States annually, so in a given year the average American citizen's chance of contracting the disease is about one in a million. In 1982, I became one of the unlucky few when I skinned an infected beaver in the Tanana River drainage, 100 miles downstream from the military's test site. After a six-month period of debilitation, the infection was finally conquered with the surgical removal of a portion of my lymph.

316   "In one incident uncovered by . . . minutes": R. Fineberg, "The Chemical War Tests in Alaska," two part series in *Anchorage Daily News,* 6 and 7 June 1970; see also Fineberg (1971), pp. 90–96; P. Epler, "A Potentially Deadly Area," *Anchorage Daily News,* 16 Nov. 1986.

316   "when the program . . . disposed of": Nielson (1988), p. 210.

317   "If the logistical . . . site": ibid., p. 211.

317   "In the Kara Sea . . . fuel": "The Problem of Russian Arctic Pollution," Office of Senator Frank H. Murkowski, (updated) 3 July 1992.

317   "A person who merely stands . . . radiation": M. Hertsgaard, "From Here to Chelyabinsk," *Mother Jones,* Jan./Feb. 1992, p. 71.

317   "CHASE . . . corrode": J. Roderick, "The Case of the *Robert Louis Stevenson,*" *Cook Inlet Vigil,* March 1991, p. 3.

317   "Alaska's Aleutian . . . Attu": J. Roderick, "Chemical Weapons in Alaskan Waters," *Peace Illustrated: The Newsletter of SANE/Alaska,* Spring 1991.

318   "The military has expanded . . . exercises": U.S. Army Engineering
       District, Alaska, "Joint Military Training Exercises Environmental Impact
       Statement: Public Scoping Process," undated (1992), Army Corps of
       Engineers, Anchorage, Alaska.
318   "Vowing that all . . . non-military": Fritz (2002), p. 142.
318   "A privately-owned . . . launches": ibid, p. 146.
319   "The DOE continues to employ . . . image": Titus (1989), pp. 25–26.
319   "In fact . . . relations": quoted in "In Brief," *Bulletin of the Atomic
       Scientists,* Nov. 1993, p. 4.
319   "It's still the same . . . door": quoted in Titus (1989), p. 25.

Afterword

321   "A scientist's allegiance . . . government": L. Viereck to W. Wood, 29 Dec.
       1960.
322   "But throughout . . . America": see R. Krejci OHI in O'Neill (1989), p.
       460.
322   Udall quotes: quoted in K. Schneider, "A Longtime Pillar of the
       Government Now Aids Those Hurt by Its Bombs," *The New York Times,*
       9 June 1993.

Methodology

333   "The Bush administration . . . attacks": Roberts, Alasdair, "Battle for
       transparency fought on evolving fronts," *Bellingham Herald,* 13 March
       2007, p. A-7.
333   "At the time . . . your decisions": U.S. House of Representatives,
       Committee on Government Reform—Minority Staff, Special
       Investigations Division, prepared for Rep. Henry A. Waxman, *Secrecy in
       the Bush Administration,* September 14, 2004, pp. 3–6. See also:
       Johnson, C., p. 247; Shakir, F., et al, "The Dark Ages," at Center for
       American Progress Action Fund, 13 March 2007,
       http://www.americanprogressaction.org/progressreport/2007/03/dark
       _ages.html
333   "Delaying actions . . . 2005": Shakir, F., et al, "The Dark Ages," at Center
       for American Progress Action Fund, 13 March 2007,
       http://www.americanprogressaction.org/progressreport/2007/03/dark
       _ages.html
333   "As University of Massachusetts . . . history": U.S. House of
       Representatives, Committee on Government Reform—Minority Staff,
       Special Investigations Division, prepared for Rep. Henry A. Waxman,
       Secrecy in the Bush Administration, 14 September 2004, p. 4.

334   "The Clinton administration . . . quadrupled": from "More Government Secrecy," a graph prepared by McClatchy-Tribune Information Services, published in *Bellingham Herald*, 12 March 2007, p. A-9.

334   "Since the Freedom . . . instantly available": Roberts, Alasdair, "Battle for transparency fought on evolving fronts," *Bellingham Herald*, 13 March 2007, p. A-7.

334   "Despite these trends . . . people": U.S. House of Representatives, Committee on Government Reform—Minority Staff, Special Investigations Division, prepared for Rep. Henry A. Waxman, *Secrecy in the Bush Administration*, 14 September 2004, p. iv.

BIBLIOGRAPHY

Allardice, Corbin, and Edward R. Trapnell. *The Atomic Energy Commission.* Praeger Publishers, 1974.

Ambrose, Stephen E. *Eisenhower: The President,* Vol. 2. Simon and Schuster, 1984.

American Association for the Advancement of Science, Committee on Science in the Promotion of Human Welfare. "Science and Human Welfare: The AAAS Committee on Science in the Promotion of Human Welfare states the issues and calls for action," *Science* (8 July 1960), pp. 68–73.

Arnold, Robert D. *Alaska Native Land Claims.* The Alaska Native Foundation, 1978.

Ball, Howard. *Justice Downwind: America's Atomic Testing Program in the 1950's.* Oxford University Press, 1986.

Beechey, Frederick W. "Narrative of a Voyage to the Pacific and Beering's Strait." *Bibliotheca Australiana,* #34, Vol. 1. Nico Israel; Da Capo Press, 1968.

Berry, Mary Clay. *The Alaska Pipeline: The Politics of Oil and Native Land Claims.* Indiana University Press, 1975.

Berry, Wendell. "Preserving Wilderness." *Home Economics,* North Point Press, 1987.

Blackman, Margaret B. *Sadie Brower Neakok: An Iñupiaq Woman.* University of Washington Press, 1989.

Blumberg, Stanley A. and Gwinn Owens. *Energy and Conflict: The Life and Times of Edward Teller.* G. P. Putnam's Sons, 1976.

Blumberg, Stanley A., and Louis G. Panos. *Edward Teller: Giant of the Golden Age of Physics.* Charles Scribner's Sons, 1990.

Bockstoce, John R. *Whales, Ice and Men: The History of Whaling in the Western Arctic.* University of Washington Press, 1986.

Bradley, David. *No Place to Hide: 1946/1984.* University Press of New England, 1983.

Broad, William J. *Teller's War: The Top Secret War Behind the Star Wars Deception.* Simon & Schuster, 1992.

Brooks, Paul, and Joseph Foote. "The Disturbing Story of Project Chariot," *Harper's* (19 April 1962), p. 60.

Brower, Charles D., ed. *Fifty Years Below Zero.* Dodd, Mead, 1943.

Burch, Ernest S., Jr. *The Traditional Eskimo Hunters of Point Hope, Alaska: 1800–1875.* North Slope Borough, 1981.

Coates, Peter. "Project Chariot: Alaskan Roots of Environmentalism." *Alaska History.* The Alaska Historical Society, Fall 1989, pp. 1–31.

———. *The Trans-Alaska Pipeline Controversy: Technology, Conservation and the Frontier.* Lehigh University Press, 1991.

Cole, Terrence. *E. T. Barnette: The Strange Story of the Man Who Founded Fairbanks, Alaska.* Alaska Northwest Publishing Co., 1984.

Commoner, Barry. *The Closing Circle: Nature, Man, and Technology.* Knopf, 1971.

Coughlan, Robert. "Dr. Edward Teller's Magnificent Obsession," *Life.* (September 6, 1954).

Daley, Patrick, and Beverly James. "An Authentic Voice in the Technocratic Wilderness: Alaska Natives and the *Tundra Times,*" *Journal of Communication* (Summer 1986).

Daley, Patrick J., and Dan O'Neill. "Howard Rock (Sikvoan Weyahok)." *Dictionary of Literary Biography,* Vol. 127; "American Newspaper Publishers, 1951–90." Perry Ashley (volume editor). Bruccoli Clark Layman Book, Gale Research, Inc., 1993.

Defense Atomic Support Agency. *The Peaceful Uses of Nuclear Explosives: Project Plowshare,* 1964.

Divine, Robert A. *Eisenhower and the Cold War.* Oxford University Press, 1981.

E. J. Longyear Company. *Report to the University of California Radiation Laboratory on the Mineral Potential and Proposed Harbor Locations in Northwestern Alaska,* 1958.

Fejes, Claire. *People of the Noatak.* Knopf, 1967.

Findlay, Trevor. *Nuclear Dynamite: The Peaceful Nuclear Explosion Fiasco.* Brassey's Australia, 1990.

Fineberg, Richard A. *The Dragon Goes North: Chemical and Biological Weapons Testing in Alaska.* Unpublished ms., 1971.

FitzGerald, Frances. *Way Out There in the Blue: Reagan, Star Wars and the End of the Cold War.* Simon & Schuster, 2000.

Foote, Don Charles. *The Economic Base and Seasonal Activities of Some Northwest Alaskan Villages: A Preliminary Study.* Prepared for the USAEC, 1959.

————. *The Eskimo Hunter at Point Hope, Alaska: September 1959 to May 1960.* Prepared for the USAEC, 1960.

————. *Project Chariot and the Eskimo People of Point Hope, Alaska.* Prepared for the USAEC, 1961.

Foote, J. *Report from Tigara.* Unpublished ms. Foote Papers, UAF Archives, 1961.

Ford, Daniel. *The Cult of the Atom.* Simon and Schuster, 1982.

Fritz, Stacey. *The Role of Alaskan Missile Defense in Environmental Security.* (Master's Thesis, University of Alaska, Fairbanks), 2002.

Giddings, J. Louis. *Ancient Men of the Arctic.* Knopf, 1967.

Giovannitti, Len, and Fred Freed. *The Decision to Drop the Bomb.* Coward-McCann, 1965.

Glasstone, Samuel. *The Effects of Nuclear Weapons.* U.S. Atomic Energy Commission, 1957.

————. *The Effects of Nuclear Weapons.* U.S. Atomic Energy Commission, 1964.

Gofman, John W. *Radiation and Human Health.* Sierra Club Books, 1981.

Gofman, John W., and Arthur R. Tamplin. *Poisoned Power: The Case Against Nuclear Power Plants.* Rodale Press, 1971.

Groves, Leslie R. *Now It Can Be Told.* Harper, 1962.

Hewlett, Richard G., and Francis Duncan. *Atomic Shield, 1947/1952: A History of the United States Atomic Energy Commission.* Pennsylvania State University Press, 1969.

Hewlett, Richard G., and Jack M. Holl. *Atoms for Peace and War.* University of California Press, 1989.

Hilgartner, Stephen, Richard C. Bell, and Rory O'Connor. *Nukespeak: The Selling of Nuclear Technology in America.* Penguin, 1983.

Hulse, James W. *The University of Nevada: A Centennial History.* University of Nevada Press, 1974.

International Atomic Energy Agency. *Peaceful Nuclear Explosions: Phenomenology and Status Report, 1970.* 1970.

Johnson, Albert W. *Science, Society and Academic Freedom.* Paper presented to American Association of University Professors. A. W. Johnson Collection, (Undated).

Johnson, Chalmers. *Nemesis: The Last Days of the American Republic.* Metropolitan Books, 2006.

Johnson, Gerald W. *The Soviet Program for Industrial Applications of Explosions.* University of Calif. Radiation Laboratory. (UCRL #5932), 1960a.

————. *Excavation with Nuclear Explosives.* University of California Radiation Laboratory, (UCRL #5917), 1960b.

Jungk, Robert. *Brighter than a Thousand Suns.* Harcourt, Brace & World, 1958.

Kiste, Robert C. *The Bikinians: A Study in Forced Migration.* Cummings, 1974.

Kramish, Arnold. *The Peaceful Atom in Foreign Policy.* Harper & Row, 1963.

Kriebel, David, ed. *Barry Commoner's Contribution to the Environmental Movement: Science and Social Action.* Baywood Publishing, 2002.

Lacey, Michael J., ed. *Government and Environmental Politics: Essays on Historical Developments Since World War Two.* The Wilson Center Press, 1989.

Lamont, Lansing. *Day of Trinity.* Atheneum, 1965.

Langone, John. "Edward Teller: Of Bombs and Brickbats." *Discover* (July 1984).

Larsen, Helge, and Froelich Rainey. *Ipiutak and the Arctic Whale Hunting Culture.* American Museum of Natural History, 1948.

Laurence, William L. *Men and Atoms.* Simon & Schuster, 1959.

Lawrence Radiation Laboratory. *Nuclear Excavation: A Status Report.* Film A85, Undated.

Laycock, George. *Alaska: The Embattled Frontier.* Audubon Society by Houghton Mifflin, 1971.

Liberatore, Daniel. *At the 30-year Mark: The Directors Look Back at Lab History.* Lawrence Livermore National Laboratory, 1982.

Lilienthal, David E. *Change, Hope and the Bomb.* Princeton University Press, 1963.

———. *The Road to Change, The Journals of David Lilienthal,* Vol. IV. Harper & Row, 1969.

Lorenz, Konrad. *The Waning of Humaneness.* Little, Brown, 1987.

Lowenstein, Tom. *Some Aspects of Sea Ice Subsistence Hunting in Point Hope, Alaska.* North Slope Borough, 1986.

Magraw, Katherine. 1988. "Teller and the 'Clean Bomb' Episode." *The Bulletin of the Atomic Scientists* (May 1988), pp. 32–37.

McKay, Alwyn. *The Making of the Atomic Age.* Oxford University Press, 1984.

McPhee, John. *The Curve of Binding Energy.* Ballantine, 1976.

Merritt, Melvin L., and R. Glen Fuller. *The Environment of Amchitka Island, Alaska.* U.S. Research and Development Administration, 1977.

Metzger, H. Peter. *The Atomic Establishment.* Simon & Schuster, 1972.

Miller, Richard L. *Under the Cloud: The Decades of Nuclear Testing.* The Free Press, 1986.

Morgan, Lael. *Art and Eskimo Power: The Life and Times of Alaskan Howard Rock.* Epicenter Press, 1988.

Neel, J. V., and W. J. Schull. *The Effect of Exposure to Atomic Bombs on Pregnancy Termination in Hiroshima and Nagasaki.* National Academy of Sciences—National Research Council (Publication No. 461), 1956.

Nielson, Johnathan M. *Armed Forces on a Northern Frontier: The Military in Alaska's History, 1867–1987.* Greenwood Press, 1988.

Nordyke, M. D. *The Soviet Program for Peaceful Uses of Nuclear Explosions.* University of Calif. Radiation Laboratory. (UCRL-ID-124410 Rev 2), 1 Sept. 2000.

O'Keefe, Bernard J. *Nuclear Hostages.* Houghton Mifflin, 1983.

O'Neill, Dan, comp. *Project Chariot: A Collection of Oral Histories.* 2 vols. Alaska Humanities Forum, 1989.

O'Neill, Dan. "Project Chariot: How Alaska Escaped Nuclear Excavation," *The Bulletin of the Atomic Scientists* 45 (December 1989): pp. 28–37, 1989a.

Orlans, Harold. *Contracting for Atoms.* The Brookings Institution, 1967.

Orth, Donald. *Dictionary of Alaska Place Names.* Geological Survey Professional Paper #567, U.S. Government Printing Office, 1971.

Patty, Ernest N. *North Country Challenge.* David McKay Co., 1969.

Pauling, Linus. *No More War.* Dodd, Mead, 1983.

Pewe, Troy L., David M. Hopkins, and Arthur H. Lachenbruch. *Engineering Geology Bearing on Harbor Site Selection Along the Northwest Coast of Alaska From Nome to Barrow.* United States Geological Survey (TEI-678), 1958.

Piper, Arthur M. *Potential Effects of Project Chariot on Local Water Supplies Northwest Alaska.* United States Geological Survey, Professional Paper 539, 1966.

Price, Jerome. *The Antinuclear Movement.* Twayne Publishers, 1982.

Pruitt, William O., Jr. "A New Caribou Problem," *The Beaver* (Winter 1962) pp. 24–25.

———. "Radioactive Contamination," *Naturalist* (Spring 1963) pp. 20–26.

Pulu, Tupou L., Ruth Ramoth-Sampson, and Angeline Newlin. *Whaling: A Way of Life.* University of Alaska Anchorage, Undated.

Rainey, Froelich G. *The Whale Hunters of Tigara.* American Museum of Natural History, 1947.

Rhodes, Richard. *The Making of the Atomic Bomb.* Touchstone, 1988.

Salinger, Pierre. *With Kennedy.* Doubleday, 1966.

Sanders, Ralph. *Project Plowshare: Development of the Peaceful Uses of Nuclear Explosions.* Public Affairs Press, 1962.

Schooler, Dean, Jr. *Science, Scientists, and Public Policy.* The Free Press, 1971.

Seaborg, Glenn T., and William R. Corliss. *Man and Atom.* E. P. Dutton, 1971.

Seaborg, Glenn T., with Benjamin S. Loeb. *Kennedy, Khrushchev and the Test Ban.* University of California Press, 1981.

Shelton, A. Vay, et al. *The Neptune Event—A Nuclear Explosive Cratering Experiment.* University of California Radiation Laboratory (UCRL #5766), 1960.

Skartvedt, Stephen. *The Plowshare Program: Environmental Perceptions and Impacts.* (Master's Thesis, San Francisco State University), 1992.

Stone, I. F. *The Haunted Fifties.* Random House, 1963.

Sylves, Richard T. *The Nuclear Oracles.* Iowa State University Press, 1987.

Teller, Edward. "We're Going to Work Miracles." *Popular Mechanics* (March 1960).

Teller, Edward, and Albert L. Latter. *Our Nuclear Future . . . Facts, Dangers and Opportunities.* Criterion, 1958.

Teller, Edward, with Allen Brown. *The Legacy of Hiroshima.* Doubleday, 1962.

Teller, Edward, Wilson K. Talley, Gary H. Higgins, and Gerald W. Johnson. *The Constructive Uses of Nuclear Explosives.* McGraw-Hill, 1968.

Titus, A. Costandina. *Bombs in the Backyard.* University of Nevada Press, 1986.

———. "Selling the Bomb: Public Relations Efforts by the Atomic Energy Commission During the 1950s and Early 1960s." *Government Publications Review* (January/February 1989).

University of California Radiation Laboratory. *Industrial Applications of Nuclear Explosions.* Film A81, 1958.

U.S. Atomic Energy Commission. *Twenty-fifth Semiannual Report of the Atomic Energy Commission.* U.S. Govt. Printing Office, January 1959.

———. *Chariot Environmental Program.* U.S. Atomic Energy Commission San Francisco Operations Office, July 1959.

———. *Bioenvironmental Features of the Ogotoruk Creek Area, Cape Thompson, Alaska: A First Summary Report by the Committee on Environmental Studies for Project Chariot.* USAEC, Office of Technical Information, TID–12439. December 1960.

———. *Bioenvironmental Features of the Ogotoruk Creek Area, Cape Thompson, Alaska: A Second Summary by the Committee on Environmental Studies for Project Chariot.* TID-17226. October 1962.

———. *Project Chariot Phases I–V: Project Manager's Summary Report.* USAEC Nevada Operations Office and Holmes & Narver Inc. NVO-7, August 1964.

———. *Engineering with Nuclear Explosives: Proceedings of the Third Plowshare Symposium.* USAEC Division of Technical Information. TID–7695, 1964a.

U.S. Defense Nuclear Agency. *Projects Gnome and Sedan: The Plowshare Program.* DNA 6029F, March 1983.

U.S. Department of Energy. *Announced United States Nuclear Tests: July 1945 Through December 1987.* NVO-209 (Rev. 8). Nevada Operations Office, 1988.

U.S. Government Accounting Office. *Nuclear Health and Safety: Sites Used for Disposal of Radioactive Waste in Alaska.* GAO/RCED-94-130FS, 6 July 1994.

U.S. House of Representatives, Committee on Government Reform—Minority Staff, Special Investigations Division, prepared for Rep. Henry A. Waxman. *Secrecy in the Bush Administration.* September 14, 2004.

Vanstone, James W. *Point Hope: An Eskimo Village in Transition.* University of Washington Press, 1962.

Wasserman, Harvey, and Norman Solomon. *Killing Our Own: The Disaster of America's Experience with Atomic Radiation.* Delacorte Press, 1982.

Weaver, Lynn E., ed. *Peaceful Uses of Nuclear Explosives.* University of Arizona Press, 1970.

Wilimovsky, Norman J., and John N. Wolfe, eds. *The Environment of the Cape Thompson Region, Alaska.* U.S. Atomic Energy Commission, 1966.

Williams, Robert C., and Philip L. Cantelon, eds. *The American Atom: A Documentary History of Nuclear Policies from the Discovery of Fission to the Present.* University of Pennsylvania Press, 1984.

York, Herbert F. *Making Weapons, Talking Peace.* Basic Books, 1987.

Zodner, Harlan, ed. *Industrial Uses of Nuclear Explosives.* University of California Radiation Laboratory (UCRL-5253), 1958.

INDEX

CPSIA information can be obtained
at www.ICGtesting.com
Printed in the USA
BVOW10s0720300417
482679BV00001B/1/P

9 780465 003488